PURINE METABOLISM IN MAN
Biochemistry and Pharmacology of Uric Acid Metabolism

ADVANCES IN EXPERIMENTAL MEDICINE AND BIOLOGY

PURINE METABOLISM IN MAN

Biochemistry and Pharmacology of Uric Acid Metabolism

Edited by

Oded Sperling and Andre De Vries

Division of Metabolic Disease
Rogoff-Wellcome Medical Research Institute
Beilinson Hospital
Petah-Tikva, Israel

and

James B. Wyngaarden

Department of Medicine
Duke University Medical Center
Durham, North Carolina

PLENUM PRESS • NEW YORK-LONDON

Library of Congress Cataloging in Publication Data

International Symposium on Purine Metabolism in Man,
Tel-Aviv, 1973.
Purine metabolism in man: biochemistry and pharmacology of uric acid metabolism.

(Advances in experimental medicine and biology, v. 41)
Vol. 1 has subtitle: Enzymes and metabolic pathways.
Includes bibliographical references.
1. Purine metabolism—Congresses. 2. Uric acid metabolism—Congresses. I. Sperling, Oded, ed.
II. De Vries, André, 1911-   ed. III. Wyngaarden, James B., ed. IV. Title. V. Series. [DNLM:
1. Purine-pyrimidine metabolism, Inborn errors—Congresses. 2. Purines—Metabolism—
Congresses. W1AD559 v. 41 / QU58 I635p 1973]
QP801.P8I56   1973   612'.39   73-21823
ISBN 0-306-39096-5 (v. 2)

Second half of the proceedings of the International Symposium on
Purine Metabolism in Man, held in Tel Aviv, Israel, June 17-22, 1973

© 1974 Plenum Press, New York
A Division of Plenum Publishing Corporation
227 West 17th Street, New York, N.Y. 10011

United Kingdom edition published by Plenum Press, London
A Division of Plenum Publishing Company, Ltd.
Davis House (4th Floor), 8 Scrubs Lane, Harlesden, London, NW10 6SE, England

Printed in the United States of America

PREFACE

Gout and uric acid lithiasis are known to have
affected mankind for thousands of years. It is only
recently, however, that great progress has been made
in the understanding of the processes involved in purine
metabolism and its disorders in man. The key enzymes
active in the various pathways of purine synthesis and
degradation have become known and their properties are
the subject of intensive study. Major contributions to
the knowledge of normal purine metabolism in man have
derived from the study of inborn errors in patients with
purine disorders, specifically complete and partial
hypoxanthine-guanine phosphoribosyltransferase deficiency.
Mutations of other enzymes involved in purine metabolism
are being discovered. A great step forward has been
made in the treatment of gout with the introduction of
uricosuric drugs and more recently of the hypoxanthine
analogue allopurinol, a synthetic xanthine oxidase
inhibitor. Furthermore, the complex nature of the
renal handling of uric acid excretion, although still
posing difficult problems, appears to approach clarification.
In view of the intensive research on purine metabolism going on in various laboratories all over the
world it was felt by several investigators in this field
that the time was appropriate to convene a symposium
dedicated to this subject. An International Symposium
on Purine Metabolism in Man, the first of its kind, was
therefore organized and held in Tel-Aviv, June 17 to
22, 1973. The meeting dealt with the various aspects of
purine metabolism and its disorders - biochemistry,
enzyme mutations, genetics, methodology, clinical aspects
and treatment. This volume contains the full reports
of all communications made at the symposium. Its publi-

cation was made possible by the joint effort of the
contributors and the organizing and advisory committees
of the congress.

The special aid by the various sponsors, the Israel
Ministry of Health, the Tel-Aviv University, the Israel
Academy of Sciences and Humanities, the Wellcome Research
Laboratories of the Burroughs Wellcome Co. USA, and the
excellent organization by Kenes, the congress organizers,
are gratefully acknowledged.

                                    The Editors

CONTENTS OF VOLUME 41B

GOUT

Etiology of Purine Overproduction in Gout

Effects of Diet, Weight, and Stress
on Purine Metabolism

Relationship Between Carbohydrate,
Lipid and Purine Metabolism

Adenine Therapy in the
Lesch-Nyhan Syndrome

Benzbromarone, 3-Hydroxy Purines,

3-Butylazathioprine

RENAL HANDLING OF URATE

Hypouricemia and the Effect of Pyrazinamide

on Renal Urate Excretion

Various Aspects of the Renal Handling of Urate

METHODOLOGY

CONTENTS OF VOLUME 41A

ENZYMES AND METABOLIC PATHWAYS

IN PURINE METABOLISM

Purine Phosphoribosyltransferases

Nucleoside and Nucleotide Metabolism

MUTATIONS AFFECTING PURINE METABOLISM

Properties of HGPRT and APRT in HGPRT

Deficient Blood Cells

Purine Metabolism in HGPRT Deficient Cells

GOUT

Etiology of
Purine Overproduction in Gout

THE KINETICS OF INTRAMOLECULAR DISTRIBUTION OF ^{15}N IN URIC ACID FOLLOWING ADMINISTRATION OF ^{15}N-GLYCINE: PREFERENTIAL LABELING OF N-(3+9) OF URIC ACID IN PRIMARY GOUT AND A REAPPRAISAL OF THE "GLUTAMINE HYPOTHESIS"

Oded Sperling,* James B. Wyngaarden, and C. Frank Starmer

(From the Department of Medicine, Duke University Medical Center, Durham, North Carolina, and The Rogoff-Wellcome Medical Research Institute, Tel Aviv University Medical School, Beilinson Hospital, Petah Tikva, Israel)

Patients with primary gout and excessive uric acid excretion fed a test dose of ^{15}N-glycine incorporate increased quantities of ^{15}N into urinary urate [1-4]. Although enrichment of all 4 nitrogen atoms of uric acid is excessive [4,5] that of N-(3+9) is disproportionately great, especially in flamboyant overexcretors of uric acid [5]. Since N-3 and N-9 of uric acid are derived from the amide-N of glutamine [6,7], Gutman and Yu [5,8] proposed a defect in glutamine metabolism in primary gout. Since urinary ammonium, which arises principally from glutamine [9,10], is reduced in many gouty subjects [11,12], they further postulated a reduction of glutaminase activity [5,8]. The hyperglutamatemia of gout has now suggested a defect of glutamate metabolism, with diversion of glutamic acid toward glutamine and purine biosynthesis [13,14].

Several rare enzymic subtypes are now recognized in gout, e.g., glucose 6-phosphatase deficiency [15] hypoxanthine-guanine phos-

*Present address: Beilinson Hospital, Petah Tikva, Israel

phoribosyltransferase deficiencies of variable severity [16-18], and superactive PP-ribose-P[1] synthetases [19-21]. In at lease the last two the demonstrably raised intracellular level of PP-ribose-P is regarded as the driving force of excessive purine biosynthesis. The findings reported in this paper indicate that preferential labeling of N-(3+9) occurs in gouty patients with PRT deficiency or superactive PP-ribose-P synthetases, as well as in overproducers with idiopathic gout. Thus this labeling pattern appears to be a consequence of kinetic factors common to several different metabolic subtypes of gout, and is not necessarily indicative of a defect of glutamine metabolism.

MATERIALS AND METHODS

[15]N-Glycine, 95.1 atom percent excess ^{15}N was synthesized according to Schoenheimer and Ratner [22], or purchased from Isomet Corp. [1-^{14}C]-Phenylacetic acid was purchased from New England Nuclear Corp.

Uric acid was isolated from urine by adsorption on Darco-G60 at pH 4.2 overnight at 4°C, eluted with hot 0.1N NaOH, and repeatedly recrystallized from Li_2CO_3 solution. Uric acid analyses were performed by the differential spectrophotometric method [23] employing purified uricase purchased from Worthington Biochemical Corporation.

Uric acid samples were degraded by procedures which yield N-7 as glycine [24] and N-(1+3+9) [24], N-(1+7) [25], N-(1+3) [26] and N-(7+9) [26] as NH_3. After alkalinization, the NH_3 was collected in 1N HCl. Hippuric acid was isolated according to Quick [27,28]. Urinary ammonia was recovered from fresh urine following alkalinization with saturated K_2CO_3 and collection in 1N HCl.

Phenylacetylglutamine was isolated from urine following the ingestion of [[1-^{14}C] phenylacetic acid by consecutive separations on cation and anion exchange resins as follows: AG-50W X 8 H[+] (200-400 mesh) was converted to its Na[+] form, placed in a chromatographic column (40 X 600 mm) and washed with citrate buffer (pH 3.42) according to Stein [29]. A urine sample of 50 to 100 ml was acidified to pH 2.3 and applied to the column, which was washed with successive 100 ml portions of citrate buffers of pH 2.5 and

---

1. Abbreviations used are: PP-ribose-P, α-5-phosphoribosyl-1-pyrophosphate; PRA, β-5-phosphoribosyl-1-amine; GAR, glycinamide-ribonucleotide; formyl-GAR, α-formyl-GAR; formyl-GAM, α-formyl glycinamidineribonucleotide; IMP, inosine 5´-monophosphate; AMP, adenosine 5´-monophosphate; GMP, guanosine 5´-monophosphate; PP$_i$, inorganic pyrophosphate; CoA-SH, reduced coenzyme A; NADP, nicotinamide adenine dinucleotide phosphate; PRT, hypoxanthine-guanine phosphoribosyltransferase.

2.9. Adsorbed compounds were eluted from the resin with citrate buffer of pH 3.42. Two radioactive peaks appeared, one preceding and one following the urea peak. The first of these represented unchanged ^{14}C-phenylacetic acid, and generally contained about 30 percent of the counts of the urine sample. The eluates comprising the second peak and containing phenylacetylglutamine (see below) were pooled and concentrated to a small volume by freeze-drying.

AG-2 X 8 Cl-(200-400 mesh) was converted to its acetate form according to Hirs et al [30] and placed in a 25 X 300 mm column. The eluate containing phenylacetylglutamine was reduced to 50 ml volume, brought to pH 7.0 and applied to the column. The resin was washed with 100 ml of 0.5 M acetic acid and then with successive 50 ml volumes of acetic acid of increasing concentrations (1.0 M, 1.5 M, 2.0 M, etc.). ^{14}C-labeled material appeared as a single peak in the 2.5 M acetic acid eluate and was collected in one sample of 100 ml. After freeze-drying, the solid material was extracted with and recrystallized from hot ethyl acetate.

Both the crystalline material thus obtained, and authentic commercial phenylacetylglutamine, gave an Rf of 0.46 on silica gel thin layer chromatography in butanol:acetic acid:water, 4:1:1; and yielded equivalent amounts of ammonia and glutamic acid after hydrolysis in 3 N HCl at 100° for 3 hours [31]. The ammonia derived from the amide-N of glutamine was separated, following alkalinization, by aeration into 1N HCl.

All compounds or fractions which were not already in the form of ammonia were digested by the micro-Kjeldahl procedure and steam-distilled to collect the ammonia.

Samples were prepared for ^{15}N analysis according to Rittenberg [32]. ^{15}N analyses were performed in duplicate on a 180° Consolidated Electrodynamic mass spectrometer, by Dr. L. A. Pogorski, Chemical Projects Limited, Ontario, Canada. ^{15}N values were obtained on uric acid, on glycine isolated from uric acid and representing N-7, and on NH_3 produced by degradation of uric acid and representing N-(1+3+9), N-(1+7), N-(1+3), or N-(7+9). From these six analytical values, ^{15}N enrichment values were calculated for other N-atoms or pairs of atoms, as follows:

$$^{15}N\text{-}1 = 2\left[^{15}N\text{-}\left(\frac{1+7}{2}\right)\right] - \left[^{15}N\text{-}7\right] \tag{1}$$

$$^{15}(N\text{-}9) = \left[2\ ^{15}N\text{-}\left(\frac{7+9}{2}\right)\right] - \left[^{15}N\text{-}7\right] \tag{2}$$

$$^{15}N-(3+9) = 4\left[^{15}N-\left(\frac{1+3+7+9}{4}\right)\right] - 2\left[^{15}N\left(\frac{1+7}{2}\right)\right] \qquad (3)$$

$$^{15}N-3 = {}^{15}N-(3+9) - {}^{15}(N-9) \qquad (4)$$

Subjects. The four gouty subjects were carefully selected to represent different subtypes of primary gout (Table I). S. G. had a 2 year history of acute gouty attacks. He excreted an average of 648 mg of uric acid per day during the study (normal 418 ± 70 mg per day on a comparable diet [33]). He is classified as a mild overexcretor with "idiopathic gout." R.Jo. was a 17-year-old boy who had carried a diagnosis of cerebral palsy until an attack of gout supervened at age 15. Erythrocyte PRT activity was 0.22 n moles of IMP formed/mg protein/hr, a value of less than 0.25 percent of normal. This patient has many features of the Lesch-Nyhan syndrome [34], but shows no self-destructive tendencies [17]. O.G., an Israeli subject, was an extraordinary overproducer with a history of uric acid lithiasis and acute gout. In later studies, both he and a hyperuricemic brother were found to have excessive levels of PP-ribose-P in erythrocytes, and to produce PP-ribose-P at excessive rates [35]. He was recently shown to have an erythrocyte PP-ribose-P synthetase more than four-fold as active as the normal enzyme at concentrations of inorganic phosphate ranging from 0.2 to 0.5 mM but of normal activity above 2 mM [19,20]. T.B. has been the subject of several reports from the N.I.H. [16,36-38]. Although first suspected of having a mutant PP-ribose-P amidotransferase with altered control features [36], he has more recently been found to have >2-fold excessive levels of PP-ribose-P synthetase activity at inorganic phosphate concentration values up to at least 32 mM [21]. Thus the defect in T.B. appears to differ from that of the prototype case of Sperling et al [20]. We are indebted to Dr. J. E. Seegmiller for gifts of ^{15}N-uric acid samples from control subject L.L., and gouty subject T.B. [3,16].

Design of Experiments. Each subject was placed on a standard low purine diet 5 days prior to the oral administration of 65 mg per kg body weight of ^{15}N-glycine, 95 atom percent excess. Urine was collected in consecutive periods of 18, 4, and 2 hours each day throughout the study. At the beginning of the 18th hour each day, 1 gm of [1-^{14}C]-phenylacetic acid (specific activity 2 μCi/gm) was administered orally in 240 ml of water. At the beginning of the 22nd hour each day, 400 mg of sodium benzoate was administered orally in 240 ml of water. Uric acid was isolated from the 18-hour urine samples; phenylacetylglutamine, and ammonia, from the 4-hour samples; and hippurate from the final 2-hour portions. In patient O.G. the experiment was started by administration of 1 gm of phenylacetic acid (2 μCi) followed one hour later by 400 mg of sodium

TABLE I

DATA ON SUBJECTS OF THIS STUDY

Subject	Age	Wt	Uric Acid		Incorporation of Glycine ^{15}N*	Comment
			Serum	Urine		
(males)	Yrs	kg	mg/100 ml	mg/24 hrs	% into urinary uric acid/7 days	
				Non-gouty subjects		
R.H.	27	84	5.4	523	0.12	Normal subject. [28]
F.L.	23	75	5.1	580	0.13	Normal subject.
L.L.	28	70	5.5	388	0.17	Normal subject. [3]
				Gouty subjects		
S.G.	46	86	8.0	648	0.17	Idiopathic gout.
R.Jo.	17	44	15.3	1100	1.02	Partial PRT deficiency. [17]
O.G.	32	90	13.5	2400	3.32	Superactive PP-ribose-P synthe- tase. [19,20,35]
T.B.	48	78	10.3	1242	1.09	Superactive PP-ribose-P synthe- tase. [21,36-38]

*Values have not been corrected for extrarenal disposal [3].

benzoate. ^{15}N-glycine was given one hour after the benzoate and
was followed by two urine collections of one hour each and one
collection of two hours. Second doses of phenylacetic acid and
benzoate (combined) were then given followed by two collection
periods of three and four hours, respectively. Third doses of
phenylacetic acid and benzoate (combined) were followed by one
collection period of 12 hours; this terminated the first day. The
second and subsequent days were started by combined administrations
of phenylacetic acid and benzoate, following which urine was
collected in sequential aliquots of 6 and 18 hours. Uric acid,
hippurate, phenylacetylglutamine and ammonia were isolated from
each of the 1-, 2-, 3-, 4-, 6- and 12-hour collections, and uric
acid only from the 18-hour samples. Two ml aliquots were taken
for collection of ammonia. Uric acid was separated from the rest
of the sample by adsorption on charcoal. Hippuric acid was precipitated from the charcoal filtrate, and phenylacetylglutamine was
then isolated from the filtrate of the hippuric acid crystals.

RESULTS

Incorporation of ^{15}N into Urinary Uric Acid. Time courses of
enrichment of urinary uric acid in the three control and four
gouty subjects are shown in Fig. 1. Cumulative incorporation
values of ^{15}N into urinary uric acid in 7 days are presented in
Table I. The values on S.G., the patient with idiopathic gout,

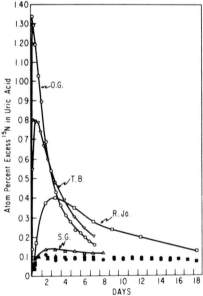

Figure 1. [15]N enrichment of urinary uric acid in three control
(solid symbols) and four gouty subjects.

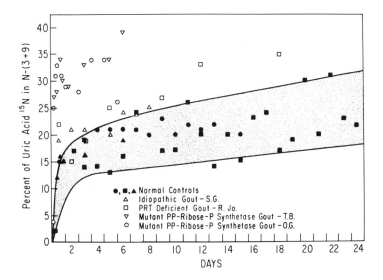

Figure 2.   Intramolecular distribution of ^{15}N in uric acid, expressed as percent of total ^{15}N in uric acid, expressed as percent of total ^{15}N found in N-(3+9). The shaded area includes essentially all values on the three control subjects.

fall at the upper limits of the normal range in both respects [39], but an enlarged miscible pool and enhanced extrarenal disposal of urate may have obscured more decisive evidence for overproduction in this subject [6].   The 3 other gouty patients show unequivocal evidence for extraordinary overincorporation of ^{15}N; and for excessive excretion and flamboyant overproduction of uric acid.

Intramolecular Distribution of ^{15}N in Uric Acid.   Fig. 2 plots the percent of total uric acid-^{15}N in N-(3+9) as a function of time. Within hours an appreciable amount of ^{15}N has appeared in N-(3+9). All of the gouty subjects show elevated percentages of ^{15}N in N-(3+9).   Those of S.G. appear slightly above normal on most days, especially on days 1 and 2, but 2 of 7 values fall within the upper range of control values.   Those of R.Jo. are slightly high initially (though not without some overlap with normal values) and become more strikingly elevated with time.   Those of O.G. and T.B. exceed normal values by a factor of 2 within the first day.

^{15}N Enrichment and Turnover of Metabolic Glycine Pool(s). Small doses of benzoic acid were administered periodically to two controls (R.H. and F.L.) and three gouty subjects (S.G., R.Jo. and O.G.), in order to sample body glycine [28]. Hippurate was iso-lated from urine collected during the next few hours.   The

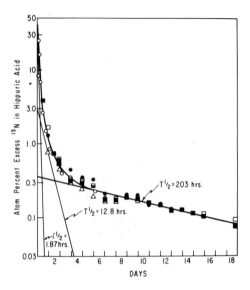

Figure 3. ^{15}N enrichment of urinary hippuric acid in control and gouty subjects (■, R.H.; ●, F.L.; ○, O.G.; □, R.Jo.; ▽, S.G.)

enrichment data (Fig. 3) form a typical polyexponential die-away curve, in which values in gouty and nongouty subjects are indistinguishable.[2]

We conclude, with others [3,5], that excessive incorporation of ^{15}N-glycine into uric acid in gout is not a reflection of abnormal enrichment or turnover of hepatic glycine, but rather that it is indicative of the utilization of a larger than normal fraction of the glycine turnover for purine biosynthesis per unit time [2]. Since even in the most extraordinary overincorporators this fraction is less than 1 percent per day, increased rates of purine biosynthesis do not produce detectable changes in the kinetics of turnover of the glycine pool.

^{15}N Enrichment and Turnover of the Metabolic Glutamine Pool(s). Gutman and Yu [5] originally proposed that the disproportionately

2.   The die-away curves were analyzed by an unweighted least squares method. Parameter variances were estimated from the inverse design matrix defined by a Taylor expansion [40] of the non-linear model and the estimated error variance. Unweighted analysis was selected because measurement errors in the mass spectrometer are constants, approximating ±0.003 atom percent excess.

great enrichment of N-(3+9) in gouty subjects was attributable to "glutamine amide nitrogen containing an unduly high concentration of N^{15}." To evaluate this possibility we have sampled body gluta-mine by the periodic administration of phenylacetic acid.

Normal fasting adult subjects excrete 250 to 500 mg of phenyl-acetylglutamine per day [29]. Amounts are increased following ingestion of phenylacetic acid [41] and in patients with phenyl-ketonuria who produce increased quantities of phenylacetic acid [29]. The reactions concerned have been defined by Moldave and Meister [42] in human liver mitochondria:

$$\text{Phenylacetic acid} + \text{ATP} \rightleftharpoons \text{Phenylacetyl-AMP} + \text{PP}_i \qquad (5)$$

$$\text{Phenylacetyl-AMP} + \text{CoA-SH} \rightleftharpoons \text{Phenylacetyl-CoA} + \text{AMP} \qquad (6)$$

$$\text{Phenylacetyl-CoA} + \text{L-glutamine} \rightleftharpoons \text{Phenylacetylglutamine}$$

$$+ \text{CoA-SH} \qquad (7)$$

In preliminary studies in 3 normal controls and 4 normal excretor patients with idiopathic gout, from 66 to 70 percent of 1 gm of $[1-^{14}\text{C}]$-phenylacetic acid (2 μCi ^{14}C/gm) was excreted in urine in 3 hours, and more than 96 percent in 12 hours. About 70 percent of the ^{14}C in the 3 hour sample was in the form of phenyl-acetylglutamine. The rates of excretion of ^{14}C-phenylacetylgluta-mine were the same in control and gouty subjects. Therefore, 1 gm of phenylacetic acid was administered periodically to sample what is presumably chiefly the hepatic glutamine pool. Urinary phenylacetylglutamine was isolated serially from one control (F.L.) and three gouty subjects (S.G., R.J., and O.G.). These four subjects span a 4-fold range of urinary uric acid excretion and a 25-fold range of ^{15}N-glycine incorporation values (Table I). Immediately following administration of ^{15}N-glycine, enrichment of urinary phenylacetylglutamine was very low. It rose to a maximum at about 3 hours (see below) and then declined rapidly. The declining limbs of the enrichment curves are plotted in Fig. 4. As in the case of urinary hippurate, the enrichment values form a single poly-exponential die-away curve, in which enrichment in the four subjects studied appears to be indistinguishable.

We conclude that the excessive incorporation of ^{15}N into N-(3+9) of uric acid in gouty overproducers is not a reflection of abnormal enrichment or turnover of hepatic glutamine, but that (as in the case of glycine) it is indicative of the utilization of a

larger than normal fraction of the precursor glutamine pool for
purine biosynthesis per unit time.

Precursor-Product Relationships. Following the experimental
labeling of a precursor pool, the concentration of isotope declines
progressively as labeled molecules are replaced by newly synthesized unlabeled compound. The isotope concentration of each
successive product of an unbranced irreversible pathway will achieve
a maximal value when its enrichment equals that of its immediate
precursor [43].

Fig. 5 presents the enrichments of N-7 of urinary uric acid
(which comes from glycine) in comparison with those of hippurate;
Fig. 6 presents the enrichments of N-3 in comparison with those of
the amide-N of phenylacetylglutamine.

In spite of several additional reactions between GAR and
formyl-GAM (the intermediates which first contain the eventual N-7
and N-3 atoms, respectively) and the ultimate product, uric acid,
the classic isotopic relationships of precursor and immediate product are closely approximated: each enrichment peak is found in
reasonably close proximity to the value of hippurate or amide-N of
glutamine at the corresponding time.

In addition, the more rapid the rate of incorporation of ^{15}N
into uric acid (Table I, Fig. 1) the earlier and higher the peak

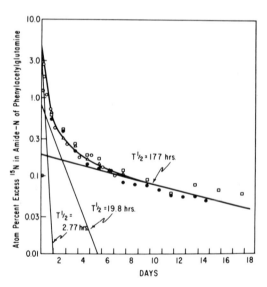

Figure 4. ^{15}N enrichment of amide-N of urinary phenylacetylglutamine in control and gouty subjects. (●, F.L.; 0, O.G.;☐, R.Jo.;
Δ, S.G.).

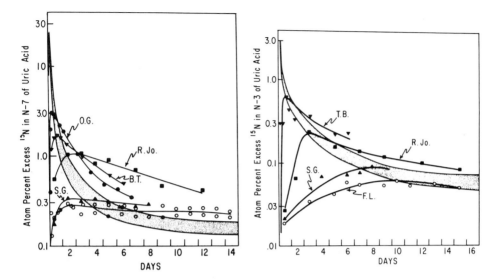

Figure 5. Time course of ^{15}N enrichment of N-7 of uric acid compared with the range of ^{15}N enrichment values of urinary hippuric
acid. (Open circles represent data on control subjects R.H. and
F.L.)

Figure 6. Time course of ^{15}N enrichment of N-3 of uric acid compared with range of ^{15}N enrichment values of the amide-N of
urinary phenylacetylglutamine.

_____

enrichment value is found in both N-7 and N-3. This result requires
that the increase in rate of conversion of precursor to product is
not balanced by a proportionate expansion of the diluting pool of
product, for commensurate expansion of product pool size would
lead to an enrichment curve that was independent of rate of synthesis or turnover of product.

   A computer model has been constructed simulating the kinetics
of the reaction sequence of Fig. 7. The isotopic enrichment of
glycine is represented by the equation for the polyexponential die-
away curve of hippurate shown in Fig. 3. That of the amide-N of
glutamine is represented by a first order product curve derived
from the enrichment of phenylacetylglutamine, which starts at
zero, reaches a maximum at 3 hours, and thereafter closely approxi-
mates the die-away curve of Fig. 4. The model also includes
an arbitrary convention which accelerates a late reaction of the
sequence (IMP → hypoxanthine) as a higher power function of [PP-
ribose-P], in order to achieve a ratio of >1 in the expression
(increase in rate of synthesis of β-phosphoribosylamine)/(increase

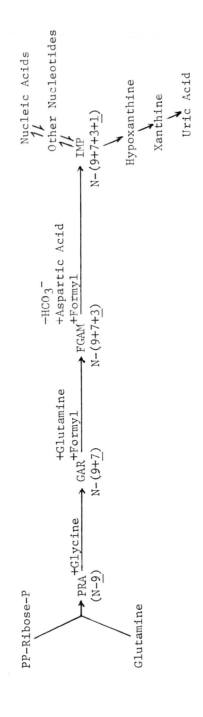

Figure 7. Reaction sequence showing order of incorporation of nitrogen atoms into purine ring.

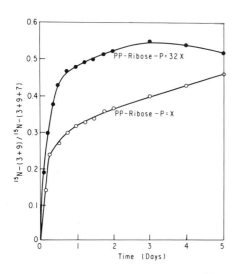
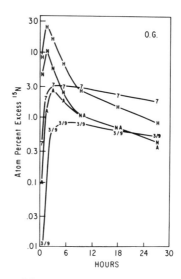

Figure 8.   Enrichment ratios ^{15}N-(3+9)/^{15}N-(3+9+7) predicted by simulation model at two levels of PP-ribose-P.   Compare with observed values in Fig. 2, and note that the simulation model omits ^{15}N-1 values from the denominator.   X = 0.015625.

Figure 9.   Time courses of ^{15}N enrichment of various nitrogenous products of urine in subject O.G. (H = hippuric acid, N = ammonia, A - amide-N of glutamine, 7 = N-7 of uric acid, 3/9 = $\frac{N-(3+9)}{2}$ of uric acid.

of product pool size).   With this model it is possible to show that an increase in concentration of PP-ribose-P will lead to a substantial increase in the ratio ^{15}N-3/^{15}N-7, or ^{15}N-(3+9)/^{15}N-(3+9+7) (see Fig. 8) within the first few hours following administration of ^{15}N-glycine.   Similar results are obtained if increased synthesis of uric acid is predicated upon an increase in concentration of glutamine, or an increase in activity of the amidotransferase, provided that compensatory expansion of product pool size is prevented by some convention in each case.   This model will be presented in detail in another publication [44].

Sequence of Enrichment of Urinary Products.   The study on the remarkable overproducer, O.G., offered an opportunity to observe the sequences of enrichments of various urinary constitutents (Fig. 9).   The earliest peaks were found in hippurate and ammonia at 1.5 hours, in agreement with findings by Wu and Bishop [45] in a normal subject.   Maximal enrichment of the amide-N of phenyl-acetylglutamine occurred at 3 hours.   Peak enrichment of N-7 was found at 3 hours, and N-(3+9) at 5.5 hours.   Glycine, the immediate

precursor of hippurate and of N-7, had presumably reached its
enrichment maximum somewhat earlier, perhaps at about 1 hour [45].
Glutamine probably reached its maximal enrichment sometime past 1.5
hours, based on the time of peak enrichment of urinary ammonia, and
the presumed operation of the glutamine synthetase reaction. The
source of the urinary ammonia highly labeled so early is probably
glycine itself. In the first hour, the enrichment of urinary
ammonia is nearly 50 times greater than that of the amide-N of
phenylacetylglutamine, whereas that of hippurate is less than twice
that of ammonia (Fig. 9). Enrichment values of ammonia declined
to equal those of the amide-N of phenylacetylglutamine in the 7-12
hour sample, and the two values were much the same thereafter
throughout the 7 days of the study. The results in other subjects
were similar.

DISCUSSION

In this paper we demonstrate disproportionately increased
labeling of N-(3+9) of uric acid from ^{15}N glycine in patients with
rare subtypes of primary gout attributable to specific enzymatic
defects unrelated to glutamine metabolism. We postulate that the
changes in ^{15}N-(3+9)/^{15}N-uric acid ratios observed with accelerated
purine production are related to 1) different time courses of
enrichment of precursor glycine and glutamine pools which are
labeled consecutively rather than simultaneously and have different
turnover characteristics; and 2) kinetic factors which reflect the
differences of early enrichment patterns of these precursor pools
most strikingly in the most rapid producers, irrespective of the
precise molecular cause of purine overproduction.

The first specific step of purine biosynthesis de novo involves
the following reaction catalyzed by glutamine phosphoribosylpyrophosphate amidotransferase [46]:

$$\alpha\text{-PP-ribose-P} + \text{glutamine} + H_2O \xrightarrow{Mg^{2+}} \beta\text{-phosphoribosylamine}$$

$$+ \text{ glutamic acid} + PP_i \qquad\qquad (8)$$

This reaction appears to be the slow and rate-controlling step
of the purine biosynthetic pathway. The amidotransferase is subject to synergistic allosteric feedback inhibition by adenyl and
guanyl ribonucleotides [46,47]. The molecular basis of this control
mechanism has recently been elucidated in our laboratory. Purine
ribonucleotides convert the human enzyme from a form with a
molecular weight of 133,000, to a larger form which has a molecular
weight of 270,000 and is enzymatically inactive. PP-ribose-P

converts the larger form of the enzyme to the smaller active form. Glutamine has no effect upon this association-dissociation reaction [48].

Increases in rates of purine production could in theory result from increased concentrations of the substrates PP-ribose-P or glutamine, increased activity of the amidotransferase, or decreased availability or non-optimal ratios of feedback inhibitors [39,46].

The Km value of human amidotransferase for PP-ribose-P is 0.25 mM [47]. The intracellular concentrations of PP-ribose-P are 0.005 to 0.05 mM (erythrocytes and fibroblasts) [49-51]. Thus, intracellular concentrations of PP-ribose-P are in the low range of the substrate-velocity curve, and factors which affect their intracellular, perhaps chiefly intrahepatic, concentrations will have a corresponding influence upon purine biosynthesis de novo. The amidotransferase reaction displays sigmoidal kinetics with respect to PP-ribose-P in the presence of nucleotide regulators such as AMP and GMP [47]. Small increases in concentrations of PP-ribose-P may therefore affect purine biosynthesis disproportionately. Indeed, in isolated systems such as Ehrlich ascites cells [52] or cultured human fibroblasts [49,53], manipulations of the concentrations of PP-ribose-P profoundly affect the rate of the first steps of purine biosynthesis.

Concentration values of PP-ribose-P are normal in erythrocytes urate excretion values are in the normal range [47,49,51]. These subjects are representative of 70 to 75 percent of the gouty population [3,33]. However, PP-ribose-P values are significantly elevated in erythrocytes of patients with partial [49] or virtually complete [49,50] deficiencies of PRT activity, and in patients with superactive PP-ribose-P synthetases [20,21], as well as in some gouty patients with idiopathic gout and excessive urate excretion values [36,54,55].

In an earlier study from this laboratory on 9 subjects with idiopathic gout, an increased labeling of the ribose moiety of urinary imidazoleacetic acid ribonucleoside (which is derived from PP-ribose-P) was found in 3 overproducers following administration of [U-^{14}C]-glucose, but not in 6 normal excretors [56]. These data strongly suggested overproduction of PP-ribose-P from glucose in the 3 gouty overproducer subjects. The kinetics of the glutamine PP-ribose-P amidotransferase reaction are such that small increases of PP-ribose-P concentration could contribute to the lesser degrees of purine overproduction believed to be present in many patients with idiopathic gout whose urate excretions fall within the upper segments of the normal range. The critical values to be determined in patients with all forms of gout are those of PP-ribose-P

concentration or production rate in liver, on which no data are as
yet available.

Plasma levels of glutamine are 0.5 to 0.7 mM in both normal and
gouty subjects [13,14,57]. These values are below the glutamine Km
value of 1.6 mM for human placental [47] or lymphoblast [58] amidotransferase. The rate of purine biosynthesis in cultured fibroblasts can be reduced by glutamine starvation [59], for these cells
do not have the highly developed enzymatic machinery for synthesis
of glutamine present in liver. We have been unable to find data
in the literature on concentrations of glutamine in human liver.
However, values of free glutamine in the protein free extracts of
quick frozen liver of fed rats [60-62], range from 2.15 to 5.03
μmoles/gm wet weight, or about 3 to 7 mM in cell water. The Km for
glutamine of PP-ribose-P amido-transferase of rat liver is <1 mM
[63]. Therefore, in the fed rat this enzyme of the purine pathway
may be nearly saturated with respect to glutamine.

If the human enzyme is not saturated with glutamine, variations
in glutamine concentrations would be expected to affect the rate of
purine biosynthesis, although less exquisitely than variations in
concentrations of PP-ribose-P. The kinetics of the amidotransferase
reaction are hyperbolic with respect to glutamine at all levels of
PP-ribose-P, in the presence or absence of nucleotide inhibitors
[47,63]. Purine ribonucleotide inhibition is non-competitive with
respect to glutamine [47,63] and glutamine does not reverse the
association of amidotransferase subunits caused by ribonucleotides
[48].

The three elements of the hypothesis of abnormal glutamine
metabolism in primary (idiopathic) gout are: 1) disproportionate
labeling of N-(3+9) in urate overproducers following ingestion of
^{15}N-glycine [5,8], 2) reduced production of urinary NH_3 at a given
acid load [11,12], and 3) hyperglutamatemia [13,14] which appears
to be independent of dietary protein level [13]. The disproportionate labeling of N-(3+9) has now been shown to be a general
feature of accelerated purine biosynthesis, and therefore this
finding does not specifically favor a defect of glutamine metabolism
in primary gout. The reduced production of urinary NH_3 in gout was
initially attributed to a defect in glutaminase activity [6,10], but
this postulated mechanism appears to have been excluded by normal
renal glutaminase assay values in vitro in 4 gouty subjects [64].
However, Pitts [65] has cited possible disparities between glutaminase activities as assessed in vitro and in vivo, and a "functional
deficit" of glutaminase activity in gout has been suggested by
Gutman and Yu [11]. Others have considered the reduced NH_3 production of gouty subjects a consequence of renal injury. The hyperglutamatemia [13,14] has led to the proposal that surplus glutamate may drive purine biosynthesis via glutamine in certain subjects.

However, even if there should be a defect of glutamate or of glutamine metabolism, such a metabolic error could not directly account
for the isotopic data, for an increase in concentration of free
glutamine in liver would result in greater dilution of newly
synthesized labeled glutamine and reduced rather than increased
labeling of N-3 and N-9 relative to N-7 or total uric acid.

The computer-simulated kinetic model offers additional support
for our interpretation that the increased fractional labeling of
N-3 and N-9 of uric acid in gouty overproducers fed ^{15}N-glycine is
a kinetic phenomenon related to the different time-courses of
enrichment of precursor glycine and glutamine pools, and to changes
in the patterns of incorporation of ^{15}N into intermediary products
which contribute to the purine ring, as rates of purine production
are increased. The results of this paper indicate that the disproportionate labeling of N-(3+9) following ingestion of ^{15}N-glycine
does not selectively favor a specific defect of glutamate or glutamine metabolism in primary idiopathic gout.

Summary. An abnormality of glutamine metabolism in primary
gout was first proposed on the basis of isotope data: when ^{15}N-
glycine was administered to gouty subjects, there was disproportionately great enrichment of N-(3+9) of uric acid which derive
from the amide-N of glutamine. An unduly high concentration of
^{15}N in glutamine was postulated, and attributed to a defect in
catabolism of glutamine. Excess glutamine was proposed as the
driving force of uric acid overproduction.

We have reexamined this proposition in four gouty subjects:
One mild overproducer of uric acid with "idiopathic gout"; one
marked overproducer with high grade but "partial" hypoxanthine-
guanine phosphoribosyltransferase deficiency; and two extraordinary
overproducers with superactive phosphoribosylpyrophosphate synthetases. In the last three, the driving force of excessive purine
biosynthesis is a known surplus of PP-ribose-P. Disproportionately
high labeling of N-(3+9) was present in all four gouty subjects,
most marked in the most flamboyant overproducers. The precursor
glycine pool was sampled by periodic administration of benzoic acid
and isolation of urinary hippuric acid. Similarly, the precursor
glutamine pool was sampled by periodic administration of phenylacetic
acid and isolation of the amide-N of urinary phenylacetylglutamine.
The time course of ^{15}N-enrichment of hippurate differed from that
of the amide-N of glutamine. Whereas initial enrichment values of
hippurate were very high, those of glutamine-amide-N were low,
increasing to a maximum at about 3 hours, and then declining less
rapidly than those of hippurate. However, enrichment values of
hippurate and of glutamine were normal in all of the gouty subjects
studied. Thus, preferential enrichment of N-(3+9) in gouty overproducers given ^{15}N-glycine does not necessarily reflect a specific

abnormality of glutamine metabolism, but rather appears to be a kinetic phenomenon associated with accelerated purine biosynthesis per se.

REFERENCES

1. Benedict, J. D., T.-F. Yu, E. J. Bien, A. B. Gutman, and DeW. Stetten, Jr. 1953. A further study of the utilization of dietary glycine nitrogen for uric acid synthesis in gout. J. Clin. Invest. 32:775.

2. Gutman, A. B., T.-F. Yu, H. Black, R. S. Yalow, and S. A. Berson. 1958. Incorporation of glycine 1-C^{14} and glycine-N^{15} into uric acid in normal and gouty subjects. Amer. J. Med. 25:917.

3. Seegmiller, J. E., A. I. Grayzel, L. Laster, and L. Liddle. 1961. Uric acid production in gout. J. Clin. Invest. 40:1304.

4. Gutman, A. B., T.-F. Yu, M. Adler, and N. B. Javitt. 1962. Intramolecular distribution of uric acid-N^{15} after administration of glycine-N^{15} and ammonium N^{15} chloride to gouty and nongouty subjects. J. Clin. Invest. 41:623.

5. Gutman, A. B., and T.-F. Yu. 1963. An abnormality of glutamine metabolism in primary gout. Amer. J. Med. 35:820.

6. Sonne, J. C., I. Lin, and J. M. Buchanan. 1956. Biosynthesis of the purines. IX. Precursors of the nitrogen atoms of the purine ring. J. Biol. Chem. 220:369.

7. Levenberg, B., S. C. Hartman, and J. M. Buchanan. 1956. Biosynthesis of the purines. X. Further studies in vitro on the metabolic origin of nitrogen atoms 1 and 3 of the purine ring. J. Biol. Chem. 220:379.

8. Gutman, A. B., and T.-F. Yu. 1963. On the nature of the inborn metabolic error(s) of primary gout. Trans. Assoc. Amer. Physicians 76:141.

9. Pitts, R. F., L. A. Pilkington, and J. C. M. DeHaas. 1965. ^{15}N tracer studies on the origin of urinary ammonia in the acidotic dog, with notes on the enzymatic synthesis of labeled glutamic acid and glutamine. J. Clin. Invest. 44:731.

10. Stone, W. J., S. Balagura, and R. F. Pitts. 1967. Diffusion equilibrium for ammonia in the kidney of the acidotic dog. J. Clin. Invest. 46:1603.

11. Gutman, A. B., and T.-F. Yu. 1965. Urinary ammonium excretion in primary gout. J. Clin. Invest. 44:1474.

12. Sperling, O., M. Frank, and A. de Vries. 1966. L'excretion d'ammoniac au cours de la goutte. Rev. Franc. Etud. Clin. Biol. 11:401.

13. Yu, T.-F., M. Adler, E. Bobrow, and A. B. Gutman. 1969. Plasma and urinary amino acids in primary gout, with special reference to glutamine. J. Clin. Invest. 48:885.

14. Pagliara, A. S., and A. D. Goodman. 1969. Elevation of plasma glutamate in gout, its possible role in the pathogenesis of hyperuricemia. New Eng. J. Med. 281:767.

15. Kelley, W. N., F. M. Rosenbloom, J. E. Seegmiller, and R. R. Howell. 1968. Excessive production of uric acid in type I glycogen storage disease. J. Pediat. 72:488.

16. Kelley, W. N., M. L. Greene, F. M. Rosenbloom, J. F. Henderson, and J. E. Seegmiller. 1969. Hypoxanthine-guanine phosphoribosyl-transferase deficiency in gout. Ann. Intern. Med. 70:155.

17. Emmerson, B. T., and J. B. Wyngaarden. 1969. Purine metabolism in heterozygous carriers of hypoxanthine guanine phosphoribosyltransferase deficiency. Science. 166:1533.

18. Sperling, O., M. Frank. R. Ophir, U. A. Liberman, A. Adam, and A. de Vries. 1970. Partial deficiency of hypoxanthine-guanine phosphoribosyltransferase associated with gout and uric acid lithiasis. Rev. Europ. Etud. Clin. Biol. 15:942.

19. Sperling, O., P. Boer, S. Persky-Brosh, E. Kanarek, and A. de Vries. 1972. Altered kinetic property of erythrocyte phosphoribosylpyrophosphate synthetase in excessive purine production. Europ. J. Clin. Biol. Res. 17:703.

20. Sperling, O., S. Persky-Brosh, P. Boer, and A. de Vries. 1973. Human erythrocyte phosphoribosylpyrophosphate synthetase mutationally altered in regulatory properties. Biochem. Med. In press.

21. Becker, M. A., L. J. Meyer, A. W. Wood, and J. E. Seegmiller. 1973. Purine overproduction in man associated with increased phosphoribosylpyrophosphate synthetase activity. Science 179:1123.

22. Schoenheimer, R., and S. Ratner. 1939. Studies in protein metabolism. III. Synthesis of amino acids containing isotopic nitrogen. J. Biol. Chem. 127:301.

23. Praetorius, E. 1949. An enzymatic method for the determination of uric acid by ultraviolet spectrophotometry. Scand. J. Clin. Lab. Invest. 1:222.

24. Shemin, D., and D. Rittenberg. 1947. On the utilization of glycine for uric acid synthesis in man. J. Biol. Chem. 167: 875.

25. Brandenberger, H. 1954. The oxidation of uric acid to oxonic acid (allantoxanic acid) and its application in tracer studies of uric acid biosynthesis. Biochim. Biophys. Acta. 15:108.

26. Sonne, J. C., I. Lin, and J. M. Buchanan. 1956. Biosynthesis of the purines. IX. Precursors of the nitrogen atoms of the purine ring. J. Biol. Chem. 220:369.

27. Quick, A. J. 1940. Clinical application of hippuric acid and prothrombin tests. Amer. J. Clin. Path. 10:222.

28. Howell, R. R., M. Speas, and J. B. Wyngaarden. 1961. A quantitative study of recycling of isotope from glycine-1-C^{14}, α-N^{15} into various subunits of the uric acid molecule in a normal subject. J. Clin. Invest. 40:2076.

29. Stein, W. H. 1953. A chromatographic investigation of the amino acid constituents of normal urine. J. Biol. Chem. 201:45.

30. Hirs, C. H., S. Moore, and W. H. Stein. 1954. The chromatography of amino acids on ion exchange resins. Use of volatile acids for elution. J. Amer. Chem. Soc. 76:6063.

31. Prescott, B. A., and H. Waelsch. 1947. Free and combined glutamic acid in human blood plasma and serum. J. Biol. Chem. 167:855.

32. Rittenberg, D. 1946. Preparation of gas samples for mass spectrographic isotope analysis. In Preparation and Measurement of Isotopic Tracers. D. W. Wilson, A. O. C. Nier, and S. P. Reimann, editors. Edwards, Ann Arbor, 31.

33. Gutman, A. B., and T.-F. Yu. 1957. Renal function in gout with a commentary on the renal regulation of urate excretion, and the role of the kidney in the pathogenesis of gout. Amer. J. Med. 23:600.

34. Lesch, M., and W. L. Nyhan. 1964. A familial disorder of uric acid metabolism and central nervous function. Amer. J. Med. 36:561.

35. Sperling, O., G. Eilam, S. Persky-Brosh, and A. de Vries. 1972. Accelerated erythrocyte 5-phosphoribosyl-1-pyrophosphate synthesis. A familial abnormality associated with excessive uric acid production and gout. Biochem. Med. 6:310.

36. Henderson, J. F., F. M. Rosenbloom, W. N. Kelley, and J. E. Seegmiller. 1968. Variations in purine metabolism of cultured skin fibroblasts from patients with gout. J. Clin. Invest. 47:1511.

37. Seegmiller, J. E., J. R. Klinenberg, J. Miller, and R. W. E. Watts. 1968. Suppression of glycine-N^{15} incorporation into urinary uric acid by adenine-8-C^{13} in normal and gouty subjects. J. Clin. Invest. 47:1193.

38. Kelley, W. N., F. M. Rosenbloom, and J. E. Seegmiller. 1967. The effect of azathioprine (Imuran) on purine synthesis in clinical disorders of purine metabolism. J. Clin. Invest. 46:1518.

39. Wyngaarden, J. B. 1966. Gout. In The Metabolic Basis of Inherited Disease, 2nd ed. J. B. Stanbury, J. B. Wyngaarden, and D. S. Fredrickson, editors. New York, McGraw-Hill, p. 667.

40. Draper, N. R., and H. Smith. 1966. Applied Regression Analysis. New York, John Wiley, Chp. 10.

41. Sherwin, C. P., M. Wolf, and W. Wolf. 1919. The maximum production of glutamine by the human body as measured by the output of phenylacetylglutamine. J. Biol. Chem. 37:113.

42. Moldave, K., and A. Meister. 1957. Synthesis of phenylacetylglutamine by human tissue. J. Biol. Chem. 229:463.

43. Zilversmit, D. B., C. Entenman, and M. C. Fishler. 1943. On the calculation of "turnover time" and "turnover rate" from experiments involving the use of labeling agents. J. Gen. Physiol. 26:325.

44. Starmer, C. F., O. Sperling, and J. B. Wyngaarden. A kinetic model for uric acid labeling in primary gout. In preparation.

45. Wu, H., and C. W. Bishop. 1959. Pattern of N^{15}- excretion in man following administration of N^{15}-labeled glycine. J. Appl. Physiol. 14:1.

46. Wyngaarden, J. B. 1972. Glutamine phosphoribosylpyrophosphate amidotransferase. In Current Topics in Cellular Regulation. B. Horecker and E. R. Stadtman, editors. New York, Academic Press, Vol. V, p. 135.

47. Holmes, E. W., J. A. McDonald, J. M. McCord, J. B. Wyngaarden, and W. N. Kelley. 1973. Human glutamine phosphoribosylpyro-phosphate amidotransferase: Kinetic and regulatory properties. J. Biol. Chem. 248:144.

48. Holmes, E. W., J. B. Wyngaarden and W. N. Kelley. 1973. Human glutamine phosphoribosylpyrophosphate amidotransferase: Two molecular forms interconvertible by purine ribonucleotides and phosphoribosylpyrophosphate. J. Biol. Chem. (In press.)

49. Greene, M. L., and J. E. Seegmiller. 1969. Elevated erythro-cyte phosphoribosylpyrophosphate in X-linked uric aciduria: Importance of PRPP concentration in the regulation of human purine biosynthesis. J. Clin. Invest. 48:32a.

50. Fox, I., and W. N. Kelley. 1971. Phosphoribosylpyrophosphate in man: Biochemical and clinical significance. Ann. Intern. Med. 74:424.

51. Fox, I. H., J. B. Wyngaarden, and W. N. Kelley. 1970. Deple-tion of erythrocyte phosphoribosylpyrophosphate in man, a newly observed effect of allopurinol. New Eng. J. Med. 283:1177.

52. Henderson, J. F., and M. K. Y. Khoo. 1965. Synthesis of 5-phosphoribosyl-1-pyrophosphate from glucose in Ehrlich ascites tumor cells in vitro. J. Biol. Chem. 240:2349.

53. Kelley, W. N., I. H. Fox, and J. B. Wyngaarden. 1970. Regu-lation of purine biosynthesis in cultured human cells. I. Effects of orotic acid. Biochim. Biophys. Acta. 215:512.

54. Hershko, A., C. Hershko, and J. Mager. 1968. Increased forma-tion of 5-phosphoribosyl-1-pyrophosphate in red blood cells of some gouty patients. Israel Med. J. 4:939.

55. Sperling, O., R. Ophir, and A. de Vries. 1971. Purine base incorporation into erythrocyte nucleotides and erythrocyte phosphoribosyltransferase activity in primary gout. Rev. Europ. Etudes Clin. et Biol. 16:147.

56. Jones, O. W., Jr., D. M. Ashton, and J. B. Wyngaarden. 1962. Accelerated turnover of phosphoribosylpyrophosphate, a purine nucleotide precursor, in certain gouty subjects. J. Clin. Invest. 41:1805.

57. Segal, S., and J. B. Wyngaarden. 1955. Plasma glutamine and oxypurine content in patients with gout. Proc. Soc. Exp. Biol. Med. 88:342.

58. Wood, A. W. and Seegmiller, J. E. 1973. Properties of 5-phosphoribosyl-1-pyrophosphate amidotransferase from human lymphoblasts. J. Biol. Chem. 248:138.

59. Raivio, K. O., and J. E. Seegmiller. 1971. Role of glutamine in purine synthesis and interconversion. Clin. Res. 19:161.

60. Williamson, D. H., O. Lopes-Vieira, and B. Walker. 1967. Concentrations of free glucogenic amino acids in livers of rats subjected to various metabolic stresses. Biochem. J. 104:497.

61. Kennan, A. L. 1962. Glutamine synthesis in rats with diabetic acidosis. Endocrinology. 71:203.

62. Bergmeyer, H. U. 1970. Methoden der enzymatischen Analyse, Verlag-Chemie. Weinheim/Bergstrasse. Book II. p. 2206.

63. Caskey, C. T., D. M. Ashton, and J. B. Wyngaarden. 1964. The enzymology of feedback inhibition of glutamine phosphoribosyl-pyrophosphate amidotransferase by purine ribonucleotides. J. Biol. Chem. 239:2570.

64. Pollak, V. E., and H. Mattenheimer. 1965. Glutaminase activity in the kidney in gout. J. Lab. Clin. Med. 66:564.

65. Pitts, R. F. 1964. Renal production and excretion of ammonia. Amer. J. Med. 36:720.

ACKNOWLEDGEMENTS

We wish to thank Dr. R. Rodney Howell for participation in the study of R.H. (1961); Dr. Bryan Emmerson for participation in the study of R.Jo. (1966); Dr. Joseph Hollander for permission to study S.G. (1966); Dr. J. Edward Seegmiller for generously contributing uric acid samples from earlier studies on control subject L.L., and gouty patient T.B. (1967); and Mrs. M. Evans and Mr. Z. Weidenfeld for excellent technical assistance.

These investigations were supported in part by Grants AM-10301, AM-12413 and 1-K 4-HL-70102 from the U. S. Public Health Service.

HYPERGLUTAMATEMIA IN PRIMARY GOUT

ALEXANDER B. GUTMAN* AND TS'AI-FAN YÜ   *DECEASED

MOUNT SINAI SCHOOL OF MEDICINE, CITY UNIVERSITY

OF NEW YORK, NEW YORK 10029

IN THE PRESENT STUDY WE FOUND THAT THE PLASMA GLUTAMATE IN PATIENTS WITH PRIMARY GOUT WAS SIGNIFICANTLY HIGHER THAN THAT IN THE NONGOUTY SUBJECTS. (TABLE I) IN 62 GOUTY SUBJECTS WITH PLASMA URATE >8.0 MG%, THE PLASMA GLUTAMATE WAS 80 ± 27 μM/L AS COMPARED TO 53 ± 16 μM/L IN 28 NORMAL CONTROLS. THERE WAS APPRECIABLE OVERLAP IN THE TWO GROUPS, BUT THE LEVEL IN 2/3 OF THE GOUTY SUBJECTS EXCEEDED THE NORMAL MEAN BY $+1$ S.D., AND IN 1/3 EXCEEDED THE NORMAL MEAN BY $+2$ S.D. IN 12 DISTINCTLY HYPERURICEMIC, BUT ASYMPTOMATIC YOUNG SONS OF OVERTLY GOUTY SUBJECTS, THE MEAN PLASMA GLUTAMATE LEVEL WAS 79 ± 14 μM/L. IN 24 SIMILAR BUT NORMOURICEMIC YOUNG SONS, THE MEAN PLASMA GLUTAMATE LEVEL WAS SIGNIFICANTLY LOWER, 56 ± 15 μM/L, NOT DISTINGUISHABLE FROM THE NORMAL CONTROLS. THE DATA INDICATE THAT HYPERGLUTAMATEMIA IS NOT ONLY A FREQUENT ACCOMPANIMENT OF HYPERURICEMIA IN PRIMARY GOUT, BUT APPEARS EARLY IN THE COURSE OF THE DISEASE, BEFORE THE ONSET OF SYMPTOMS AND APPARENTLY IN CONJUNCTION WITH THE ONSET OF HYPERURICEMIA AT ABOUT THE TIME OF PUBERTY.

KINETIC STUDIES BY PREVIOUS WORKERS SUGGESTED IMPERMEABILITY OF THE ERYTHROCYTES TO GLUTAMATE. SUBSEQUENT STUDIES BY OTHERS INDICATED THAT GLUTAMATE CONCENTRATIONS IN THE RED CELLS OF THE DOGS WERE CONSIDERABLY HIGHER THAN THAT IN PLASMA. IN NORMAL MAN, THE ERYTHROCYTE TO PLASMA RATIOS OF GLUTAMATE CONCENTRATIONS SIGNIFICANTLY EXCEEDED UNITY, INDICATING A CONCENTRATION GRADIENT ACROSS THE CELL MEMBRANE. OUR PRESENT STUDY SHOWED THAT THE ERYTHROCYTE GLUTAMATE WAS 216 ± 91 μM/KG WATER IN 5 NORMAL MEN. THE ERYTHROCYTE GLUTAMATE CONCENTRATION IN THE GOUTY MEN EITHER DISTINCTLY OR MILDLY HYPERURICEMIC WERE RESPECTIVELY 243 ± 31 AND 214 ± 67 μM/KG WATER. THUS IT SEEMS THAT THERE IS NO DEFECT FOR

TABLE 1
PLASMA AND ERYTHROCYTE GLUTAMATE IN PRIMARY GOUT, GOUTY OFFSPRING AND NONGOUTY SUBJECTS

	Nongouty		Gouty		Hyperuricemic Offspring		Normouricemic Offspring	
	No.	∿M/L	No.	∿M/L	No.	∿M/L	No.	∿M/L
Plasma	28	53+16	62	80+27	12	79+14	24	56+15
Erythrocyte	5	216+91	10	243+31*	(*P urate > 8.0 mg%)			
			7	214+67**	(**P urate 7.2 mg%)			

TABLE 2

EFFECT OF L GLUTAMIC ACID LOADING ON PLASMA GLUTAMIC ACID

Hours after Loading	Nongouty	Gouty	Hyperuricemic Offspring	Normouricemic Offspring
	(6)	(16)	(6)	(6)
	∿M/L	∿M/L	∿M/L	∿M/L
0	53 + 12	83 + 30	80 + 7	45 + 12
1	110 + 22	124 + 67	214 + 28	101 + 29
2	150 + 59	283 + 141	334 + 99	178 + 48
3	146 + 30	238 + 128	241 + 102	133 + 49

TABLE 3

EFFECT OF GLUTAMATE LOAD ON

PLASMA GLUTAMINE, ALANINE, GLYCINE, THREONINE AND SERINE

		0 HR	1 HR	2 HR	3 HR
		$\mu M/L$	$\mu M/L$	$\mu M/L$	$\mu M/L$
GLUTAMINE	NG	506+60	531+48	520+57	544+55
	G	530+79	536+61	547+62	521+64
ALANINE	NG	307+98	333+89	347+84	337+97
	G	348+88	367+77	383+72	360+47
GLYCINE	NG	234+53	247+61	206+40	222+42
	G	244+47	241+35	235+37	212+38
THREONINE	NG	131+16	147+23	136+29	131+21
	G	122+29	128+28	123+29	112+28
SERINE	NG	97+11	111+15	114+12	115+14
	G	97+21	101+30	102+20	103+22

TABLE 4

EFFECT OF GLUTAMATE LOAD

ON URINARY EXCRETION OF GLUTAMATE

HOUR	$0°$	$1°$	$2°$	$3°$
	NM/MIN	NM/MIN	NM/MIN	NM/MIN
NONGOUTY	24 + 8	23 + 9	22 + 10	18 + 8
GOUTY	28 + 19	27 + 15	33 + 16	47 + 38

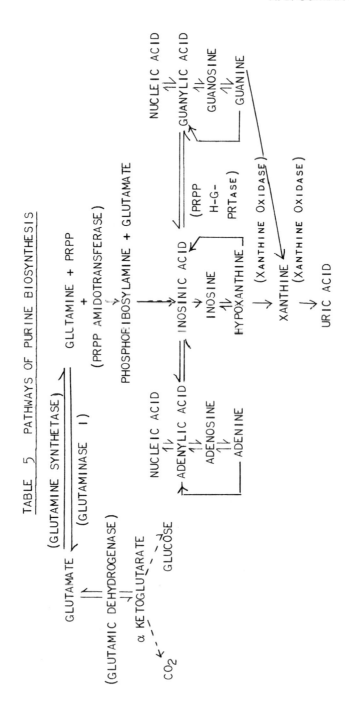

TABLE 5   PATHWAYS OF PURINE BIOSYNTHESIS

GLUTAMATE TRANSFER ACROSS THE RED CELL MEMBRANE IN PRIMARY GOUT.

IN ORAL LOADING AFTER GIVING L GLUTAMIC ACID 1.3 MM/KG BODY WEIGHT IN 16 GOUTY SUBJECTS, 6 HYPERURICEMIC AND 6 NORMOURICEMIC GOUTY OFFSPRINGS AND 5 NONGOUTY CONTROLS, WE FOUND DISPROPORTIONATE ACCENTUATION IN HYPERGLUTAMATEMIA IN THE HYPERURICEMIC SUBJECTS. (TABLE 2) IN BOTH THE GOUTY AND THE NONGOUTY SUBJECTS THE PEAK INCREMENTS IN PLASMA GLUTAMATE WERE ATTAINED IN THE SECOND HOUR AFTER THE LOADING. THE MEAN PEAK PLASMA GLUTAMATE REACHED 283 ± 141 ᴜM/L IN THE GOUTY, WHILE IT WAS 150 ± 59 ᴜM/L IN THE NONGOUTY. THUS THE MEAN PEAK PLASMA GLUTAMATE INCREMENT IN THE GOUTY WAS ABOUT TWICE THAT IN THE CONTROLS. A SIMILAR RESPONSE WAS NOTED IN 6 HYPERURICEMIC BUT ASYMPTOMATIC YOUNG SONS OF THE GOUTY SUBJECTS. THE PLASMA GLUTAMATE CONCENTRATION WAS INCREASED FROM 80 ± 7 ᴜM/L TO A PEAK OF 334 ± 99 ᴜM/L IN THE SECOND HOUR. THE RESULTS WERE COMPARABLE TO THOSE IN THE GOUTY SUBJECTS. FOR THE 6 NORMOURICEMIC YOUNG SONS, THE PLASMA GLUTAMATE CONCENTRATIONS WERE FROM 45 ± 12 TO 178 ± 48 ᴜM/L, COMPARABLE TO THAT OF THE NONGOUTY SUBJECTS.

(TABLE 3) NO SIGNIFICANT INCREASES IN PLASMA GLUTAMINE, ALANINE AND OTHER AMINO ACIDS WERE OBSERVED IN BOTH GOUTY AND NONGOUTY SUBJECTS. (TABLE 4) THE INCREASED PLASMA GLUTAMATE CONCENTRATIONS WERE NOT ACCOMPANIED BY ANY SIGNIFICANT INCREASE IN THE URINARY GLUTAMIC ACID. IN FACT, THE AMOUNT WAS NEGLIGIBLE IN THE URINE UNDER THE CONTROL AS WELL AS AFTER THE LOADING, INDICATING EFFICIENT RENAL TUBULAR REABSORPTION OF THE FILTERED GLUTAMATE.

OUR DATA INDICATE THE HIGH PLASMA GLUTAMATE IN GOUT IS NOT DUE TO ANY BLOCK IN THE ENTRY OF GLUTAMATE INTO THE CELLS. TUBULAR REABSORPTION OF THE FILTERED GLUTAMATE APPEARS TO BE AS RAPID AND COMPLETE IN GOUT AS IN NONGOUT. ABSORPTION OF GLUTAMATE IN THE INTESTINAL TRACT IS NOT DELAYED NOR ACCELERATED IN GOUT. THE DISPROPORTIONATE RISE IN PLASMA GLUTAMATE AFTER ORAL LOADING IS MOST LIKELY A REFLECTION OF DISTURBED GLUTAMATE CATABOLISM, POSSIBLY AS A RESULT OF DIMINISHED GLUTAMIC DEHYDROGENASE ACTIVITY, AS SUGGESTED BY PAGLIARA AND GOODMAN. THE REDUCED ACTIVITY OF GLUTAMIC ACID DEHYDROGENASE PRESUMABLY WOULD MAKE MORE INTRACELLULAR GLUTAMATE AVAILABLE FOR CONVERSION TO GLUTAMINE BY GLUTAMINE SYNTHETASE, A PRECURSOR IN THE RATE LIMITING REACTION OF DE NOVO PURINE BIOSYNETHESIS. (TABLE 5)

REDUCED RENAL AMMONIAGENESIS IN PRIMARY GOUT

ALEXANDER B. GUTMAN* AND TS'AI-FAN YÜ   *DECEASED

MOUNT SINAI SCHOOL OF MEDICINE, CITY UNIVERSITY

OF NEW YORK, NEW YORK   10029

THE RATE OF NH_3 DIFFUSION IS INFLUENCED BY (1) THE TRANS-
TUBULAR PH GRADIENT, (2) THE AVAILABILITY OF TITRATABLE ACID PRE-
CURSORS, AND (3) THE URINE FLOW RATES. IN OUR SUBJECTS FOR THE
PRESENT STUDY OF URINARY EXCRETION OF TITRATABLE ACID, NH_4^+, AND
URIC ACID, URINE SAMPLES WERE COLLECTED AS A RULE FROM 6:30 A. M.
TO 8:30 A. M. AFTER AN OVERNIGHT FAST, BOTH THE GOUTY AND THE NON-
GOUTY SUBJECTS ALIKE. THE URINE FLOW RATES WERE MORE OR LESS IN
THE SAME RANGE, APPROXIMATELY 1.0 ± 0.3 ML/MIN.

(TABLE 1) THE URINARY NH_4^+ AND TITRATABLE ACID IN 119 PRI-
MARY GOUT AND 73 NONGOUTY SUBJECTS ARE LISTED HERE. NONE OF THE
SUBJECTS WITH GOUT HAD HISTORY OF RENAL CALCULI. ABOUT 2/3 OF THE
119 PRIMARY GOUT HAD URINE PH IN THE RANGE 4.8 - 5.3, DISTINCTLY
MORE ACID THAN THE URINE PH OF THE NONGOUTY CONTROLS, AND YET THERE
IS SIGNIFICANT REDUCTION IN NH_4^+ OUTPUT WITH NO DIFFERENCE IN TI-
TRATABLE ACIDITIES.

(TABLE 2) UNDOUBTEDLY, MEASUREMENTS OF RENAL FUNCTIONS ARE
REDUCED WITH ADVANCING YEARS, EVEN IN THE ABSENCE OF OVERT RENAL
DISEASE. TABLE 2 SHOWS THE RESULTS OF THE STUDY IN 38 GOUTY AND
19 NONGOUTY SUBJECTS, NONE EXCEEDED 40 YEARS OF AGE. ALL HAD RE-
LATIVELY LOWER URINE PH. THE CREATININE AND TOTAL NITROGEN EX-
CRETIONS WERE ALSO QUITE COMPARABLE FOR BOTH GROUPS. IN 19 NON-
GOUTY SUBJECTS, THE MEAN PLASMA URATE WAS 5.5 ± 0.8 MG%, AND THE
MEAN URINE PH WAS 5.4 ± 0.2, TITRATABLE ACID WAS 24.8 ± 6.1 μEQ/
MIN, NH_4^+ WAS 35.0 ± 7.3 μEQ/MIN, URINARY URIC ACID WAS 0.5 MG $\pm$
0.11 MG/MIN, CREATININE 1.46 MG/MIN, AND TOTAL NITROGEN 10.5 MG/
MIN.

TABLE I

URINARY TITRATABLE ACIDITIES AND NH_4^+ EXCRETION

IN 119 PRIMARY GOUT AND 73 NONGOUTY SUBJECTS

	NONGOUTY				GOUTY			
	No	%	TA	NH_4^+	No	%	TA	NH_4^+
			µEQ/MIN	µEQ/MIN			µEQ/MIN	µEQ/MIN
4.8-5.0	11	15	25.8±4.0	31.2±5.1	35	39	26.4±7.6	24.3±6.3
5.1-5.3	25	24	26.1±5.7	33.2±8.2	45	38	26.1±5.8	26.8±8.6
5.4-5.6	18	25	23.5±9.0	32.5±7.5	23	19	23.1±5.5	26.2±6.7
5.7-5.9	8	11	17.9±2.3	25.5±5.9	11	9	21.6±7.7	23.2±5.2
6.0-6.7	11	15	12.3±3.9	27.8±6.3	5	4	13.3±3.2	15.2±4.7

OF THE 38 GOUTY SUBJECTS, 14 HAD A MEAN PLASMA URATE OF 8.0 + 0.6 MG%. THE URINE PH, TA, NH_4^+ AND UA VALUES WERE NOT DIFFERENT FROM THE CORRESPONDING VALUES FOR THE NONGOUTY CONTROLS. IN 10 OTHER GOUTY SUBJECTS, WHOSE MEAN PLASMA URATE WAS 9.6 + 0.4 MG%, THERE WAS A SIGNIFICANT LOWERING IN NH_4^+ EXCRETION, AND IT WAS FURTHER DECREASED WHEN THE MEAN PLASMA URATE OF ANOTHER GROUP OF 14 GOUTY SUBJECTS WERE UP TO 11.7 + 1.3 MG%. AS THE URINARY NH_4^+ EXCRETION BECAME STEADILY LOWER, THE URINARY URIC ACID WAS STEADILY HIGHER. (TABLE 3) THE TOTAL METABOLIC ACID WAS LOWER IN THE GOUTY, PARTICULARLY WHEN THE PLASMA URATE BECAME HIGHER, NH_4^+/TA RATIOS CORRESPONDINGLY FELL FROM 1.5 + 0.5 TO 1.0 + 0.2. AS THE NH_4^+/TN RATIOS FELL, AN INVERSE RELATIONSHIP IN UA-N/TN WAS NOTED. TA WAS NOT DIFFERENT IN VARIOUS GROUPS.

(TABLE 4) THE RESPONSE TO ACID LOADING IN THE GOUTY AND NONGOUTY SUBJECTS LIKEWISE WAS DIFFERENT. SUBJECTS IN BOTH GROUPS WERE ON SIMILAR DIETS. NH_4 CL WAS GIVEN ORALLY AT DOSAGES 8 GM THE FIRST DAY, AND 4 GM EACH DAY FOR 3 SUBSEQUENT DAYS. TWENTY-FOUR HOUR URINES WERE COLLECTED. THE FOLLOWING VALUES REPRESENTED THE MEANS OF DAYS 3 AND 4. THE MEAN AGE OF THE 6 NORMOURICEMIC NONGOUTY SUBJECTS WAS 38, AND THAT OF THE 19 HYPERURICEMIC GOUTY SUBJECTS 41. IN THE NONGOUTY SUBJECTS, NH_4^+ EXCRETION WAS 24.9

TABLE 2

URINARY TITRATABLE ACIDITIES, NH_4^+ AND URIC ACID EXCRETION

IN 38 GOUTY AND 19 NONGOUTY SUBJECTS

(AGE NOT MORE THAN 40)

SUBJECTS	No.	P URATE MG%	PH	TA µEQ/MIN	NH_4^+ µEQ/MIN	NH_4^+ MG/MIN	UA MG/MIN
Nongouty	19	5.5±0.8	5.4±0.2	24.8±6.1	35.0±7.3	0.49±0.10	0.50±0.11
Gouty	14	8.0±0.6	5.4±0.3	26.7±9.0	32.3±7.9	0.45±0.11	0.61±0.20
	10	9.6±0.4	5.3±0.2	25.4±3.7	27.3±4.4	0.38±0.06	0.61±0.13
	14	11.7±1.3	5.2±0.3	25.5±6.2	25.4±6.8	0.36±0.10	0.67±0.19

TABLE 3

TOTAL ACID, NH_4^+/TA, NH_4^+/TN AND UA-N/TN

IN 38 GOUTY AND 19 NONGOUTY SUBJECTS

SUBJECTS	NO.	TA + NH_4^+ μEQ/MIN	NH_4^+/TA	NH_4^+/TN %	UA-N/TN %
NONGOUTY	19	60.0+10.0	1.50+0.52	4.7+1.2	1.6+0.3
GOUTY	14	59.0+15.8	1.30+0.45	4.2+0.8	1.8+0.6
	10	52.7+ 5.6	1.10+0.25	4.1+0.9	2.2+0.6
	14	50.9+11.6	1.01+0.23	3.8+0.9	2.4+0.7

TABLE 4

RESPONSE OF ACID LOADING IN THE GOUTY AND NONGOUTY SUBJECTS

		URINE PH	NH_4^+ MEQ/GM CR	TA MEQ/GM CR	NH_4^+ + TA MEQ/GM CR	NH_4^+/TA
NONGOUTY	A.	5.5	24.9+ 4.9	21.0+ 3.3	45.9	1.2
(6)	B.	4.8	47.7+10.1	31.5+ 5.4	79.2	1.5
GOUTY	A.	5.3	19.0+ 4.6	21.1+ 4.9	40.1	0.9
(19)	B.	4.8	34.2+ 6.1	27.7+ 3.9	61.9	1.2

A. CONTROL

B. AFTER ACID LOADING.

+ 4.9 MEQ AND TITRATABLE ACID 21.0 + 3.3 MEQ/GM CREATININE WITH A CONTROL PH 5.5. THE MEAN NH_4^+/TA RATIO WAS 1.2 AND TOTAL ACID EX-CRETION WAS 45.9 MEQ/GM CREATININE. AFTER NH_4 CL LOADING URINE PH DECREASED TO 4.8, NH_4^+ INCREASED TO 47.7 + 10.1 MEQ/GM CRE-ATININE AND TA 31.5 + 5.4 MEQ/GM CREATININE. TOTAL CREATININE EX-CRETION WAS PRACTICALLY UNCHANGED, BUT TOTAL NITROGEN WAS INCREAS-ED FROM 13.3 TO 15.6 GM PER DAY. IN THE 19 HYPERURICEMIC GOUTY SUBJECTS, THE INITIAL URINE PH WAS 5.3, AND AFTER NH_4 CL LOADING, THE URINE PH WAS DECREASED TO 4.8, NOT DIFFERENT FROM THAT OF THE CONTROL SUBJECTS. ALTHOUGH THE URINE PH RANGES WERE QUITE COM-PARABLE, THE NH_4^+ EXCRETIONS WERE SIGNIFICANTLY LESS. IT WAS 19.0 + 4.6 MEQ/GM CREATININE BEFORE NH_4 CL LOADING, AND INCREASED TO 34.2 + 6.1 AFTER THE LOADING. BOTH VALUES WERE SIGNIFICANTLY LOWER THAN THOSE IN THE NONGOUTY CONTROLS. TITRATABLE ACIDITIES ON THE OTHER HAND WERE QUITE SIMILAR COMPARING WITH THOSE OF THE NONGOUTY CONTROLS. NH_4^+/TA RATIOS WERE 0.9 AND 1.2, AND TOTAL ACIDITIES WERE 40.1 AND 61.9 MEQ BEFORE AND AFTER NH_4 CL LOADING RESPECTIVELY.

FALLS ALSO OBSERVED A LOW NH_4^+/TA RATIOS IN THE GOUTY SUB-JECTS, PARTICULARLY WHEN COMBINED WITH A LOW PO_4 DIET TO REDUCE TITRATABLE ACID. APPARENTLY THE GOUTY SUBJECTS AFTER ACID LOAD-ING SHOWED A SIGNIFICANT INCREASE IN TA AND TOTAL ACID, WITH ONLY A MINIMAL INCREASE IN NH_4^+ EXCRETION.

THE DATA INDICATE THAT THE SELECTIVE IMPAIRMENT IN RENAL PRO-DUCTION AND EXCRETION OF NH_4^+ IN MANY PATIENTS WITH PRIMARY GOUT COULD NOT BE RELATED TO RENAL INSUFFICIENCY, AGE OF THE PATIENTS, OR DURATION OF THE MANIFEST GOUT. WHEN THERE IS DISTINCT REDUCT-ION OF RENAL AMMONIAGENESIS, THERE IS AN INDICATION OF INCREASED URIC ACID EXCRETION, AS DEMONSTRATED BY OUR EARLIER WORK NOT ONLY IN THE TOTAL URIC ACID-^{15}N ENRICHMENT BUT ALSO THE DISPROPORTION-ATE INTRAMOLECULAR DISTRIBUTION OF URIC ACID-^{15}N IN N-9 AND N-3. THE REGULATION OF RENAL NH_3 PRODUCTION FROM GLUTAMINE, CHIEFLY THE AMIDE N, INVOLVES MANY FACTORS, OF WHICH THE MOST IMMEDIATE IS RENAL GLUTAMINASE I ACTIVITY. ENZYME ASSAYS APPEAR TO RULE OUT REPRESSION OF GLUTAMINASE I SYNTHESIS IN PRIMARY GOUT, OR SYN-THESIS OF A CATALYTICALLY LESS ACTIVE ENZYME PRODUCTION. HOWEVER, SUCH IN VITRO ASSAYS DO NOT EXCLUDE THE POSSIBILITY OF SOMEWHAT REDUCED IN VIVO ACTIVITY. GLUTAMATE IS ALSO A POTENT INHIBITOR OF RENAL GLUTAMINASE I.

THE UPTAKE OF GLYCINE - ^{14}C INTO THE ADENINE AND GUANINE OF DNA AND INSOLUBLE RNA OF HUMAN LEUCOCYTES

G-N. Chang, A. Fam, A.H. Little and A. Malkin

Departments of Clinical Biochemistry and Medicine

Sunnybrook Hospital, University of Toronto, Toronto, Canada

SUMMARY

Kaplan et al. (1) have recently reported that glycine-^{14}C uptake into the adenine and guanine of DNA and 'insoluble' RNA of human leucocytes was found to be significantly greater in patients with primary gout than in normal controls. These observations have been confirmed in this investigation and extended to include other subjects. It was noted that patients with asymptomatic hyperuricemia showed a normal glycine uptake in this in vitro system. On the other hand, in patients with chronic renal disease and secondary hyperuricemia, the glycine uptake was found to be enhanced, indicating that excessive purine formation may contribute to the hyperuricemia which accompanies chronic renal failure. In addition, leucocytes from normal females showed an increased glycine uptake as compared with those obtained from normal males, despite the presence of significantly lower serum uric acid levels in the female subjects. Furthermore, while a positive but insignificant correlation was observed between the glycine uptake and serum uric acid in the normal males studied (r=+0.577), the correlation was negative in the females (r=-0.585; P < 0.05). These results indicate that there may be feedback mechanisms controlling purine synthesis in females that may not be present in males. The in vitro findings in the patients with primary gout appear to parallel the observations made by earlier investigators on glycine incorporation into uric acid in the intact subject. It is suggested, therefore, that the isolated human leucocyte may serve as a useful and convenient model for studying some of the manifold aspects of purine metabolism in man. Nevertheless, additional studies will be required in order to confirm that overall changes in purine metabolism in a variety of clinical situations is accurately reflected in the simple intact cell system described in this investigation.

INTRODUCTION

In 1969, Diamond, Friedland, Halberstam and Kaplan (1), in attempting to develop a model for the study of purine metabolism in vitro in an intact human cell system, chose to study leucocytes because of the ease with which they can be obtained and their relative resistance to non-physiological conditions (2). Glycine-^{14}C incorporation into the adenine and guanine of DNA and 'insoluble' RNA* in leucocytes was found to be greater in gouty subjects with hyperuricemia than normouricemic controls. We undertook to confirm the findings resulting from these investigations, to extend them to include patients with asymptomatic and secondary hyperuricemia and to compare a normal male control group with normal females.

METHODS

The techniques used to study glycine incorporation into human leucocytes were those described by Diamond et al (1), with minor modifications.

White blood cell suspensions were prepared as follows: Three tubes of blood were collected in heparinized vacutainers (100 X 16 mm; Becton, Dickinson and Co.). A 20 ml aliquot of the heparinized blood was transferred into a 25 ml graduated cylinder together with 5 ml of 5% dextran (M.W. 200,000 to 275,000) in 0.85% sodium chloride. The mixture was allowed to stand at 4°C for 40 to 45 minutes. At the end of this period, the supernatant was transferred to a 50 ml round-bottom tube and centrifuged at 50 g for 15 minutes at 4°C. The supernatant was discarded and the cell pellet was re-suspended in 5 ml of phosphate buffer pH 7.40 (3). Ten ml of distilled water was added, to facilitate red cell lysis, and the mixture centrifuged as before. The supernatant was discarded and the cell pellet resuspended in approximately 2 ml phosphate buffer. Duplicate cell counts were made, and the final white blood cell concentration adjusted to 15 X 10^6 cells/ml with phosphate buffer.

A 2 ml aliquot of the leucocyte suspension was then added to a test-tube (1.5 X 11 cm.), pre-loaded with the glycine[+] substrate (1 ml containing 2μCi of uniformly labelled glycine-^{14}C and 6.6 mgm

---

* 'insoluble' RNA refers to the RNA which is insoluble in 2% perchloric acid.

[+] glycine-^{14}C was obtained from Schwarz/Mann. Specific activity of the material was 104-106 mCi/mmole and specific concentration 50 μCi/ml.

glucose in phosphate buffer). 100% oxygen was bubbled through the
mixture for 20 seconds. The mixture was then incubated at 37°C
for 4 hours, with gentle shaking, in a Dubnoff metabolic
shaking incubator (Precision Scientific Co.). At the end of
incubation, the tube was placed in ice-water to inhibit further
reaction.

Adenine and guanine were extracted from the insoluble nucleic
acids by the following modification of the technique described by
Cooper and Rubin (4) and Diamond et al(1):

Following incubation, the sample was centrifuged at 600 g for
10 minutes at 4°C. The supernatant was discarded, and the cells
washed with 10 ml of cold phosphate buffer. After recentrifugation
for 10 minutes, the supernatant was removed and 5 mls of fixative
containing absolute ethyl alcohol-glacial acetic acid (3:1) added
to the cell pellet. After 10 minutes at room temperature, the
sample was recentrifuged, 10 ml of 70% ethanol was added to the
cell pellet and the mixture stored at 4°C overnight. The following
day, the cells were recovered by centrifugation, the supernatant
discarded and the pellet suspended in 5 ml of 2% perchloric acid
for 50 minutes at 4°C. The mixture was recentrifuged for 5 minutes,
the supernatant removed, and the cells washed with 10 mls of 2%
perchloric acid. 0.5 ml of 10% perchloric acid was then added to
the cells and the mixture heated at 65°C for 3 hours. Following
centrifugation, 0.4 ml of the supernatant was now transferred to
a pyrex tube (17 X 150 mm), 0.4 ml of 70% perchloric acid added
and the tube placed in a boiling water bath for 2 hours. The
solution was desalted by neutralizing with 0.45 ml of 8.4 N
potassium hydroxide, the mixture centrifuged and the supernatant
saved for quantitation.

Adenine and guanine (0.05 mg of each) were added to the desalted
extract, and 100 microliters (measured in a disposable Clay-Adams
micropet) spotted as a band on a strip (1½" X 19") of Whatman No.1
chromatography paper. Five such strips were prepared for each
sample. The strips were developed by descending chromatography
using an isopropyl alcohol - concentrated hydrochloric acid-water
media (130:33:37) (5). After drying, adenine and guanine bands
were located under a UV lamp, the bands cut out and each placed
vertically in a polyvinyl counting vial. About 15 ml of Insta-Gel
(Packard Co.) was then added. Each sample was counted for 50
minutes in a Packard Tri-Carb liquid scintillation counter. The
activities of adenine and guanine in the five strips was summated
and the results expressed as total dpm of adenine and guanine/30 X
10^6 white blood cells.

Serum uric acids were determined by the standard Autoanalyzer
method.

CLINICAL MATERIAL

Normal Subjects

Nine normal males were included in the study. Their serum uric acid levels fell in the range of 5.0-7.2 mgm/100ml (Table 1), while the range for the 13 normal females was found to be 3.5-5.9 mgm/100ml (Table 4). These values were within the normal limits established for our laboratory.

TABLE 1

GLYCINE UPTAKE INTO DNA AND RNA OF HUMAN LEUCOCYTES

	NORMAL MALES			GOUT (all males)	
	Uric Acid (mg%)	Glycine Uptake *		Uric Acid (mg%)	Glycine Uptake *
JS	6.4	653	JP	5.9	542
DR	6.2	714	SG	12.0	1056
RH	5.7	742	FS	6.8	1153
VA	5.0	836	JB	7.0	1161
JT	6.9	851	FM	6.8	1170
PB	6.2	866	JM	7.5	1526
BK	6.7	876	AA	7.7	1618
GC	6.9	1050	WP	8.9	1631
TF	7.2	1424	AH	8.7	2349
			JR	5.8	3014
MEAN $\pm$ SD	6.4 $\pm$ 0.7	890 $\pm$ 230		7.7 $\pm$ 1.8	1522 $\pm$ 707

* Expressed as DPM/30x10^6 WBC/4 hrs. incubation
Gout vs. Normal Males: $P < 0.025$

Gouty Arthritis

The 10 patients with gout (Table 1) had episodes of hyper-uricemia in association with a typical history of recurrent acute arthritis and/or demonstration of monosodium urate monohydrate crystals on synovianalysis or examination of tophaceous deposits. Patients J.P., A.H. and J.R. were on allopurinol at the time their leucocytes were taken for investigation. Patient J.M. was on phenylbutazone for 4 days before his studies were done. All the other patients in this group were off medication for 3 days prior to investigation.

Asymptomatic Hyperuricemia

The 14 patients with asymptomatic hyperuricemia all had serum uric acid levels over 7.8 mgm/100ml (Table 2). Half the subjects were normal hospital staff and the remainder were orthopedic patients. With the exception of one or two patients who were on small doses of salicylates, there was no clear explanation for the elevated serum uric acids in this group.

TABLE 2

GLYCINE UPTAKE INTO DNA AND RNA OF HUMAN LEUCOCYTES

NORMAL MALES			ASYMPTOMATIC HYPERURICEMIA (all males)		
	Uric Acid (mg%)	Glycine Uptake *		Uric Acid (mg%)	Glycine Uptake *
JS	6.4	653	JJ	8.1	477
DR	6.2	714	JM	10.5	672
RH	5.7	742	CB	8.9	713
VA	5.0	836	DE	8.2	728
JT	6.9	851	LS	8.4	736
PB	6.2	866	EL	10.4	827
BK	6.7	876	GH	8.3	863
GC	6.9	1050	HL	8.5	881
TF	7.2	1424	JW	7.9	971
			JG	9.6	1021
			AF	9.2	1077
			JS	8.6	1193
			RM	9.4	1262
			JS	8.7	2314
MEAN	6.4	890		8.9	981
+ SD	+ 0.7	+ 230		+ 0.8	+ 439

* Expressed as DPM/30x10⁶ WBC/4 hrs. incubation
Asymptomatic Hyperuricemia vs Normal Males: N.S.

Chronic Renal Failure

The majority of the 14 patients in the chronic renal failure group were on a hemodialysis program. Serum uric acids were generally elevated, although normal values were occasionally observed following treatment (Table 3).

TABLE 3

GLYCINE UPTAKE INTO DNA AND RNA OF HUMAN LEUCOCYTES

NORMAL MALES			CHRONIC RENAL FAILURE (all males)		
	Uric Acid (mg%)	Glycine Uptake *		Uric Acid (mg%)	Glycine Uptake *
JS	6.4	653	LW	9.8	936
DR	6.2	714	EH	11.5	938
RH	5.7	742	HB	6.0	1206
VA	5.0	836	GS	8.7	1318
JT	6.9	851	RJ	12.0	1865
PB	6.2	866	GM	5.3	1945
BK	6.7	876	JS	7.8	1976
GC	6.9	1050	EM	8.6	2072
TF	7.2	1424	JD	11.6	2137
			RM	8.2	2179
			WB	11.3	2575
			GR	6.2	2763
			RH	11.1	2784
			JC	12.0	4476
MEAN ± SD	6.4 ± 0.7	890 ± 230		9.3 ± 2.3	2084 ± 920

* Expressed as $DPM/30 \times 10^6$ WBC/4 hr. incubation
Chronic Renal Failure vs Normal Males: $P < 0.005$

RESULTS

The results obtained in the gouty subjects are summarized in Table 1. While these patients showed considerable variation in the uptake of glycine into the adenine and guanine of DNA and insoluble RNA of their leucocytes, on the whole the uptake was greater than normal, the difference being highly significant ($P < 0.025$). On the other hand, while the patients with asymptomatic hyperuricemia were obviously comprised of a very heterogeneous group, their glycine uptake showed no significant difference from normal (Table 2).

In patients with chronic renal failure with secondary hyper-uricemia the elevated serum uric acid is generally attributed to a diminished glomerular filtration. Surprisingly, the glycine uptake was found to be appreciably increased ($P < 0.005$) (Table 3).

The results in normal females were also rather unexpected. It is well-known that females have lower serum uric acid levels than males (6). In our subjects the difference was very highly significant ($P < 0.0005$). Therefore, one might have anticipated a priori a lower glycine uptake than in males. However, the reverse was true; glycine uptake into adenine and guanine of DNA and insoluble RNA in leucocytes in females is considerably greater than in males ($P < 0.005$) (Table 4).

In the normal males, a positive correlation was observed between serum uric acid levels and glycine uptake ($r=+0.557$; N.S.). In the normal females, the opposite was noted ($r=-0.585$; $P < 0.05$). In the other groups studied, no particular correlation between these two parameters of purine metabolism could be found.

TABLE 4

GLYCINE UPTAKE INTO DNA AND RNA OF HUMAN LEUCOCYTES

NORMAL MALES			NORMAL FEMALES		
	Uric Acid (mg%)	Glycine Uptake *		Uric Acid (mg%)	Glycine Uptake *
JS	6.4	653	TH	5.4	869
DR	6.2	714	HE	5.9	871
RH	5.7	742	JJ	5.3	998
VA	5.0	836	BW	5.4	1010
JT	6.9	851	MO	4.5	1055
PD	6.2	866	SF	5.6	1483
BK	6.7	876	KB	4.4	1594
GC	6.9	1050	LT	4.9	1749
TF	7.2	1424	JP	3.5	1852
			SG	5.1	2202
			BB	4.2	2510
			SH	4.6	2526
			CC	4.4	2768
MEAN + SD	6.4 ±0.7	890 + 230		4.9 + 0.7	1653 + 680

* Expressed as DPM/30x10⁶ WBC/4 hr. incubation
   Normal Males vs Normal Females: $P < 0.005$
 Males: Uric Acid vs Glycine Uptake: $r=+0.557$; N.S.
 Females: Uric Acid vs Glycine Uptake: $r=-0.585$; $P < 0.05$

DISCUSSION

It has long been established that glycine incorporation into uric acid, in vivo, is enhanced in certain patients with primary or idiopathic gout (7,8). Since increased glycine incorporation into uric acid precursors can also be demonstrated in vitro in the leucocytes of patients with primary gout, a readily accessible model may be available for the study of purine metabolism in these patients, and other groups as well.

Considerable variation in the uptake of glycine was noted in our gouty patients. Diamond's results are similar (1). This finding is compatible with observations made in vivo (8,9), that only about two-thirds of patients with gouty arthritis show enhanced incorporation of glycine into uric acid.

The hyperuricemia of chronic renal failure is generally attributed to a diminution of glomerular filtration. Since glycine uptake into the adenine and guanine of DNA and insoluble RNA in leucocytes in these patients is increased, excessive synthesis of purines appears to be a factor contributing to the hyperuricemia. In this connection, it is interesting to note that in the early in vivo studies on glycine incorporation into uric acid (8), the two patients with azotemia due to chronic renal disease who were included in the investigation showed an enhanced incorporation of labelled glycine into their urinary uric acid.

Since young females are known to have lower serum uric acids than young males (6), it was paradoxical to find that the glycine uptake in leucocytes from young females was much greater than in their male counterparts. When attempts were made to correlate glycine uptake with levels of serum uric acid, the situation became even more complicated. In normal males, in the small group investigated, while the correlation between glycine uptake and serum uric acid was not quite significant, it was, nevertheless, positive ($r=+0.557$). In the female subjects, the correlation was both negative and significant ($r=-0.585$; $P<0.05$). In other words, it would appear at first glance that there are feedback mechanisms controlling purine synthesis in leucocytes from human females which may not be present in males. The nature of these mechanisms are, at the moment, rather obscure. In addition, while the urinary urate clearance is evidently the same in the two sexes (10), the overall handling of purine metabolites may nevertheless be quite different.

The group of patients with asymptomatic hyperuricemia showed no difference in glycine uptake as compared to the normal subjects, although, like the patients with gout, the variability was rather marked. Nevertheless, all of the patients in this group, with one exception, had glycine uptakes that fell within the range noted for the group of normal males. Here again, as with the normal females,

our knowledge of factors controlling purine metabolism is still
rather incomplete.

One difficulty inherent in these investigations is the fact
that while care was taken to ensure that the total number of leucocytes was the same in each incubation, differential cell counts
were not done, and it is possible that there may have been some
differences in the proportions of the different types of leucocytes
in the groups of individuals investigated. Selection of a homo-
geneous population of leucocytes might have yielded somewhat different results. Matching for cell age as well as cell type would
not seem possible, however, and the differences observed could
just as easily be due to this factor. In this connection, it may
be noted that while increased leucocyte turnover has not been
demonstrated in patients with primary gout, there is evidence that
increased platelet turnover does occur in this condition (11).

In patients with chronic renal failure, the bone marrow is
generally hypercellular with a shift to the left - as in toxic
conditions. The neutrophilic leucocytosis in the peripheral blood
is accompanied by a shift to the right in the granulocytic series
to include many segmented neutrophils (12). Whether this would
account for the increased glycine incorporation into purines in
leucocytes of patients with chronic renal failure is entirely
speculative.

The fact that glycine uptake in leucocytes in patients with
primary gout is excessive is merely symptomatic of the disease,
just as the hyperuricemia is an epiphenomenon. The factor(s)
leading to enhanced glycine uptake remain completely unresolved
just as the reasons for the increased incorporation of glycine into
uric acid in vivo in most cases of primary gout remain obscure.
It is suggested that human leucocytes may lend themselves very well
to the in vitro study of purine metabolism in man. However,
additional fundamental information should be obtained in order to
be certain of the adequacy of the model. For example, normal
excretors of uric acid may be expected to have either normal or
enhanced glycine uptakes into their leucocytes. Hyperexcretors,
on the other hand, should show only an increased incorporation of
glycine into their leucocytic purines (8,9). Similarly, the effects
of allopurinol in inhibiting purine synthesis (13) should be reflected
in a diminished glycine uptake in the isolated leucocytes of
individuals taking this medication. These questions are now under
active investigation.

ACKNOWLEDGEMENTS
      This work was supported by the Canadian Arthritis and
Rheumatism Society, Grant # 7-152-71. We should like to express
our thanks to Miss Barbara Bruser for her invaluable technical
assistance.

REFERENCES:

1. Diamond, H.S., Friedland, M., Halberstam, D. and Kaplan, D. Ann. Rheum. Dis. 28:275,1969.

2. Fallon, H.J., Frei, E. III, Davidson, J.D., Trier, J.S., and Burk, D.   J. Lab. Clin. Med. 59:779,1962.

3. Munroe, J.F. and Shipp, J.C.   Diabetes 14:584,1965.

4. Cooper, H.L. and Rubin, A.   Blood. 25:1014,1965.

5. Thomson, R.Y.   Purines and pyrimidines and their derivatives. In Chromatographic and Electrophoretic Techniques.   Smith, I. editor.   Heinemann, London. 3rd edition. 1:295,1969.

6. Gutman, A.B. and Yü, T-F.   N.Engl.J.Med. 273:252,1965.

7. Benedict, J.D., Roche, M., Yü, T-F., Bien, E.J., Gutman, A.B. and Stetten, D., Jr.   Metabolism 1:3,1952.

8. Gutman, A.B., Yü, T-F., Black, H., Yalow, R.S. and Berson,S.A. Amer.J.Med. 25:917,1958.

9. Seegmiller, J.E., Grayzel, A.I., Laster, L. and Liddle, L. J.Clin.Invest. 40:1304,1961.

10. Gutman, A.B. and Yü, T-F.   Amer.J.Med. 23:600,1957.

11. Mustard, J.F., Murphy, E.A., Ogryzlo, M.A. and Smythe, H. Can.Med.Ass.J. 89:1207,1963.

12. Callen, I.R. and Limarzi, L.R.   Amer.J.Clin.Path. 20:3,1950.

13. Fox,I.H., Wyngaarden, J.B. and Kelley, W.N.   N.Engl.J.Med. 283:1177,1970.

Effects of Diet, Weight, and Stress on Purine Metabolism

IN CERTAIN PHYSIOLOGICAL STRESS CONDITIONS
ON THE PROBLEM OF HYPERURICEMIA

I. Machtey and A. Meer

Rheumatology Clinic and Department of Medicine

Hasharon Hospital, Petah-Tiqva, Israel

It is well known that diseases, other than gout, may be associated
with elavated serum uric acid levels. Myeloproliferative diseases are
the main examples of such a condition. Certain drugs, especially
diuretics, are also known to produce hyperuricemia and, eventually, an
acute attack of gout. Further-more, attacks of gout have been noted
following stress, as, for example, after surgery. In the acute stage
of myocardial infarction (MI), hyperuricemia may also be found. There
is some discussion about whether this phenomen on is connected with
the stress caused by the MI, or is it due to an extensive tissue
breakdown.
Gout, and, of course, the hyperuricemia which is connected with it, is
apparently a hereditary disease usually affecting only young men and
is extremely rare among young women. That is why we have chosen for
our study young women.

The subjects in our study were mostly student–nurses between the
ages 18 – 25. Fasting blood samples were taken during ovulation,
menstruation and in the period between. In those who were not certain
of their ovulation date, we assumed it to be on the fourteenth day
before the expected menstruation.
In the second group, which included pregnant women aged 18 – 30, blood
samples were taken during labour at the peak of their pains, and in
most cases on the other 3rd day after delivery. All blood samples were
examined by an autoanalizer for uric acid, creatinin and a few other
laboratory parameters. The results are given in the following slides.

TABLE **1**

SERUM URIC ACID

	NUMBER OF CASES	RANGE MG%	MEAN MG%	ST. DEV.
OVULATION	14	3.8–6.4	4.8	0.8
MENSTRUATION	17	3.4–6.0	4.7	0.8
OUTSIDE BOTH PERIODS	23	3.8–6.6	4.8	0.5
TOTAL	54	3.4–6.6	4.8	0.7

In the first table the serum uric acid levels during ovulation, menstruation and between both periods are given. It is obvious that there is no difference in values. Should stress really cause hyperuricemia, than it is debatable, whether ovulation or menstruation should be regarded as inadequate stress or, as suggested by others, the results are masked by estrogens which may lower serum uric acid levels (1).

TABLE 2

SERUM URIC ACID

	NUMBER OF CASES	RANGE MG%	MEAN MG%	ST. DEV.
a) DURING LABOUR	25	5.2–7.6	6.1	0.8
b) 3 DAYS LATER	20	4.4–8.8	5.9	1.0
c) NON–PREGNANT CONTROLS	54	3.4–6.6	4.8	0.7

Significance of difference between a & c:

$t = 7.97 \qquad p < 0.01$

The uric acid levels during labour just before delivery and 3 days after delivery, as compared with those from women who are not pregnant are given in the second table. There is no question, that the women during labour and shortly after delivery had in our series, much higher serum uric acid levels than the controls.

This difference is statistically highly significant ($p < 0.01$).

The real cause of the increase in serum uric acid during labour is undetermined. In another report, published many years ago, the author had found very low serum uric acid levels during the entire pregnancy (2.9 - 3.6 mg%) (2). This values increased during labour to 5 mg%, but returned to previous levels in puerperium. This possibly could exclude an inadequate renal handling of urate. It is therefore an atractive assumption, that the marked stress connected with labour is the possible cause of the hyperuricemia found in the reported cases. It should be pointed out that all pregnant women in our series had no other diseases and especially no hypertension or renal disease. They took no drugs during pregnancy as far as we were able to ascertain it. The serum creatinine values were normal in all cases, with a mean value of 0.7 mg%.

In conclussion, we were unable to find any difference between the serum uric acid levels in young women during menstruation and ovulation as compared with the period between these two points. On the other hand, the uric acid levels were much higher during labour and immediately following delivery as compared with non pregnant women. In view of the strikingly low uric acid levels prevailing during the entire pregnancy, which have been reported earlier, there is a distinct possibility that the elevated serum uric acid during labour is largly due to the physical and emotional stress connected with it. We are fully aware of the necessity for further large scale studies in-order to ascertain this point.

References

1. Nicholls A. et al. Brit. Med. J. 1:449, 1973.

2. Steenstrup O.R. Scandinav. J. Clin, Lab. Investig. 8:263, 1956.

URIC ACID METABOLISM FOLLOWING ACUTE MYOCARDIAL INFARCTION

J.A. Dosman, J.C. Crawhall and G.A. Klassen

Divisions of Cardiology and Clinical Biochemistry, McGill University Clinic, Royal Victoria
Hospital, Montreal, Quebec, CANADA

The purpose of this investigation was to quantitate
the rate of uric acid production in patients following
acute myocardial infarction. In previous reports (Spring
et al. 1960) it was observed that serum uric acid levels
could be elevated following acute myocardial infarction.
In view of other metabolic changes which have been observed such as the increased excretion of catecholamines and
their metabolites and changes of glucose tolerance and
lipid metabolism, it seemed adviseable to investigate abnormalities of uric acid metabolism in terms of pool size
and turnover rates. 7 male patients who had had uncomplicated myocardial infarctions were admitted to the study following informed consent. The criteria of myocardial infarction were dependent upon clinical, electrocar-
diographic and biochemical enzyme changes. The patients
were selected to exclude those with pre-existing disease
and were restricted to those patients whose post-operative
course did not require the administration of drugs particularly diuretics. The patients were admitted to an in-
vestigative ward from the coronary care unit between the
2nd and 5th day following infarction. Throughout the period their total caloric intake was adequate but their
diet was essentially free of purines. At the beginning
of the study uric acid-2-C^{14} (10µc) was injected intravenously, urine was collected into glass jars containing
lithium carbonate. Urines were collected in 12 hour bat-
ches for the first 6 samples and then at 24 hour periods
for the next 6 days. Uric acid was isolated from the
urine by conventional crystallization procedures without

423

the addition of carrier uric acid. The purity of the
uric acid was confirmed by analysis with uricase and by
its characteristic UV absorption spectrum. Radioactivity
was determined by a liquid scintillation counting technique using Carb-o-Sil gel. The total uric acid of each
urine specimen was measured by the uricase assay and the
final data obtained as radioactivity per total weight of
uric acid excreted in each period of time. Typical results of one such study are shown in figure 1 which indicates that the rate of decrease of radioactivity in the
urinary uric acid followed an exponential decay from
which it was possible to calculate the initial pool size
of uric acid in these subjects and the turnover rate of
uric acid. Table 1 shows a summary of the results obtained on 6 of the male subjects in this group. The 7th
subject, known to have had a previous history of gout,
was included in the study to compare the significance of
the increase of uric acid pool size and turnover with the
data obtained from a typical gouty subject. Two normal
subjects were studied who were hospitalized for medical
conditions not associated with disorders of uric acid
metabolism. It can be seen from Table 1 that the excretion of uric acid in the post-myocardial infarct subjects
was not significantly greater than that found in the normal controls but the increase of uric acid pool size was
approximately 4 times that found in normal controls. The
rate of turnover of uric acid was $2\frac{1}{2}$ to 3 times that of
the normal controls.

These findings in themselves do not lead to an explanation of the etiology of this abnormality of purine
metabolism and it is pertinent to examine the various
possibilities that might be related to this disorder.
Exogenous purine intake was minimal and no drugs were
administered which would interfere with the normal pathways of excretion of uric acid. Some other relevant findings as listed in Table 2 indicate that renal function
as measured by creatinine clearance was not grossly impaired and urate clearance was in the range of normality.
There was no evidence of increased circulating serum lactate and the serum nor-adrenaline levels and urinary VMA's
were within normal limits in those patients from whom data were obtained. In view of this rather negative evidence a possible mechanism suggested as an alternative
hypothesis would be that some more general disorder of
metabolism had occurred and that one aspect of this was
related to the feed-back control of purine synthesis.
Holmes et al., (1973) have recently shown that the subunit structure of the enzyme glutamine phosphoribosyl

TABLE 1

URIC ACID KINETIC STUDIES IN 7 PATIENTS POST ACUTE MYOCAR-
DIAL INFARCTION AND IN 2 NORMAL SUBJECTS

Name	Wt. in kg.	Mean Serum Uric Acid mg%	Mean 24hr. Urinary Uric Acid mg. (±S.D.)	Uric Acid Pool mg.	mg/kg.	Uric Acid Turnover mg/24hr	mg/kg/24hr
SUBJECTS							
J.C.	58	9.8	571 ± 135	4090	70.5	1889	32.6
D.C.	70	8.7	407 ± 81	5571	79.6	2642	37.7
D.V.	73	8.5	459 ± 65	4976	68.2	1934	26.5
G.A.	81	5.4	505 ± 93	2069	25.5	1503	18.6
L.L.	41	4.0	353 ± 48	1511	36.9	1036	25.3
N.L.	64	2.9	623 ± 63	4400	68.8	2772	43.3
Mean±1SD		6.6±2.8	487 ± 100	3769±1624	58.3±22	1962±662	30.6±9
T.W.	80	6.8	1217 ± 179	8322	104.0	4274	53.4
NORMALS							
E.K.	60	6.0	245 ± 94	753	12.6	612	10.2
A.Z.	60	4.4	503 ± 97	1009	16.8	872	14.6

TABLE 2

CREATININE, URATE CLEARANCE AND ADDITIONAL DATA ON CONTROL SUBJECTS

Pat.	Creatinine Clearance ml/min	Urate Clearance ml/min.	Urate Clearance/ Creatinine Clearance x 100	Serum Lactate μM/ml	Serum Nor- Adrenaline μg/ml	Urine VMA mg/gm crea- tinine
J.C.	77.9	3.18	4.1	0.70	–	2.8
D.C.	85.0	5.85	6.9	0.50	0.31	–
D.V.	79.3	7.57	9.5	1.03	1.25	–
G.A.	85.6	6.44	7.5	1.77	0.50	0.4
L.L.	79.4	7.35	9.3	0.77	0.50	0.6
N.L.	112.0	7.00	6.3	1.00	0.10	0.0
T.W.	139.0	11.15	8.0	1.11	1.01	0.0
Normals						
A.K.	86.0	3.82	4.4	–	–	–
A.Z.	75.6	6.91	9.1	–	–	–

FIGURE 1

URINARY URIC ACID RADIOACTIVITY AS DE-
TERMINED IN ONE OF THE POST-MYOCARDIAL
INFARCTION PATIENT

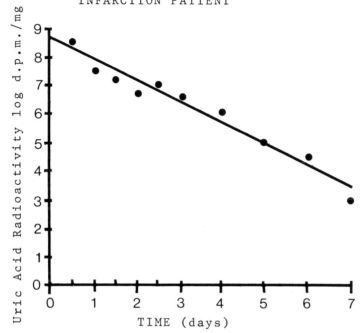

pyrophosphate amidotransferase can be brought together
in the presence of purine nucleotides resulting in a
loss of enzyme activity or can be dissociated in the presence of phosphoribosyl pyrophosphate giving rise to an
increase of enzyme activity. Whether this change of
feed-back control is specifically related to the loss of
intermediates by ischemic myocardial tissue or whether
it is partly a response to a change of circulatory haemodynamics remains to be investigated but further studies of this type might throw some light on the complex
mixture of metabolic changes which occur in the post-myocardial infarction period.

    Spring, M., Cavusoglu, M., Chiu, Y.C., Artymowska,
C.; Circulation, 22, 817, 1960.
    Holmes, F.W., McDonald, J.A., McCord, J.M., Wyngaar-
den, J.B., and Kelley, W.; J. Biol. Chem., 248, 144, 1973.

(Supported by Canadian M.R.C. grants No. MA-3331
and MT-3238).

THE EFFECT OF WEIGHT REDUCTION ON URATE METABOLISM

B.T. EMMERSON

University of Queensland Department of Medicine

Princess Alexandra Hospital, Brisbane, Australia

The common view of the etiology of gout has related it to obesity and over-indulgence in food. Contrariwise, gout was seen much less frequently in undernourished European countries during World War II. Our observations of patients with primary gout have shown that many weigh considerably more than their desirable weight for height (1). Little is understood however, concerning possible pathogenetic mechanisms involved in this association. One approach would be to characterize urate metabolism in an obese subject who had suffered from gout and then to define it on a second occasion by an identical technique after he had lost weight.

The patient studied had been found to be hypertensive when he presented at the age of 55 years with transient hemiparesis and was treated with methyldopa and phenytoin. He had suffered from frequent attacks of gout over a 15 year period, was a regular drinker and had been obese for many years. He subsequently reduced his alcohol consumption and, his methyldopa having been withdrawn, his urate metabolism was studied some 12 months later, as shown in the first study in Table 1. After this, moderate caloric restriction was instituted which led to gradual weight reduction. Over an 18 month period, his weight fell by 18 kilograms, although he still did not achieve desirable weight for height. He continued to be troubled by occasional attacks of gouty arthritis despite prophylactic colchicine. However, his blood pressure and serum urate concentration gradually fell to normal. On a second occasion, 18 months after the first study, his urate metabolism was again defined on an identical purine-free diet for the 5 pre-study days and the 7 study days. Identical investigational techniques were used for the two studies. Uric acid crystals

429

TABLE I

Studies of Urate Metabolism in a Gouty Subject on Two Occasions,
Firstly while Obese and Secondly after Weight Reduction

	First Study	Second Study	Significance of Difference
Weight (kilogrammes)	99	81	
Surface area (square metres)	2.12	1.94	
Blood pressure (mm. Hg)	170/110	140/80	
Serum urate (mg/100 ml)	7.0 ± 0.2	5.3 ± 0.3	p < .001
Urine urate (mg/24 hrs)	494 ± 46	487 ± 46	N.S.
Renal clearance urate (ml/min)	5.3 ± 0.5	8.0 ± 0.9	p < .01
Urine creatinine (mg/24 hrs)	1734 ± 92	1634 ± 111	N.S.
Renal clearance creatinine (ml/min)	141 ± 13	132 ± 10	N.S.
Miscible urate pool (mg)	2050	1419	p < .001
(95% range)	1956-2148	1304-1545	
Urate turnover (pools/24 hrs)	0.43	0.49	p < .001
Urate production (mg/24 hrs)	885	696	
I.V. urate excreted 7 days (per cent)	51.7	72.3	
24 hr urinary urate excretion / 24 hr urate production %	55.8	70.0	
Extra-renal disposal (mg/24 hrs)	391	209	
^{14}C-glycine incorporation into urinary urate 7 days (% dose)	0.29	0.25	
^{14}C-glycine incorporation into produced urate 7 days (% dose)	0.56	0.34	

were isolated from the urine by adsorption and elution from a
Dowex anion exchange resin (2) and urate was measured by the
spectrophotometric uricase method (3). The miscible urate pool,
turnover rate and daily urate production were assessed after
intravenous administration of 22 mg of lithium urate containing
15 atoms% ^{15}N by measurement of the ^{15}N enrichment of pure uric
acid crystals isolated from sequential 24 hour urine collections
over a 7 day period (4). The per cent of glycine incorporated into
urinary urate was measured following the simultaneous administration
of 10 μCi of ^{14}C glycine by the ^{14}C content of the uric acid
crystals isolated on to the Dowex resin (5). This was then
corrected to the per cent of glycine incorporated into produced
urate by reference to the per cent of intravenous ^{15}N urate which
was excreted into the urine in a 7 day period (6). Renal clearances
of urate and creatinine were calculated from three consecutive 30
minute clearance periods.

The results of the two studies are shown in Table 1. The
main difference in the patient between the studies was that he
weighed 18 kg less at the time of the second study. Along with
this weight loss, his blood pressure had fallen significantly to a
level which was within the normal range, a phenomenon which is well
recognized as occurring in some obese subjects with weight reduction
(7). The serum urate, which had been at the upper limit of normal
during a purine free diet while the first study was being undertaken
had fallen to a significantly lower level during the second study,
but no significant difference in the mean 24 hour urinary urate
excretion occurred. Correspondingly, the urate clearance was
significantly greater at the time of the second study than it had
been during the first study. The creatinine clearances and the 24
hour urinary creatinine excretions were completely comparable on
the two occasions.

A significant fall in the urate pool was seen on the second
occasion. A 95% reliability value of each pool estimate is shown
and these demonstrate no overlap whatsoever between the two assess-
ments. The urate turnover, in pools/day, was significantly
greater on the second occasion and this is in keeping with the
improvement in his urate clearance. The urate production, however,
assessed as the product of pool and turnover, was significantly less
on the second occasion. If the normal range of urate production is
taken as being 343 ± 36 mg/m^2/day, as suggested by Rieselbach et
alii (8), then the urate production has fallen to just within the
normal range on the second occasion. The per cent of labelled
urate excreted by the kidney in the 7 days after its administration
had increased from 51.7 to 72.3% on the second occasion and again,
this is probably in keeping with the improved urate clearance
during the second study. These figures are nicely matched with the
ratio of the urinary urate excretion to urate production. Extra-

renal disposal of urate was less when the urate clearance was
higher. Glycine incorporation into urinary urate was within normal
limits on both occasions although it was less on the second
occasion than on the first. This reduction is even more marked when
the glycine incorporation into urinary urate is corrected by the
per cent renal excretion of labelled urate to give the per cent
glycine incorporation into produced urate.

     In summary, on the first occasion when the patient was obese
and hypertensive, he had a definitely abnormal urate metabolism,
but on the second occasion, after weight loss, when his blood
pressure had returned to normal and his urate clearance had
improved, his urate metabolism was completely within the normal
range.

     The remission of hypertension which was associated with weight
loss in this patient is probably an important factor in the
improvement of the renal excretory capacity for urate and an
impairment of urate clearance by hypertension is well documented
(9, 10, 11). In addition, his daily production of urate per unit
surface area fell considerably with weight reduction, which implies
that urate production in an individual may be greater in absolute
terms when he is obese than at normal body size. The study also
shows that the daily production of urate in an individual is not a
constant function, but is susceptible to a variety of exogenous
inferences.

     Because it is limited to only one patient, interpretation of
this study must be cautious. However, a wide variety of
aetiological factors are now recognized as being important in the
syndrome of gout, so that one does not expect hyperuricaemia and
gout to be due to the same cause in all patients, or that all
hyperuricaemic patients will respond in the same way to changes in
environment or therapy. It should therefore not be expected that
all obese subjects with gout would respond to weight reduction by
a return to normal of an abnormal urate metabolism and clearly this
would not occur in obese subjects whose hyperuricaemia was due to
an intrinsic overproduction of urate, in whom urate metabolism
would be abnormal whether the patient was obese or not. On the
other hand, it is unlikely that this patient is unique, so that
other patients are likely to be found in whom obesity is an
important contributing factor to their hyperuricaemia and in whom
obesity makes the difference between normal and abnormal urate
metabolism. The difficulty is in predicting which obese subjects
will respond to weight reduction as this particular patient has.
The practical difficulties in finding further subjects who could be
studied in this way are, primarily, to find obese subjects who are
prepared to cooperate in the investigations and secondly, to
motivate such subjects to reduce their weight between the studies.

Weight reduction is a difficult problem in any obese subject, but
it is particularly so in subjects with gout in whom ketosis will
further aggravate hyperuricaemia or precipitate acute attacks of
gouty arthritis. However, if successful, weight reduction could
be expected to be more beneficial in the long term than the
administration of drugs which would either reduce uric acid
production or promote its excretion.

REFERENCES

1. EMMERSON, B.T. and KNOWLES, B.R. (1971). Triglyceride
    concentrations in primary gout and gout of chronic lead
    nephropathy. Metabolism 20:721-729.

2. JOHNSON, L.A. and EMMERSON, B.T. (1972). The isolation of
    crystalline uric acid from urine, for urate pool and
    turnover measurements. Clin. Chim. Acta 41:389-393.

3. LIDDLE, L., SEEGMILLER, J.E. and LASTER, L. (1959). The
    enzymatic spectrophotometric method for determination of
    uric acid. J. Lab. Clin. Med. 54:903-913.

4. BENEDICT, J.D., FORSHAM, P.H. and STETTEN, DeW. (1949). The
    metabolism of uric acid in the normal and gouty human studied
    with the aid of isotopic uric acid. J. Biol. Chem. 181:183-193.

5. WYNGAARDEN, J.B. (1957). Overproduction of uric acid as the
    cause of hyperuricaemia in primary gout. J. Clin. Invest.
    36:1508-1515.

6. SEEGMILLER, J.E., GRAYZEL, A.T., LASTER, L. and LIDDLE, L.
    (1961). Uric acid production in gout. J. Clin. Invest.
    40:1304-1314.

7. CHIANG, B.N., PERLMAN, L.V. and EPSTEIN, F.H. (1969). Over-
    weight and hypertension. Circulation 39:403-421.

8. RIESELBACH, R.E., SORENSEN, L.B., SHELP, W.D. and STEELE, T.H.
    (1970). Diminished renal urate excretion per nephron as a
    basis for primary gout. Ann. Intern. Med. 73:359-366.

9. BRECKENRIDGE, A. (1966). Hypertension and hyperuricaemia.
    Lancet 1:15-18.

10. CANNON, P.J., STASON, W.B., DEMARTINI, F.E., SOMMERS, S.C.
    and LARAGH, J.H. (1966). Hyperuricemia in primary and renal
    hypertension. New Engl J Med 275:457-464.

11. GARRICK, R., BAUER, G.E., EWAN, C.E. and NEALE, F.C. (1972).
    Serum uric acid in normal and hypertensive Australian subjects.
    Aust. N.Z. J. Med. 4:351-356.

DIET AND GOUT

N. Zöllner and A. Griebsch

Medical Polyclinic, University of Munich

D-8 Munich 2, Pettenkoferstrasse 8a (Germany)

Gout, a disease of the wealthy, disappears in times of need. Although a few gouty patients had to be cared for during the last war, the disease had become exceedingly rare in Germany by 1948. At the end of that year food supply began to become normal again; by 1950 there was a noticeable incidence of new cases of gout. This was not a particular observation in Munich but was general in all Germany. Later we found out that other European Countries who had shared the low supply of food with the Germans had made the same experience.

The clinical manifestations of gout, i.e. articular and renal disease, are intimately connected with hyperuricemia and/or particularities of renal handling of uric acid. It is well established that normalization of plasma uric levels will cure the joint disease. The diminution of renal uric acid excretion by administration of allopurinol will cure nephrolithiasis in its uncomplicated forms. One may presume that the other manifestations of the gouty kidney (i.e. parenchymal renal disease, hypertension and azotemia) will also be influenced by a therapy reducing urinary uric acid; however no reliable reports exist as yet on this point.

The finding of hyperuricemia always means an increase of uric acid in the body: In early and uncomplicated cases of gout, as well as in the normal, the plasma level of uric acid is proportional to the

Table 1   Comparison of the uric acid pools of normal
           persons and hyperuricemic persons determined
           by isotope dilution (A) and calculated from
           plasma uric acid and bromide space (minimum
           pool)(from ZÖLLNER, 1960) (B).

sex	weight (kg)	plasma uric acid (mg/100 ml)	pool (mg)	
			A	B
m	73	6.0	1340	1110
m	62	6.2	1170	980
m	76	4.4	1150	850
m	88	4.6	940	1030
m	75	5.4	1270	1030
w	51	4.3	650	560
average values for men	75	5.5	1170	1000

size of the uric acid pool (table 1). If hyperuricemia
is prolonged or extensive, uric acid is deposited in
solid form and the pool size increases out of
proportion to the uric acid level of the plasma. At the
same time the mathematical assumptions on which the
calculation of pool size is based do no longer hold.

Inflow mechanisms into the uric acid pool are two,
de novo synthesis of purines from small fragments and
uptake of purines in the diet. In man, there is only
one outflow mechanism, excretion, since the human body
has no capacity for breaking down uric acid to any
degree.

There are sufficient clinical examples on hand
which prove that increased inflow as well as impaired
outflow alone may produce hyperuricemia and gout. Such
examples are overproduction in myeloproliferative
disorders, disturbed excretion in polycystic kidney
disease or under the administration of saluretics. If
gout occurs under these circumstances it is called

secondary. The influence of dietary purines, the
subject of this review, will be discussed in extenso.

Until recently, attempts at determining the
influence of diet on plasma uric acid and uric acid
excretion have been made with calculated diets prepared
from conventional foodstuffs. The results have been
inconsistent and conflicting. This is partly due to the
difficulties of determining the purine content of "low
purine foods" with accuracy. The large relative ranges
of low purine concentrations multiply to large absolute
ranges when the purine content is calculated per unit
of utilizable energy. Indeed, some vegetables contain
more purines per calorie than some meats (table 2).

In order to investigate the influence of dietary
purines, we used diets composed of "no purine foods" or
formula diets supplemented with ribonucleic or desoxy-
ribonucleic acid (RNA and DNA). Technical details of
these experiments have been described elsewhere
(ZÖLLNER and GRÖBNER, 1971; ZÖLLNER et al., 1972; see
also the following paper by GRIEBSCH and ZÖLLNER).

Table 2   Average purine nitrogen content of some foods
          calculated per unit weight and per unit energy

Food	Purine nitrogen (mg/100 g)	Calories (per 100 g)	Purine nitrogen (mg/100 kcl)
Steak (raw)	58	177	33
Chicken (boiled)	61	203	30
Cauliflower (raw)	12.5	25	50
Cabbage (raw)	7.5	25	30
Green peas average (raw) range	72 25 - 145	64 60 - 80	112 41 - 242
Spinach (raw)	70	24	

    In subjects put on these diets, plasma uric acid
and urinary uric acid excretion fell immediately. The
fall continued for seven to ten days, at which time new
constant levels were approached. These levels were
around 3.1 mg/100 ml uric acid in the plasma, while the
uric acid excretion was around 300 mg/day. The latter
value corresponds rather well with the "endogene Harn-
säure" of BURIAN and SCHUR who estimated endogenous
purine nitrogen to be 100 - 200 mg/day. When RNA was
added to the basic diet, plasma uric acid levels rose
until they reached a new equilibrium after eight to ten
days. Concomitantly, uric acid excretion rose. The
increase was the same in all subjects who were normouricemic before starting on the no purine diets. If the
amount of nucleic acid supplementation was increased,
there were further rises of plasma level and excretion
of uric acid. Over a range of 0 - 4 g of daily supplementation with RNA (a range that is larger than the
usual range of dietary purine uptake) the rises were
proportional to the amounts of the nucleic acid added.
The r-values for the regression lines were 0.95. When
the experiments were repeated with DNA, identical
amounts containing the same quantities of purines
produced not only smaller rises in plasma uric acid but
also smaller rises in urinary uric acid excretion
(fig. 1).

    The different responses to the administration of
RNA and DNA demonstrate that not all dietary purines
have the same effect on purine metabolism. Therefore,
the values for total purines in conventional food tables
are of restricted usefulness for dietary purposes. If
further experiments should extend the finding that
purines from different nucleic acids exert quantitatively different effects on plasma uric acid and uric
acid excretion, a new and more detailed set of food
tables would have to be constructed. Until then, the
success of dietary prescriptions must be controlled by
repeated determinations of plasma uric acid and urinary
uric acid excretion, even if there is no doubt about
adherence to diet.

    If persons who are hyperuricemic on conventional
diets (plasma uric acid $\geq$ 6.5 mg/100 ml), but not
gouty, are used for the same type of experiments, the
response of plasma uric acid to RNA supplementations is
much larger, and the slope of the regression line is
nearly 50 % steeper than in normal subjects (fig. 2).

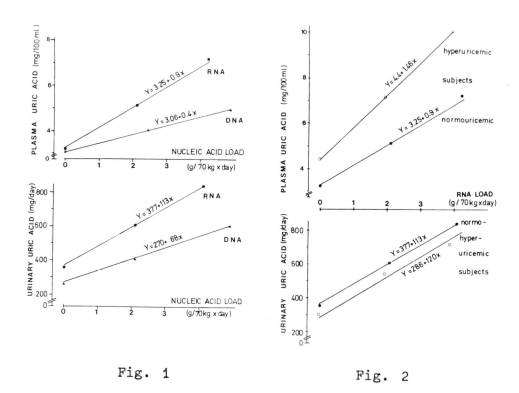

Fig. 1   Fig. 2

Fig. 1 Response of plasma uric acid level and
        urinary uric acid excretion to loads of
        RNA and DNA added to a purine free
        liquid formula diet. In the experiments
        with RNA the basal uric acid excretion
        on the purine free formula diet was
        larger than is generally our experience.

Fig. 2 Response of plasma uric acid and urinary
        uric acid excretion to loads of RNA
        added to a purine free liquid formula
        diet in persons, who were normouricemic
        or hyperuricemic (plasma uric acid above
        6.5 mg/100 ml) on conventional diets.

On the other hand, the response of urinary uric acid
excretion to the load is identical with that of the
normal controls.Repetition of the experiment with DNA
produces corresponding results, i.e., an exaggerated
increase of plasma uric acid and a normal increase of
uric acid excretion. These results show that hyperuricemia may be the consequence of a reduced capacity of
the kidneys to excrete uric acid (fig. 3). They
correlate well with earlier tabulations of WYNGAARDEN
1965 in which plasma uric acid levels and renal uric
acid excretion during the determination of renal uric
acid clearance (short time, water diuresis) were
compared.

     Under the general conditions of balance techniques,
identical outflow and constant pool size at the time of
measurement mean identical inflow into the pool. Thus,
there is no difference in purine absorption between
normal and hyperuricemic persons. However, under the
experimental conditions used, hyperuricemic persons
clearly show a lower than normal renal elimination of
uric acid at any given level of plasma uric acid. This
is further evidence in favor of a renal defect as the

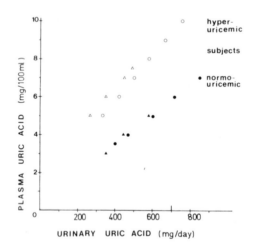

Fig. 3   Correlation of plasma uric acid levels
           and urinary uric acid excretion during
           administration of RNA (oo) and DNA (∆∆)
           in persons, who were normouricemic
           (filled symbols) or hyperuricemic(plasma
           uric acid above 6.5 mg/100ml)(open
           symbols) on conventional diets.

pathogenetic mechanism in most cases of hyperuricemia
and presumably gout. Possibly, the same mechanism
permits stone formation at a normal renal uric acid
excretion.

With respect to dietary therapy our results confirm
the clinical and epidemiological experiences that diets
of poverty and hunger cannot sustain hyperuricemia.
Therefore, therapy of the majority of cases of gout by
diet alone is theoretically feasible. However, in my
hometown nobody would accept such a diet, low in
everything people in Germany love.

On the other hand, the increase in plasma uric acid
with comparatively small purine loads shows why dietary
therapy aiming at normalizing plasma uric acid usually
meets with little success: the purines contained in
"low purine diets" suffice to raise plasma uric acid
above the limits of normal. With respect to lowering
plasma uric acid, drug therapy will in most circumstances
remain the treatment of choice.

In order to lower uric acid excretion, oral purine
uptake must be reduced unless one administers allopurinol. In view of the known tendency of patients with gout
to develop renal stones one of these two measures is
indicated. For patients treated with a uricosuric drug
a low purine diet is certainly advisable. In order to be
effective, such a diet should be supervised by repeated
24 hour collections of urine and determination of urinary acid excretion, as explained above.

One of the most important measures in dietary
therapy of gout consists in lowering total food uptake
of obese gouty patients. However, normalizing body
weight must be done slowly since the increase of ketone
bodies associated with strict fasting produces hyperuricemia, due to a rise in ß-hydroxybutyrate. It would
appear that a change in the NAD/NADH ratio is the mediator for a decrease of uric acid excretion which occurs
in ketonemia of all kinds. The same mechanism is
probably also working when uric acid excretion is
decreased after excessive intake of ethanol which
produces lactacidemia. The old advice to stay away from
alcohol is still a good one to be given to the gouty.
To be fair it should be added that moderate drinking
does not have an appreciable effect on uric acid levels.

A comment should be made with respect to combined

therapy with a drug and diet. As pointed out, drug
therapy alone is usually sufficient for the management
of hyperuricemia. If uricosuric drugs are used for this
management additional institution of a low purine diet
is wise in order to reduce the uric acid load on the
kidney. If allopurinol is given no dietary advice with
respect to purines seems necessary. Allopurinol not
only lowers plasma uric acid but also leads to a
decrease in urinary uric excretion. Indeed it can be
shown that practically all exogenous purine is taken
care of by allopurinol (ZÖLLNER and GRÖBNER, 1970).

Since dietary purines need not to be considered in
patients treated with allopurinol, there is room for
new dietetic speculations. As shown by GUDZENT in 1929
and amply confirmed recently a large proportion of
patients with gout also shows hyperlipidemia.
On typing, hyperlipoproteinemia type IV is usually
found. In our experience this hyperlipidemia is not
influenced by allopurinol. Since it is not established
what makes gout a coronary risk factor and since it is
possible that not only the gouty kidney with subsequent
hypertension but also hyperlipidemia contributes to this
risk, studies should be initiated with respect to
therapy of this aspect of the disease.

References

1   ZÖLLNER, N. and W. GRÖBNER: Der unterschiedliche
     Einfluß von Allopurinol auf die endogene und exo-
     gene Uratquote. Europ. J. clin. Pharmacol. 3, 56
     (1970)

2   ZÖLLNER, N., A. GRIEBSCH and W. GRÖBNER: Einfluß
     verschiedener Purine auf den Harnsäurestoffwechsel.
     Ernährungs-Umschau 3, 79 (1972)

3   BURIAN, R. and H. SCHUR: Über die Stellung der
     Purinkörper im menschlichen Stoffwechsel.
     Pflügers Arch. ges. Physiol. 80, 24 (1900)

4   BURIAN, R. and H. SCHUR: Das quantitative Verhalten
     der menschlichen Harn-Purin-Ausscheidung.
     Pflügers Arch. ges. Physiol. 94, 273 (1903)

5   WYNGAARDEN, J. B.: Gout. In Advances in Metabolic
     Disorders, Vol. 2 (R. LEVINE and R. LUFT edit.),
     Academic Press, New York - London (1965)

EFFECT OF RIBOMONONUCLEOTIDES GIVEN ORALLY ON

URIC ACID PRODUCTION IN MAN

A. Griebsch and N. Zöllner

Medical Polyclinic, University of Munich

D-8 Munich 2, Pettenkoferstrasse 8a [Germany]

During recent years WASLIEN et al. as well as our group have studied the influence of dietary purines on plasma levels and urinary excretion of uric acid in man, using modern feeding techniques. In these experiments defined quantities of RNA and DNA were added to an isocaloric, purine free formula diet. [For a review see ZÖLLNER et al., 1972].

From these studies several new facts emerged: The effect of dietary purines proved to be much larger than expected from the literature; e.g. nearly 50 percent of the purines in RNA being absorbed and converted to uric acid. There is a straight line relationship between the amount of purines added to the basal diet and the rise in plasma uric acid and uric acid excretion. These rises are nearly twice as large after RNA than after DNA.

So far no modern data existed on the response of uric acid parameters to the administration of mononucleotides, although these data are much needed to unterstand the absorption of uric acid precursors. In the following paper we will report on them.

ESTABLISHMENT OF BASE LINE

Formula Diet

In order to establish a reproducible base line of
endogenous uric acid production purine free liquid
semisynthetic formula diets were used. They were isocaloric and contained carbohydrates, fat and protein in
a caloric relation of 55 : 30 : I5 per cent. Carbohydrates were given as a scarcely sweet mixture of
oligosaccharides [Maltodextrin[R]], produced by partial
hydrolysis of corn starch, fat as sun flower seed oil.
Protein sources were skim milk powder containing 30 per
cent, lactalbumine with a content of 80 per cent or a
new protein mixture Hyperprotidine[R] containing more
than 90 per cent protein. Vitamins and minerals were
added.

An essential feature of the experiments were careful attention to body weight and total dietary uptake
in order to prevent undernutrition, since ketoacidosis
leads to renal retention of uric acid. Thus the average
change of body weight was never higher than minus
0.7 kg during 33 days.

Analysis

Blood samples were taken Monday, Wednesday and
Friday in the morning before breakfast. Urine samples
were collected during 24 hours daily. To check for
errors in the urine collection endogenous creatinine
in blood and urine were determined. Uric acid was
measured by our modification [ZÖLLNER,I963] of the
uricase method.

Endogenous Uric Acid Production

During the administration of the purine free diet
uric acid excretion reached a minimum of 336 $\pm$ 39 mg
per day [average $\pm$ S.E.M.] in eight healthy young
volunteers, to whom the mononucleotides were administered later. Fig. I shows results of endogenous uric acid
excretion in a total of I6 subjects, including the
above 8 persons. Plasma levels of uric acid decreased
from values between 4.6 - 4.9 mg per I00 ml during
purine free diet to average values of 3.I0 $\pm$ 0.4 mg

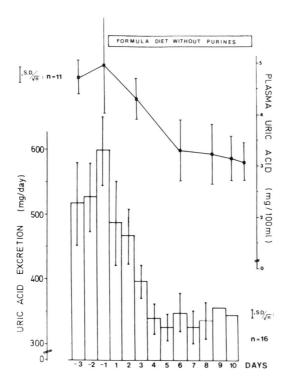

Fig. I   Endogenous uric acid production:
           plasma levels and urinary excretion of
           uric acid during isocaloric purine free
           formula diet

per I00 ml in eleven subjects after I0 days, a time
necessary to reach a steady state.

EFFECT OF RIBOMONONUCLEOTIDES

Materials and Amounts

        Crystalline 5'-AMP [$C_{10}H_{14}N_5O_7P \cdot H_2O$], pure "pro
analysi" with a content of 99.8 per cent, free from the
2'-and 3'-isomeres, produced by enzymatic phosphorylation of adenosine, molecular weight 347.2 was used. The
content of lead [Pb] and other heavy metals was less
than I ppm. 5'- GMP was given as crystalline sodium salt
[$C_{10}H_{12}N_5O_8P\ Na_2 \cdot 4H_2O$], molecular weight 479.2, lead
content less than 5 ppm. The amounts utilized were
equimolecular and corresponding to the amounts of AMP

or GMP contained in 2 g respectively 4 g RNA. The daily
intake at the lower dose level was 8.74 mg per kg body
weight or an average of 553.9 mg per day and per person.
In case of 5'- AMP.I2.06 mg per kg body weight of the
sodium salt of GMP were given, which corresponded to
8.8 mg per kg of 5'-GMP pure or to an average uptake
of 574.9 mg per day and per person. These amounts were
applied during a first period of II days, followed by
amounts twice as high during a second period of IO days.
Corresponding amounts of RNA were applied to 8 other
volunteers comparable in age [24 years instead of 25
years in experiments with mononucleotides] and in body
weight [65.7 kg instead of 65.0 kg].

Plasma Levels of Uric Acid

The results of the experiment are shown in fig. 2.
Plasma uric acid levels rose during application of

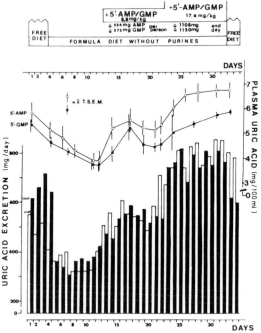

Fig. 2   Plasma levels [above] and urinary excretion
          [below] of uric acid during purin free formu-
          ladiet and during administration of different
          amounts of 5'-AMP and 5'-GMP

ribomononucleotides from values below, 4.0 mg per I00 ml
to averages of 5.3 mg per I00 ml in the group of 5'-
AMP and to 4.5 mg per I00 ml in the 5'- GMP group. In
the second period with doubled supplementation the
averages were 6.7 mg per I00 ml and 5.85 mg per I00 ml
respectively. The rise was in proportion to the
amounts administered.

Urinary Uric Acid Excretion

The effects of mononucleotides supplementation on
uric acid excretion were comparable to those on plasma
levels. During the first period the 24-hours excretion
rose from 336 ± 39 mg per day to 579 ± 28 mg per day in
the 5' - AMP group and to 537 ± 62 mg per day in the
5'- GMP group. During the application of the doubled
amounts the values reached were 847 ± 64 mg per day
and 757 ± 2I2 mg per day respectively. Effects of RNA
are shown in fig. 3. Plasma levels and renal uric acid
excretion also rose proportional to the amounts
administered.

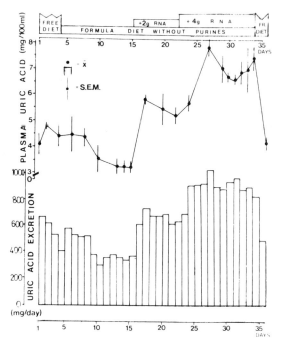

Fig. 3   Plasma levels [above] and renal excretion
           [below] of uric acid during administration of
           amounts of 2 g and 4 g RNA

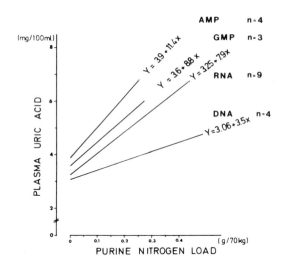

Fig. 4   Comparison on basis of the purine nitrogen
         content:Correlation between purine nitrogen
         load [from purines in different sources as in
         5'-AMP, 5'-GMP, RNA and DNA] and plasma levels
         of uric acid

Comparison on the Basis of Purine Nitrogen Content

     The rises of uric acid levels and renal excretion,
effected by the mononucleotides were compared to the
values reached during administration of the nucleic
acids RNA and DNA an the basis of equal purine load.
The purine nitrogen content is II.5 per cent in RNA
and II.44 per cent in DNA; but 5'-AMP contains 20.16
per cent and 5'-GMP I9.27 per cent of purine nitrogen.
The comparison is shown in fig. 4 for the plasma
levels. The purines from the ribomononucleotides
effect a significantly larger increase of plasma uric
acid than those of nucleic acids. The same is true for
the urinary uric acid excretion [fig. 5]. Thus e.g.
nearly all purine nitrogen administered as 5'-AMP can
be accounted for by the increase of uric acid excretion
while only about 25 per cent of the purines in DNA
appear as uric acid in the urine.

     From these results on may conclude that purines
in mononucleotides are absorbed to a larger extent than
purines in nucleic acids;in the case of 5'-AMP this
absorption would appear to be nearly quantitative.

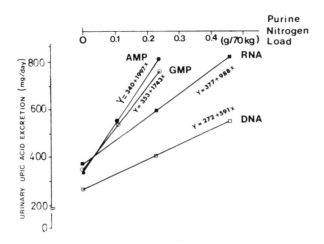

Fig. 5   Correlation between purine nitrogen load
          [g/person/day] as in fig. 4 and urinary uric
          acid excretion [mg/person/day] to compare
          effects of purines from different sources

Therefore it may be assumed, that the limiting factor
in the absorption of dietary purines from their major
source, i.e. the nucleic acids in foods is the rate of
hydrolysis, since the products of this nucleic acid
hydrolysis, AMP and GMP, are absorbed to a very large
degree.

References

I   WASLIEN, C. I., D. H. CALLOWAY, S. MARGEN and
     F. COSTA:Uric acid levels in men fed algae and
     yeast as protein sources. J. Food Sci. 35, 294
     [1970]

2   ZÖLLNER, N., A. GRIEBSCH and W. GRÖBNER:Einfluss
     verschiedener Purine auf den Harnsäurestoffwechsel.
     Ernährungs-Umschau 3, 79-82 [1972]

3   ZÖLLNER, N.:Eine einfache Modifikation der enzy-
     matischen Harnsäurebestimmung. Normalwerte in der
     deutschen Bevölkerung. Z. Klin. Chem. I, 178 [1963]

RELATIONSHIP BETWEEN GOUT AND ARTERIAL HYPERTENSION

Aurelio Rapado, M.D.

Laboratorio de Urolitiasis y Enfermedades Metabólicas

Fundación Jiménez Díaz. Madrid. Spain

The incidence of arterial hypertension (A.H.) in gout has been observed for many years, the rate being about 36 per cent in gouty patients (Table I) (1-4), which is well above the percentage in normal population (about 10 per cent in our country).

A higher incidence of hyperuricemia in hypertensive patients has been also described (5, 6) though data are contradictory (7). On the other hand, there is a nephropathy of familial basis where hypertension and gout are associated (8, 9), which confirms previous studies on the harmful action of hyperuricemia upon the kidney (10).

In a survey of 750 cases of gout studied during the last ten years, we have analyzed both diseases according to diverse parameters.

TABLE I

INCIDENCE OF ARTERIAL HYPERTENSION IN GOUT

AUTHOR	YEAR	CASES OF GOUT	A.H. %
Barlow & Beilin	1968	1410	36
Rotés & Muñoz	1968	380	30
Grahame & Scott	1970	354	52
Barceló et al.	1971	534	37
Present series	1973	750	36

TABLE II

CLINICAL CORRELATION IN GOUT
WITH AND WITHOUT ARTERIAL HYPERTENSION

	NORMAL B.P.	HIGH B.P.
NUMBER OF CASES	480	270
	%	%
Female sex	7.3	17.4
Clinical type:		
Acute	35	26
Chronic	53	54
Tophaceous	12	20
Renal Lithiasis	55	37
Uric Acid Lithiasis	20	17
Thiazide-induced Gout	9	32
Steroid-dependent Gout	6	12
Lead contact	4	12
Family history of Gout	29	26
Family history of A.H.	12	29
Obesity	65	68
Diabetes	3.5	3.7
Visible Tophi	42	51

MATERIAL AND METHODS

Every patient was studied by a special case history record and thorough physical examination. Blood pressure was measured at least twice under rest conditions. Creatinine and uric acid clearance in samples of 24-hour urine was done after a low purine diet had been administered for three previous days and all drug had been withdrawn. X-ray examination of joins was also carried out. Results were assembled in a special filing card and statistically analyzed.

RESULTS

Table II shows the clinical findings in 480 cases of gout with normal blood pressure and 270 cases in whom A.H. was associated (the values represent the percentage of cases).

A higher incidence of females in the group of arterial hypertension was observed, which agrees with the higher incidence of hypertension in the feminine sex (11).

There were no significant differences in the clinical type of gout though the acute type was more frequent in the normotensive group and tophi were more frequent in the cases with A.H.

No relationship between renal lithiasis (either in general or uric acid lithiasis) and high blood pressure was found (12).

In the group of hypertensive patients we have observed a high incidence of thiazide-induced attacks of gout as well as the so-called steroid-dependent gout (13).

Thiazides are employed in the treatment of A.H. and their inhibition of the secretion of uric acid favours the acute attack of gout. On the other hand, the sodium retentive action of corticosteroids may explain the high blood pressure (14). The reversible character of A.H. as well as of renal insufficiency in this group of patients would support this assumption.

Drivers and people in contact with lead, to which we have attributed an aetiopathogenic factor in the hyperuricemias (15), were frequent in our series, the percentage being higher in cases with A.H.

The familial history of gout was proportionally alike in both groups (29 and 26 per cent respectively), which shows the hereditary basis of this disease. By contrast, the familial history of A.H. was markedly higher among hypertensive patients which confirms its genetic trait.

The association of obesity and diabetes did not difer in either group and therefore the "risk factor" attributed to both parameters, seems to be questionable in A.H., at least in our series (16).

The presence of visible tophi did not contributed to a higher incidence of A.H. in our gouty patients.

The age at the time of consulting (Table III) showed statistically significant differences, which implies that A.H. is more related to this parameter than to gout disease. However in the whole the duration of gout exerted a marked influence in the incidence of A.H. We have not studied whether this fact was due to better or worse therapy or to the patient's age.

TABLE III

CLINICAL CORRELATION IN GOUT
WITH AND WITHOUT ARTERIAL HYPERTENSION

	NORMAL B.P.	HIGH B.P.
NUMBER OF CASES	480	270
Age (years)	50.2±11.9	55.5±10.2
Duration of Gout (years)	7.4±7.8	10.3± 9.4
Serum Uric Acid (mg/100 ml)	8.2±1.64	8.9± 2.26
Urinary Uric Acid (above 800 mg/day) %	21	10
Primary Renal Gout %	6	21
Secondary Renal Gout %	4	19

Hyperuricemia presented higher mean values in patients with A.H. At the same time, the hyperexcretion of urates (above 800 mg of urinary uric acid/24 hours) was significantly more frequent in in the normotensive gouty patients. This hyperuricemia seems to be a consequence more than a cause of A.H., including the renal involvement, both organic and drug-induced.

The cases of gout with renal involvement (primary renal gout) were more frequent in the hypertensive group. The renal diseases that develop gouty attacks during their evolution (secondary renal gout) also showed a higher incidence in the group with A.H. The finding that even in cases of severe renal failure, the blood pressure could be normal (17) was rather striking.

The survey of cases of gout with renal involvement in relation to blood pressure (Table IV) revealed, in the hypertensive secondary renal gout, an earlier age of onset; a higher incidence of females; a predominance of the acute clinical type and a higher incidence of gouty attacks caused by the administration of diuretic drugs. On the contrary, in the normotensive group, uric acid renal lithiasis and diabetes were more frequent, which confirms the non-specificity of these parameters.

TABLE IV

GOUT WITH RENAL INVOLVEMENT
(number of cases)

	Primary Renal Gout		Secondary Renal Gout	
	NORMAL B.P.	HIGH B.P.	NORMAL B.P.	HIGH B.P.
Number of cases	30	56	20	51
Female Sex	2	6	5	12
Acute Gout	4	6	8	31
Uric Acid Lithiasis	7	9	3	2
Thiazide-induced G.	6	11	7	25
Steroid-dependent G.	14	27	8	17
Lead Contact	7	10	5	5
Family history of G.	8	13	3	10
Family history of A.H.	3	16	2	14
Obesity	13	31	7	21
Diabetes	2	3	3	0
Visible Tophi	17	38	6	13

T A B L E   V

GOUT WITH RENAL INVOLVEMENT

(mean values)

	Primary Renal Gout		Secondary Renal Gout	
	NORMAL B.P.	HIGH B.P.	NORMAL B.P.	HIGH B.P.
NUMBER OF CASES	30	56	20	51
Age (years)	57.5 ± 13.2	56.4 ± 12.4	56.9 ± 10.1	46.1 ± 13.6
Duration of Gout (years)	11.1 ± 9.7	12.1 ± 11.7	4.4 ± 4.5	3.0 ± 3.6
Serum Uric Acid (mg/100 ml)	9.68 ± 2.03	10.20 ± 2.91	9.77 ± 1.75	10.90 ± 3.48

Kidney transplant was performed in 8 patients of the normotensive group, in whom the hyperuricemia and gouty attacks disappeared.

Primary renal gout with normal as well as high blood pressure
did not show striking differences save a higher incidence of the
female sex in the hypertensive group.

There were no marked differences in age, duration of the gout
or serum uric acid values in the two groups (Table V).

In those cases in which the cause of death could be studied
(Table VI), we found a higher incidence of renal and cardiovascular
causes in the hypertensive than in the normotensive group, while
cerebral strokes were alike. Four cases died because of complications
of the renal transplant in patients with secondary renal gout. The
other causes of death were diverse, being striking those due to the
administration·of drugs used to control the gout.

TABLE VI

CAUSES OF DEATH IN GOUT

	NORMAL B.P.	HIGH B.P.
NUMBER OF CASES	36	40
Renal	10	20
Cardiovascular	4	10
Cerebral	6	6
Others	16	4 (transplant)

CONCLUSIONS

The higher incidence of A.H. in gout is related to female sex,
age, duration of the gouty process, certain drugs (particularly
steroids), contact with lead and familial history of A.H.

No relation to renal lithiasis, visible tophi, obesity, diabetes or familial history of gout was found. Neither to the values
of uric acid in blood or to the urinary excretion of uric acid.

Renal involvement confers a severe prognosis to A.H. as well as to gout disease, since on one hand, the former worsens the articular picture (with higher incidence of attacks caused by the treatment with thiazides) and, on the other hand, rises the incidence of death owed to renal or cardiovascular causes.

Secondary renal gout bears a worse prognosis and the higher incidence of females and juvenile age indicates the existence of a larvate gouty gene which manifests in the persistent hyperuricemia. On the other hand, kidney transplant solves the hyperuricemia as well as the gout.

We have not studied the prognosis of primary renal gout, but it seems that this process is a chronic disease which is well tolerated, particularly when hyperuricemia is properly controlled by the new drugs available (18). On the other hand, accelerated A.H. was a rare finding in our cases.

It will be interesting to have more experience on the follow-up of these patients who undergo early treatment for their hyperuricemia in order to study whether this parameter contributes to A.H. as well as to the renal involvement.

REFERENCES

1.- Barlow, K.A. and Beilin, L.J.: "Renal diseases in primary gout" Quar.J.Med., 37, 79-1968

2.- Rotés Querol, J. and Muñoz, J.: "La gota". Ed. Toray. Barcelona. 1968

3.- Grahame, R. and Scott, J.T.: "Clinical survey of 354 patients with gout". Ann.rheum.Dis., 29, 461-1970

4.- Barceló, P. et al.:"Estudio estadístico sobre la gota". Rev. Esp.Reumat., 14, 1-1971

5.- Breckenridge, A.: "Hypertension and hyperuricemia". Lancet, 1, 15-1966

6.- Cannon, J. et al.: "Hyperuricemia in primary and renal hypertension". New Eng.J.Med., 275, 457-1966

7.- Myers, A.R. et al.: "The relationship of serum uric acid to risck factors in coronary heart disease". Am.J.Med., 45, 520-1968

8.- Duncan, H. and Dixon, A.St.J.: "Gout, familial hyperuricemia and renal disease". Quar.J.Med., 29, 127-1960

9.- Van Goor, W., Kooiker, C.J. and Mees, E.J.D.: "An unusual form of renal disease associated with gout and hypertension". J.Clin. Patho., 24, 354-1971

10.- Duncan, H., Wakin, K.G. and Ward, L.E.: "Renal lesions resulting from induced hyperuricemia in animals". Staff Meet.Mayo Clin., 38, 411-1963

11.- Rapado, A. and Zevallos, E.: "La hipertensión arterial esencial. Análisis de 1000 casos". Medic.Clin., 2, 73-1971

12.- Cifuentes Delatte, L. et al.: "Uric acid lithiasis and gout". Renal Stone Research Symposium. Karger. Basel. 1973

13.- Rapado, A. et al.: "Cushing iatrogénico y gota corticodependiente". Rev.Clin.Esp., 98, 275-1965

14.- Holmes, E.W., Kelley, W.N. and Wyngaarden, J.B.: "The kidney and uric acid excretion in man". Kidney Intern., 2, 115-1972

15.- Rapado, A.: "Gout and saturnism". New Eng.J.Med., 281, 851-1969

16.- Rodriguez Miñón, J.L., Rapado, A. and Conde, M.P.: "Interrelations between diabetes and arterial hypertension". Hormones, 2, 289-1971

17.- Hall, A.P.: "Arterial hypertension and primary gout". Proc.Roy. Soc.Med., 59, 317-1966

18.- Rapado, A.: "Allopurinol in thiazide-induced hyperuricemia". Ann.rheum.Dis., 25, 660-1966

Relationship Between Carbohydrate, Lipid, and Purine Metabolism

STUDIES ON THE MECHANISM OF FRUCTOSE-INDUCED HYPERURICEMIA IN MAN

I. H. Fox and W. N. Kelley

University of Toronto, Canada and Duke University Medical

Center, Durham, North Carolina 27710

The precise biochemical basis for many instances of hyperuricemia in man are not clearly understood. Elucidation of the mechanism by which certain normal intermediates and their structural analogs alter the serum uric acid may be useful in delineating potential pathophysiological alterations leading to hyperuricemia. The infusion of fructose in man precipitates a number of biochemical changes including hyperlacticacidemia, decrease in serum inorganic phosphate, and decrease in serum glucose. This results from the phosphorylation of fructose to fructose-1-P and the entrance of this compound into the glycolytic pathway. An increase in serum uric acid concentration following the infusion of fructose was initially reported by Perheentupa and Raivio in 1967., although this has remained a controversial observation.

Four fasting gouty patients were infused with 0.5 gm/Kg of a 20% fructose solution over a 10 minute period. Two of these patients were restudied after pretreatment with allopurinol, 600 mg per day. Figure 1 illustrates the changes in the serum urate concentration in the 180 minutes after the start of the fructose infusion. The serum urate showed a 33% increase over control values at 30 minutes, while this change was not seen in the two patients taking allopurinol. These data confirm the initial observation that the rapid infusion of fructose does produce a substantial increase in the serum urate concentration.

The mechanism responsible for the fructose-induced hyperuricemia in man has been unclear. In Figure 2 a schematic diagram summarizes some of the steps of fructose metabolism relevant to this discussion. In the liver, the phosphorylation of fructose to fructose-1-P utilizes the conversion of one ATP to ADP. Fructose-1-P

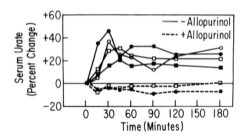

Fig. 1. Effect of fructose on serum urate. (Modified from Fox and Kelley, 1972).

FRUCTOSE METABOLISM

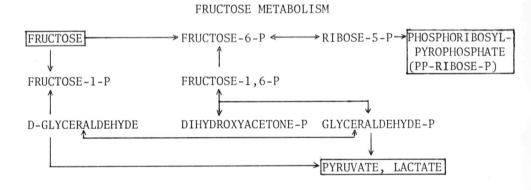

Fig. 2. Abbreviated outline of fructose metabolism.

is cleaved to D-glyceraldehyde and dihydroxyacetone-P. Both of
these compounds may be either eventually converted to lactate or
else can enter the interconversion steps of the glycolytic pathway.
Fructose-6-P is usually formed indirectly from these intermediates
except in erythrocytes, white blood cells and adipose tissue where
it is formed by direct phosphorylation of fructose. The conversion
of fructose-6-P to ribose-5-P, a substrate necessary for phosphoribosylpyrophosphate (PP-ribose-P) synthesis, provides one link between
fructose metabolism and purine biosynthesis . PP-ribose-P is an
essential substrate for purine synthesis de novo as well as the
purine salvage pathways.

Based on our understanding of fructose metabolism, we have
examined four potential mechanisms to account for fructose-induced
hyperuricemia in man. 1) Shift in the uric acid pool, 2)
decreased renal clearance of uric acid, 3) increased purine synthesis de novo by stimulating PP-ribose-P production, and 4) accelerated degradation of purine ribonucleotides. Our studies were designed
to distinguish which of these mechanisms in man could account for
the hyperuricemia observed after fructose infusion.

Figure 3 illustrates the total reducing sugars, glucose, and
phosphate in plasma during the 180 minutes after the start of
fructose infusion. The total reducing substrates, which represent
the sum of glucose and fructose concentrations, showed a rise at
15 minutes. The plasma glucose and plasma phosphate showed no
substantial change.

Blood lactate levels were determined before and 30 minutes
after the start of fructose infusion (Table 1). Hyperlacticacidemia
with a mean increase of 256% of control values was observed 30
minutes after the start of the fructose infusion. The hyperlacticacidemia also occurred in the patients treated with allopurinol
even though the serum urate concentration did not change with the
infusion of fructose under these conditions.

The urinary uric acid excretion and the urinary oxypurine
excretion were measured up to 180 minutes after the start of the
fructose infusion (Figure 4). A mean increase in the urinary
uric acid excretion to 144% of control values and in urinary
oxypurine excretion to 397% of control values occurred in the
first hour after infusion. The marked rise in urinary oxypurines
resulted primarily from a rise in hypoxanthine excretion. A
striking increase in the excretion of inosine was also noted. No
change was observed in the fractional clearance of uric acid.
Pretreatment with allopurinol enhanced the absolute increase in
urinary oxypurine excretion. These observations suggest that
fructose-induced hyperuricemia is related to stimulation of uric

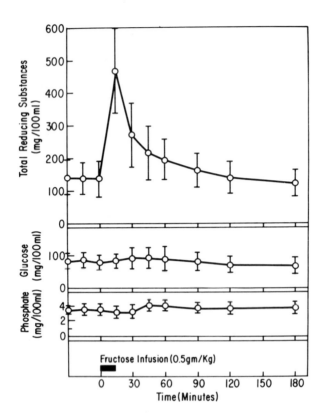

Fig. 3. Effect of fructose infusion on plasma total reducing substances, glucose and phosphate. Results are expressed in mg/100 ml. (From Fox and Kelley, 1972).

acid synthesis. A simple shift in the uric acid pool could not account for the fructose-induced increase in uric acid synthesis and the inhibition of this hyperuricemia by allopurinol. The increase in urinary uric acid virtually eliminates the possibility that renal retention of uric acid is a significant factor.

The fructose-induced increase in uric acid synthesis observed can potentially occur either by 1) increased purine biosynthesis de novo or 2) increased nucleotide catabolism. We have shown that fructose can stimulate PP-ribose-P production in human erythrocytes incubated in vitro. Since increased intracellular PP-ribose-P

TABLE 1

EFFECT OF FRUCTOSE ON BLOOD LACTATE CONCENTRATION

	Blood Lactate Control (mg/100 ml)	30 Min After Infusion (mg/100 ml)	(% Change)
No treatment			
A.B.	6.8	28.0	+312
H.L.M.	8.9	22.0	+147
J.W.	4.4	18.0	+309
Allopurinol			
A.B.	6.5	21.0	+223
J.B.	8.0	30.0	+275

(From Fox and Kelley, 1972)

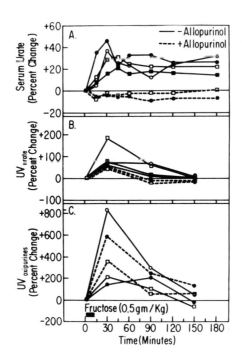

Fig. 4. Effect of fructose on serum urate and urinary excretion of uric acid and oxypurines. (From Fox and Kelley, 1972).

levels can stimulate purine biosynthesis de novo, one could
propose this mechanism to explain the stimulation of uric acid
synthesis observed. Our observations of erythrocyte PP-ribose-P
levels in vivo after fructose infusion did not substantiate this
hypothesis. Figure 5 depicts the changes in erythrocyte PP-ribose-P,
ribose-5-P and ATP levels following fructose infusion. A modest
but significant mean decrease in erythrocyte PP-ribose-P levels to
43% of control values was observed at 15 and 30 minutes. In addition

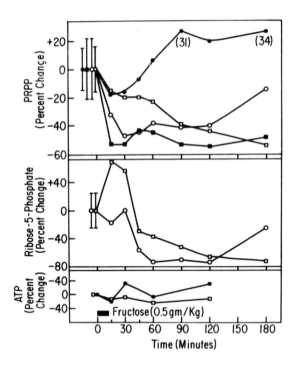

Fig. 5. Effect of fructose infusion on concentration of PP-ribose-P,
ribose-5-phosphate, and ATP in erythrocytes. Results are expressed
as per cent of control values. Control values represent mean ± S.D.
(vertical bars) of three different samples obtained during a 60
min. period immediately preceeding infusion of fructose. Open
squares, patient J.W.; closed squares, patient L.L.; open circles,
patient H.M.; and closed circles, patient A.B. (From Fox and Kelley,
1972).

intracellular ribose-5-PO_4 levels began to decrease 45 minutes after the infusion. No consistant change in ATP levels was observed. We conclude from these studies that the infusion of fructose does not increase the formation of ribose-5-P or PP-ribose-P in circulating erythrocytes in vivo.

Previous observations in hereditary fructose intolerance also suggest that fructose-induced hyperuricemia cannot be attributed to increased levels of PP-ribose-P. The infusion of fructose produces hyperuricemia in this disorder despite a block in the further metabolism of fructose-1-P in these patients.

Our observations are most consistent with the hypothesis that the infusion of fructose leads to a rapid degradation of purine nucleotides with the consequent formation of inosine, hypoxanthine, xanthine, and finally uric acid (Fig. 6). Recent studies in the rat in vivo (Raivio, Kekomaki, and Maenpaa, 1968), in perfused rat liver (Woods, Eggleston, and Krebs, 1970), and in human liver in vivo (Bode, et al., 1971), have demonstrated that fructose rapidly decreases the concentration of adenine nucleotides, especially ATP, and inorganic phosphate within the cell. Since ATP normally inhibits 5'-nucleotidase and inorganic phosphate inhibits AMP deaminase, these changes would be expected to stimulate the catabolism of AMP to inosine. This hypothesis would most readily account for the rapid rise in serum urate concentration, the increased urinary excretion of inosine, oxypurines and uric acid, and the lack of an increase in intracellular PP-ribose-P levels following fructose infusion.

PROBABLE MECHANISM OF FRUCTOSE-INDUCED HYPERURICEMIA IN MAN

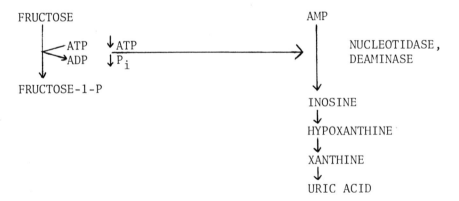

Fig. 6. Probable mechanism of fructose-induced hyperuricemia in man. The phosphorylation of fructose decreases ATP and P_i. This leads to increased activity of the purine ribonucleotide catabolic pathway which is normally inhibited by these compounds.

To summarize, the rapid infusion of large doses of fructose in man produces a marked hyperuricemia within 30 minutes. The hyperuricemia results from the increased synthesis of uric acid, which most likely occurs by stimulating the degradation of purine nucleotides.

REFERENCES

Bode, L., Schumacher, H., Goebell, H., Zelder, O. and Pelzel, H. 1971. Fructose-induced depletion of liver adenine nucleotides in man. Hormone Metab. Res. 3: 71-75.

Fox, I. H. and Kelley, W. N. 1972. Studies on the mechanisn of fructose-induced hyperuricemia in man. Metabolism. 21: 713-721.

Perheentupa, J. and Raivio, K. 1967. Fructose-induced hyperuricemia. Lancet. 2: 528-531.

Raivio, K. O., Kekomaki, M. P. and Maenpaa, P. H. 1968. Depletion of liver adenine nucleotides induced by D-fructose. Biol. Hem. Pharmacol. 18: 2615-2624.

Woods, H. F., Eggleston, L. V. and Krebs, H. A. 1970. The cause of hepatic accumulation of fructose-1-phosphate on fructose loading. J. Biol. Chem. 119: 501-510.

OBSERVATIONS OF ALTERED INTRACELLULAR PHOSPHORIBOSYLPYROPHOSPHATE

(PP-RIBOSE-P) IN HUMAN DISEASE

I. H. Fox and W. N. Kelley

University of Toronto, Canada and Duke University Medical

Center, Durham, North Carolina   27710

An intricate system of interrelated control mechanisms regulate biochemical reaction sequences. Metabolic pathways are controlled not only by specific activity and inherent kinetic properties of enzymes in the pathway but also by the intracellular concentration of certain essential substrates, activators or inhibitors. PP-ribose-P is an essential substrate of purine, pyrimidine and pyridine biosynthesis . The intracellular concentration of PP-ribose-P represents a balance between its synthesis by PP-ribose-P synthetase and its utilization which is catalyzed by several different phosphoribosyltransferase (PRT) enzymes as well as non-specific phosphatases. Alterations in the rate of synthesis or degradation of PP-ribose-P, whether drug induced or secondary to an inborn error, could potentially change the intracellular concentration of this compound.

Studies in vivo as well as in cultured human fibroblasts suggest that the intracellular concentration of PP-ribose-P plays an essential role in the regulation of purine biosynthesis de novo (Kelley, et al., 1970; Kelley, Fox and Wyngaarden, 1970).   The normal intracellular concentration of PP-ribose-P has been found to be less than the saturation of the first enzyme in the pathway.   In five different strains of cultured diploid fibroblasts the mean intracellular PP-ribose-P was 5.0 uM with the highest value being 13 μM while in normal erythrocytes the mean PP-ribose-P was found to be 4.4 ± 1.8 μM. These values are substantially less than the Km for PP-ribose-P of the human amidotransferase which is 480 μM (Holmes, et al., 1973).

The intracellular content of PP-ribose-P in fibroblasts was altered and the effect of these alterations on the rate of purine biosynthesis was investigated. The rate of the early steps of

471

purine biosynthesis was assessed by determining the incorporation
of formate-^{14}C into formylglycinamide ribonucleotide (FGAR) in the
presence of azaserine. Orotic acid at concentrations ranging from
80 to 320 μM decreased intracellular PP-ribose-P (Fig. 1.). This
was associated with a similar decrease in the rate of FGAR synthesis
by these cells. Azaorotate, which blocks the conversion of orotic
acid to OMP, inhibits the effect of orotic acid on either PP-ribose-P levels or on FGAR synthesis. Thus, the inhibitory effect of
orotic acid on FGAR synthesis requires the conversion of orotic acid
to OMP, a reaction which consumes PP-ribose-P. Fibroblasts lacking
hypoxanthine-guanine phosphoribosyltransferase (HGPRT) have
PP-ribose-P levels in fibroblasts which are 2 to 3 times higher than
normal. Orotic acid reduces the PP-ribose-P levels in these cells,
but normal levels are not achieved. Under these conditions orotic
acid has no effect on FGAR synthesis. Methylene blue, 100 μM,

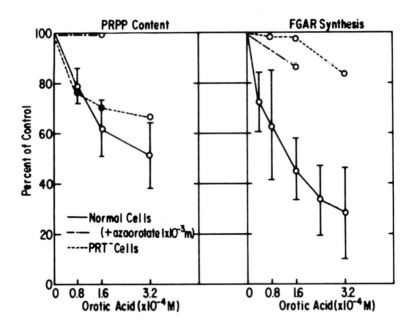

Fig. 1. Effects of varying concentrations of orotic acid on the
synthesis of FGAR and intracellular PP-ribose-P concentration.
PP-ribose-P content is indicated on the panel to the left and FGAR
synthesis on the panel to the right.o—o, mean (± S.D.) in six
different strains with normal hypoxanthine-guanine phosphoribosyltransferase;o---omean in two different cell strains deficient in
hypoxanthine-guanine phosphoribosyltransferase;o---onormal cell
strain preincubated with 1 mM azoorotate.

increases intracellular PP-ribose-P by stimulating the hexose monophosphate shunt. The increased levels of PP-ribose-P are associated
with an increased rate of FGAR synthesis (Fig. 2). This stimulatory
effect of methylene blue on FGAR synthesis is abolished if PP-ribose-P
levels are kept the same by the concomittant addition of orotic acid
indicating that methylene blue does not have a direct effect on FGAR
synthesis.

     Studies in vivo confirm these in vitro observations. Orotic acid
6 gm/day inhibited purine biosynthesis de novo by causing a 21% and
38% decrease in incorporation of glycine-1-^{14}C into urinary uric acid
in 2 normal men but had no inhibitory effect in two patients who
were deficient in HGPRT (Kelley, et. al, 1970). In two normal
patients a single 2 gram oral dose of orotic acid caused a 45%
decrease in intracellular PP-ribose-P 4 hours after ingestion

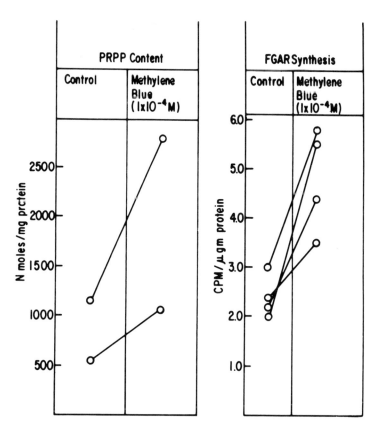

Fig. 2. Effects on methylene blue on intracellular PP-ribose-P
levels and FGAR synthesis.

(Kelley, et al., 1970).   Although orotic acid reduced the levels
of PP-ribose-P in 2 patients deficient in HGPRT the lowest levels
achieved were still substantially higher than normal.   Thus the
intracellular concentration of PP-ribose-P is an important factor
in the control of purine biosynthesis de novo.

Increased intracellular levels of PP-ribose-P have been
implicated in the cause of certain hyperuricemic states associated
with uric acid overproduction.   Fibroblasts from two patients with
the Lesch-Nyhan syndrome were found previously to have an elevated
intracellular concentration of PP-ribose-P with a normal rate of
PP-ribose-P production (Rosenbloom, et al., 1968).   Green and
Seegmiller (1969) subsequently reported a mean PP-ribose-P value of
47.1 μM in erythrocytes from seven patients with HGPRT deficiency.
We have confirmed these elevated PP-ribose-P levels in three
additional patients with the Lesch-Nyhan syndrome with values of
20.5, 39.4 and 49.5 μM (Table 1).   The mothers of these patients
are obligate heterozygotes and have normal PP-ribose-P levels.   Two
diseases associated with a deficiency of other PRT enzymes are not
associated with altered erythrocyte PP-ribose-P levels (Table 1).
PP-ribose-P levels were in the normal range in one patient with a
partial deficiency of adenine phosphoribosyltransferase (APRT) and
in one patient with orotic aciduria, which is due to a deficiency

TABLE 1

ERYTHROCYTE PP-RIBOSE-P LEVELS AND PHOSPHORIBOSYLTRANSFERASE
ACTIVITIES IN PATIENTS WITH PHOSPHORIBOSYLTRANSFERASE DEFICIENCY
STATES

	Erythrocyte PP-ribose-P (nmole/ml)
Normal values	4.4±1.8
Hypoxanthine-guanine phosphoribosyltransferase deficiency	
Hemizygote	
E. S.	39.4
J. K.	20.5
W. E.	49.5
Heterozygote	
M. S.	6.5
M. E.	4.5
H. K.	1.5
Adenine phosphoribosyltransferase deficiency	
Heterozygote	
L. L.	2.6
Orotate phosphoribosyltransferase deficiency	
Homozygote	
T. H.	2.7

(From Fox and Kelley, 1971).

of orotate phosphoribosyltransferase (OPRT) and orotidylic decar-
bosylase.

In primary gout increased erythrocyte PP-ribose-P levels have
been reported in patients with a partial deficiency of HGPRT (Greene
and Seegmiller, 1969). Several studies have provided evidence for
an increased capacity to synthesize PP-ribose-P in other patients
with gout not related to a deficiency of HGPRT. We have determined
the intracellular PP-ribose-P concentration in erythrocytes from
34 patients with hyperuricemia and normal HGPRT and in erythrocytes
from 33 normouricemic control subjects. In these studies there was
no apparent relationship between intracellular erythrocyte PP-ribose-P levels and serum urate concentration or urinary uric acid
excretion (Fig. 3). In addition, there was no relationship between
capacity for PP-ribose-P synthesis in erythrocytes and serum urate
concentration in 47 patients; 22 are illustrated in Fig. 4.

Two reports of a new genetic abnormality in primary gout
associated with elevated intracellular PP-ribose-P levels and
increased PP-ribose-P synthetase have been described in detail at
this meeting. The hyperuricemia of glycogen storage disease (Type
1) caused by deficient glucose-6-phosphatase, has been shown by a

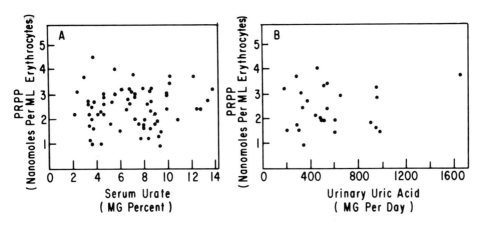

Fig. 3A and 3B. The relation of intracellular erythrocyte phosphoribosylpyrophosphate (PRPP) levels to serum urate and urinary uric
acid in patients with normal hypoxanthine-guanine phosphoribosyltransferase. (From Fox and Kelley, 1971).

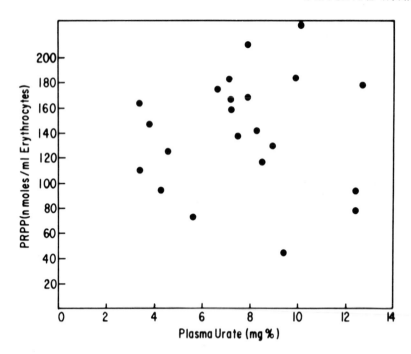

Fig. 4.   Relation of erythrocyte PP-ribose-P production to plasma
urate after 120 minutes incubation.

number of investigators to be associated with excessive uric acid
production from accelerated purine synthesis de novo as well as
with a decreased renal clearance of uric acid. Increased purine biosynthesis in this disease has been attributed theoretically to
increased PP-ribose-P production but this has remained undocumented.

Finally, Valentine, et al. (1972) has recently described a
patient with a partial deficiency of erythrocyte PP-ribose-P
synthetase and non-sperocytic hemolytic anemia. It is possible
that erythrocytes in this disorder have diminished PP-ribose-P
concentrations.

Table 2 summarizes disease states associated altered intracellular PP-ribose-P levels. This essential substrate has become
important for understanding certain types of disordered purine
metabolism and possibly other diseases in man.

TABLE 2

ALTERED INTRACELLULAR PP-RIBOSE-P IN HUMAN DISEASE

I.   Increased intracellular PP-ribose-P
     A. Decreased degradation
         1) Lesch-Nyhan syndrome
         2) Gout due to partial HGPRT deficiency
     B. Increased formation of PP-ribose-P
         1) Gout with increased PP-ribose-P synthetase
         2) Glycogen Storage Disease Type 1 (not proven)
II. Increased availability of PP-ribose-P (mechansims undetermined)
         Subtypes of gout
III. Decreased formation of PP-ribose-P
         Hemolytic anemia with decreased PP-ribose-P synthetase

REFERENCES

Fox, I. H. and Kelley, W. N. 1971. Phophoribosylpyrophosphate in
    man: Biochemical and clinical signficiance. 74: 424-433.

Greene, M. L. and Seegmiller, J. E. 1969. Erythrocyte 5-phosphor-
    ibosyl-1-pyrophosphate (PRPP) in gout: Importance of PRPP in
    the regulation of human purine synthesis (Abstract). Arth.
    Rheum. 12: 666-667.

Holmes, E. W., McDonald, J. A., McCord, J. M., Wyngaarden, J. B. and
    Kelley, W. N. 1973. Human glutamine phosphoribosylpyrophos-
    phate amidotransferase. Kinetic and regulatory properties.
    J. Biol. Chem. 248: 144-150.

Kelley, W. N., Fox, I. H. and Wyngaarden, J. B. 1970. Regulation
    of purine biosynthesis in cultured human cells. I. Effects
    of orotic acid. Biochim. Biophys. Acta. 215: 512-516.

Kelley, W. N., Greene, M. L., Fox, I. H., Rosenbloom, F. M., Levy,
    R. I. and Seegmiller, J. E. 1970. Effect of orotic acid on
    purine and lipoprotein metabolism in man. Metabolism 19:
    1025-1035.

Rosenbloom, F. M., Henderson, J. F., Caldwell, I. C., Kelley, W. N.
    and Seegmiller, J. E. 1968. Biochemical basis of accelerated
    purine biosynthesis de novo in human fibroblasts lacking
    hypoxanthine-guanine phosphoribosyltransferase. J. Biol. Chem.
    243: 1166-1173.

Valentine, W. N., Anderson, H. M., Paglia, D. E., Jaffe, E. R.
    Konrad, P. N. and Harris, S. R. 1972. Studies on human erythro-
    cyte nucleotide metabolism. II. Nonspherocytic hemolytic
    anemia, high red cell ATP and ribosephosphate phosphoribokinase
    (RPK, E.C. 2.7.6.1) deficiency. Blood. 39: 674-684.

HUMAN ERYTHROCYTE PHOSPHORIBOSYLPYROPHOSPHATE CONTENT AND "GENERATION" DURING A CONSTANT RATE INFUSION OF XYLITOL

W.Kaiser,K.Stocker and P.-U.Heuckenkamp

Medical Polyclinic, University of Munich

D-8 Munich 2, Pettenkoferstrasse 8a (Germany)

Marked increases in serum uric acid concentrations have been observed following rapid intravenous infusions of fructose (2,6). With constant rate intravenous infusions of o.5 g fructose/kg body weight/hr over 5 hours however we cannot demonstrate remarkable changes in serum uric acid levels (3,4). With the same dose xylitol, with a rate of 0.5 g xylitol/kg body weight/hr, we are able to demonstrate even higher increases in serum uric acid concentrations than with 1.5 g fructose/kg body weight/hr over 5 hours (3,4). This hyperuricemic effect of xylitol could be due in part- comparable to that of fructose- to an increased rate of purine synthesis de novo or an accelerated degradation of purine ribonucleotides. Since intracellular 5-phosphoribosyl-1-pyrophosphate (PRPP) concentration limits purine biosynthesis de novo and is also a substrate in the reutilization pathways for purine bases we assessed PRPP levels and PRPP "generation" in erythrocytes after oral administration and during constant rate infusion of varying xylitol doses.

METHODS

For the enzymatic determination of PRPP we used OPRTase and ODCase from brewers yeast. The produced $^{14}CO_2$ from $7-^{14}C$-orotic acid during the incubation period was measured. Washed erythrocytes were lysed

mechanically in the incubation medium. All values for
PRPP/ml packed cells are corrected by the determination of hemoglobin in the hemolysates. Deprotein-
ization by heating did not appear to us quite favourable because varying but significant amounts of PRPP
were lost, even after adding ethylenediaminetetra-
acetate.To prevent PRPP synthesis in vitro we followed the method of Sperling et al. (7) by adding 2,3-di-
phosphoglyceric acid (DPG) immediately after washing
the cells.

Furthermore, PRPP "generation" within the erythrocytes was assessed, i.g. the PRPP being produced
-without DPG addition- during the incubation period.
This assay comprises all changes of concentrations in
ATP,ribose-5-phosphate, inorganic phosphate and inhibitors for PRPP synthetase within the living cell.

RESULTS

Oral administration of 5o gm xylitol (within 30
minutes) leads to a significant and rapid increase in
serum uric acid concentrations with a peak one to two
hours after the beginning, followed by a gradual decrease. Seven hours after the start of the experiment,
uric acid levels still are markedly elevated. This
rise in serum uric acid levels is accompanied by a
varying but marked decrease of intracellular PRPP
concentrations without reaching the initial value within 7 hours. In contrast to the PRPP content the "generation" of PRPP within the hemolysates increases significantly, in one case the initial value was exceeded
fourfold.

During constant rate infusion of xylitol in varying doses there are significant increases in serum
uric acid levels within the first 75 minutes (within
the range of 1 - 3 mg percent). No clear-cut relationship between the administration dose and uric acid
response is observed. A plateau is reached after 75
minutes, the serum uric acid levels remaining almost
constant throughout the infusion period.

With the infusion rate of o,3 g xylitol/kg body
weight/hr the intracellular PRPP concentration remains constant in one person whereas in the other we
get a slight increase. Using the higher dose (o,4 g
xylitol/kg body weight/hr,over 5 hours) there is in
one case a slight increase in PRPP concentration
after 15 minutes followed by a marked decrease of
about 25 percent throughout the infusion period,
whereas in the other case there are no significant

changes. With the highest dose used (o,5 g xylitol/
kg body weight/hr, over 5 hours) we find a rapid decrease in intracellular PRPP concentration throughout the infusion period, in the order of 2o to 5o percent as compared to the initial values. The person
exhibiting the greatest increase in serum uric acid
level had the most pronounced fall in intracellular
PRPP concentration. Furthermore, vomiting and nausea
occured accompanied by elevation of transaminases
above normal values.

For the "generation" of PRPP - as assayed under
our conditions - we get only a slight increase using
the low dose, whereas with o,4 g xylitol/kg/hr the
rise is twofold throughout the period and decreases
rapidly after the end of infusion. With the highest
dose used there is a modest but constant decline in
PRPP "generation".

DISCUSSION

Xylitol as a direct precursor of ribose-5-phosphate
leads to an elevated PRPP "generation" in some of our
experiments. At the highest concentrations however
there is even a slight decrease in PRPP "generation".
Xylitol infusions with rates of o,5 g/kg/hr are follo-
wod by depletion of inorganic phosphate and/or ATP
influencing PRPP synthetase. Inhibition of PRPP synthetase could be also exerted by some newly generated
anticipated inhibitors (for example AMP). ATP concentrations in blood remain constant during the entire
infusion period. Thus rising serum uric acid concentration during xylitol infusion cannot be explained
sufficiently on the basis of purine biosynthesis de
novo, for PRPP "generation" is only small. Furthermore, the rapid rise of serum uric acid concentration
after oral xylitol consumption can hardly be explained
with de novo purine synthesis alone.

Intracellular PRPP, the rate limiting substrate
for de novo synthesis is lower in most of our experiments during administration as compared to the initial
values. These lowered PRPP concentrations would point
to a high consumption, for example. The observed influences on PRPP "generation" could be a consequence
of the rapid degradation of purine nucleotides, for
example through increased inhibitor concentrations for
PRPP synthetase.

According to our results the xylitol-induced
hyperuricemia can be best explained by a rapid degradation of purine nucleotides, a situation similar to
the observations during fructose infusions (1,2,5).

PRPP "generation" most likely is influenced by formation of inhibitors for PRPP synthetase, like AMP.
This could further favour our hypothesis of a rapid
breakdown of purine nucleotides being the cause of
xylitol-induced uric acid increase.

REFERENCES

1. Bode,Ch.,Schumacher,H.,Goebell,H.,Zelder,O.and
   Pelzel,H.
   Horm.Metab.Res. 3,289(1971)
2. Fox,I.H.,and Kelley,W.N.
   Metabolism 21,713(1972)
3. Heuckenkamp,P.-U.,Zöllner,N.
   Lancet I,8o8(1971)
4. Heuckenkamp,P.-U.,Schill,K.,Zöllner,N.
   Verh.dtsch.Ges.inn.Med. 77,177(1971)
5. Mäenpää,P.H.,Raivio,K.o.,and Kekomaki,M.P.
   Science 161,1253(1968)
6. Perheentupa,J.,Raivio,K.O.
   Lancet II,528(1967)
7. Sperling,o.,Eilam,G.,Persky-Brosh,S.,and DeVries,A.
   J.Lab.Clin.Med.79,1o21(1972)

DIABETES AND URIC ACID - A RELATIONSHIP INVESTIGATED

BY THE EPIDEMIOLOGIGAL METHOD

Herman, Joseph B. , Medalie, J.H. , Goldbourt, U

Dept. Int. Med. C, Hadassah Univ. Hosp. , Jerusalem

Dept. Family Med. , Sheba Hosp. , Tel Hashomer

The following data were obtained from a long term Epidemiological study by the Israel Ischemic Heart Disease Unit on 10,000 men aged 40 years and over which was started in 1963. A diabetes survey was carried out within the framework of this study. In the 1963 survey, 498 were found to be diabetic, of whom 296 were aware of diabetes, and 202 were newly diagnosed at the survey. Two years later, in 1965, another 144 new diabetics were discovered.

It was shown that the diabetics had lower uric acid values than the rest of the study population.

The percent distribution by serum uric acid values is shown in a figure for 490 diabetics and 9,238 non-diabetic men. The men with diabetes had distinctly lower serum uric acid values than other men ($P < 0.001$). Lower uric acid values were again found in the diabetics discovered in 1965 than in the rest of the study population.

The duration of existence of diabetes was related to the uric acid level. Thus, the 1963 diabetics could be divided into (1) those discovered prior to the survey: "previously diagnosed", and (2) those discovered during the survey: "newly diagnosed diabetics".

Uric acid was lower in the previously diagnosed than in the newly diagnosed diabetics. It was lower in newly diagnosed diabetics than in the non-diabetics.

All these findings were statistically significant.

Further support for the observation that the uric acid level was related to the duration of the disease was provided by the 1965 uric acid values. The 1965 uric acid values of diabetics diagnosed in 1963 were lower than those of the 1965 diabetics.

However, the 1963 uric acid values of diabetics diagnosed for the first time in 1965 (that is, at the time when they were "pre-diabetics") was higher than that of the 1963 non-diabetic population.

The significance of these findings is discussed.

METABOLIC AND GLUCOSE LOAD STUDIES IN URIC ACID, OXALIC AND

HYPERPARATHYROID STONE FORMERS

P.O.Schwille, D.Scholz, G.Hagemann and A.Sigel

Hormone and Metabolism Laboratory, Department of

Surgery and Urology, Faculty of Medicine, University

of Erlangen-Nuernberg (F. R. G.)

Introduction

There is as yet little knowledge about the metabolic origin of idiopathic hypercalciuria (HC) and associated hyperuricosuria frequently accompanied by formation of calcium containing renal stones. Following a carbohydrate-rich meal, in 1969 Lemann and coworkers (1) observed higher urinary calcium in oxalic stone formers and their relatives than in healthy controls. Endogenous resistance to insulin and its consecutive overproduction was recently reported from patients with primary hyperparathyroidism (2), and a marked loss of insulin via urine was objectived by Ching and his group (3) in patients producing calcium stones. They failed to find disturbed glucose tolerance, but their data gave no information as to the amount of glucose administered. In earlier studies (unpublished) undertaken to screen stone people by oral glucose load (100g) we found pathological plasma glucose (2 hours) in a rather great number of stone patients without symptoms of diabetes and /or overt obesity. Also there was apparently no relation to either sex or age or, most important, the type of stone, i.e. oxalic or uric acid.

These observations and foregoing reports led us to the assumption of a basic connection that might exist between metabolic stone formation and disorders of carbohydrate metabolism not as yet clarified. Our present studies were designed to investigate the role of pancreatic glucagon, an old hormone recently coming into discussion with respect to its effects upon urinary electrolytes (4, 5).

Patients and Methods

1. Determination of daily and fasting urinary uric acid and
electrolytes (endogenous creatinine clearance): controls, P,
n = 20; oxalic stones, i.e. hypercalciuria, HC, n = 30; uric
acid stones, HS, n = 16; primary hyperparathyroidism, HPT,
n = 15.
2. Fasting plasma concentration of pancreatic glucagon: same
groups as with 1.
3. Glucose load studies (intravenous: 0.5 g/kg; oral:1.75 g/
kg):P, n = 10; HC, n = 10; HS, n = 10; HPT, n = 7(i.v.) and
n = 3 (oral) resp.
4. Determination of urinary substances as in 1. excreted during both i.v. and oral load (groups as in 3.).

Apart from stone disease all patients were in a clinically
well state without signs of catabolic illness (diabetes, malignancy etc.). They belonged to the surgical board and out-
patient unit of the University Hospital. Age: 23 - 38 years
(controls) and 26 - 63 (patients). Ratio male/female = appr.
2 : 1.

Measurement of insulin and glucagon (6) radioimmunologically
using Sephadex-G 25 adsorption method (Insulin: Pharmacia,
Sweden) and the specific antibody 30 K against pancreatic
glucagon resp. (kindly donated by Dr. R. H. Unger, Dallas,
USA). Uric acid and glucose by enzymatic assay procedures
free fatty acids by mikro-titration, electrolytes according
to conventional methods (AAS).

Results and Discussion

1. The reevaluation of normalized (pro 1.73 m^2) daily uric
acid during ordinary dietary regimens confirmed increased
urinary amounts in all stone types and a decreased urinary
citrate in oxalic stones (fig. 1). In the latter group also
elevated uric acid, calcium, magnesium and phosphate during
fasting can be observed, whereas the uric stone type is regularly accompanied by more calcium, phosphate but not uric
acid and magnesium (fig.2). There are differences between
groups as to glomerular filtration rate being still within
the normal range. We conclude therefore that tubular factors from a different intensity might be operative upon uric
acid and electrolyte transport in various stone types.
2. Fasting glucagon (fig.3) confirms the normal range for
controls (P) earlier reported by Unger et al. (6). In HPT

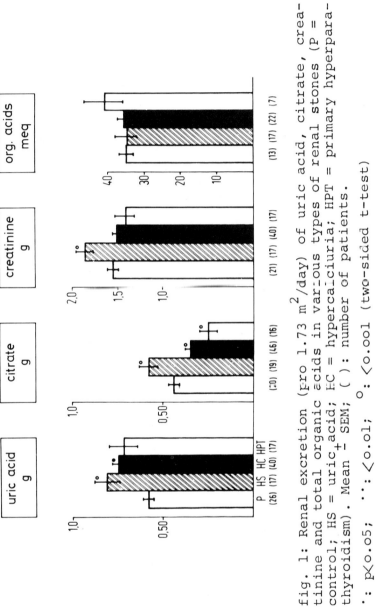

fig. 1: Renal excretion (pro 1.73 m²/day) of uric acid, citrate, creatinine and total organic acids in various types of renal stones (P = control; HS = hypercalciuria; HC = hypercalciuria; HPT = primary hyperparathyroidism). Mean ± SEM; ( ): number of patients.

•: p<0.05; ••: <0.01; °: <0.001 (two-sided t-test)

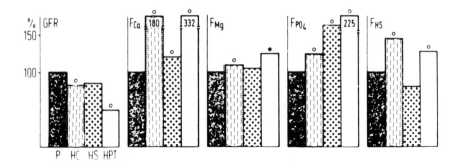

fig. 2: Medians of fractional (F) electrolytes (calcium, magnesium, inorg. phosphate), uric acid (HS) and glomerular filtration rate (GFR) during fasting in various stone types (see fig. 1). Controls (P) = 1oo per cent.

•: p < o.o5;   °: < o.oo2 (sign rank-test)

Table 1 Glucose utilisation coefficients in controls, oxalic and uric acid stone formers, primary hyperparathyroidism.

	n	range	ID_{80}	median	p
control	1o	1.27 – 4.55	3.65 – 1.27	2.26	
oxalic	1o	o.82 – 2.44	1.87 – o.82	1.33	<o.o2
uric acid	1o	o.98 – 2.04	1.98 – o.98	1.415	<o.o2
pHPT	7	o.71 – 2.66	2.66 – o.71	1.57	n.s.

glucagon is significantly elevated (p<0.01) showing a striking overlap with oxalic stones. In uric acid stones glucagon
is obviously unsuspicious indicating that some other conditions might be also involved finally resulting in tubular dys-
function.From the fact that these patients were regularly
more obese (7) we assume a state of slightly expanded extracellular volume and subsequent dilutional effects upon circulating hormones. Whatever the true nature of underlying mechanism glucose load studies could be a reasonable tool in
clarifying the role of glucagon as a possible mediator of tubular transport capacity.

3. Although insulin rises in a sufficient manner in all patients it declines slowly toward baseline (fig.4) and the
rate of glucose disappearance is in the range of latent diabetes (table 1). Glucose ingestion (fig 5) is followed by ex-
treme overproduction of insulin in uric acid (HS: 90-120 min,
111 µU/ml) and oxalic stones directing attention upon inte-

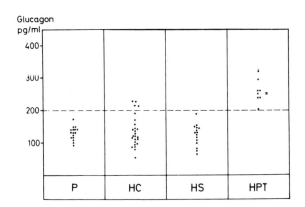

fig. 3: Fasting pancreatic glucagon in controls (P)
and patients with renal stones. Broken line: upper
range of normal.

●: symptoms of primary HPT;

x: long-term abusus of dihydrotachysterol.

Table 2 Urinary calcium, magnesium, inorg. phosphate and uric acid
(u.a.) in μg/min/1 mg creatinine following oral or intravenous (i.v.)
glucose load in controls, oxalic and uric acid stone patients.
Mean ± SEM; n = number of patients; [x]: $p > 0.05 < 0.10$

glucose load	n	U_{Ca}	U_{Mg}	U_{PO_4}	$U_{u.a.}$
oral					
controls	9	145 ± 20	75 ± 7	250 ± 59	401 ± 42
oxalic	11	207 ± 22	79 ± 12	386 ± 64	658 ± 93[x]
uric	10	128 ± 15	67 ± 9	356 ± 57	542 ± 92
i.v.					
controls	9	120 ± 24	79 ± 11	293 ± 54	500 ± 102
oxalic	11	162 ± 22	73 ± 11	448 ± 67	702 ± 45
uric	11	87 ± 13	57 ± 13	414 ± 46	603 ± 69

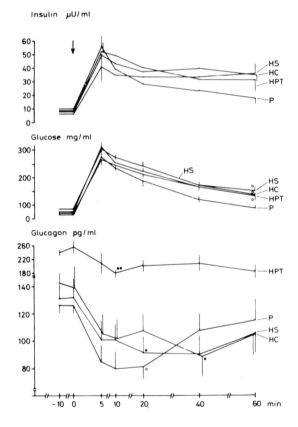

fig. 4: Intravenous glucose load (↓). Symbols for
groups and probability see fig. 1. Glucagon: t-test
for paired data; glucose: two-sided t-test.

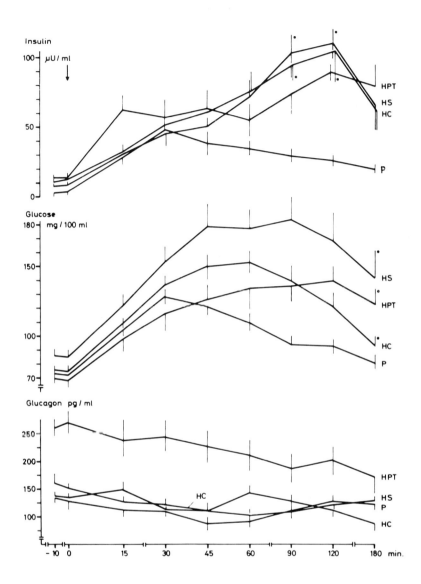

fig. 5: Oral glucose load. Symbols see fig. 1 and 4.
Insulin: significance according to variance analysis.

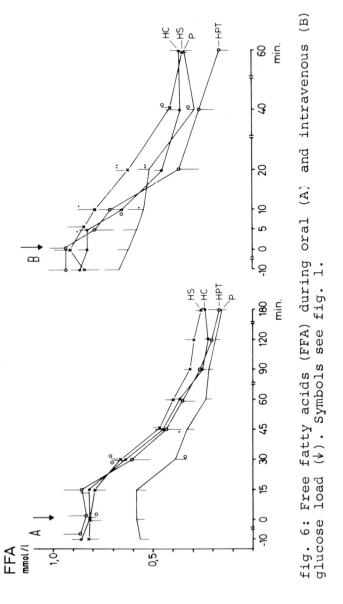

fig. 6: Free fatty acids (FFA) during oral (A) and intravenous (B) glucose load (↓). Symbols see fig. 1.

stinal insulinogenic factors (8). A prompt and steep fall in free fatty acids can be accepted as reliable indicator of biological activity of measured insulin (fig.6).

In controls glucagon declines rapidly (fig.4,P) but not so in stones: minimum values are not established before 40 min post inj. and with HPT glucagon remains strongly elevated. If the aforementioned hypothesis of a tubular effect of glucagon holds true then different patterns of uric acid and electrolytes could well be the major result of a glucose load (table 2). Although fasting hypercalciuria/-magnesiuria were near totally relieved, phosphaturia and uricosuria are likely to persist at a higher level in both oxalic and uric stones.

This study sheds some light upon disturbed mineral and purine metabolism now coming into discussion (9,10). The results add another aspect to the concept that relief of glucagon would in deed be responsible for abolishing fasting natriuresis (11) and, as demonstrated here, some other minerals. Long-term administration of pharmacological doses of glucagon evoked a state of glucose intolerance (12), but the concept of gradually increased resistance to insulin such as described for hyperparathyroid patients (2) would be much more likely to occur. At present one might only speculate upon the true nature of relative hyperglucagonaemia, hyperinsulinism and a delay in glucose disappearance in stones but there is no doubt that they must be in some way interrelated.

Conclusions
1. Hyperuricosuria in uric and oxalic stone patients is not a single symptom;
2. Fasting urinary calcium, magnesium, phosphate agree with the assumption of a factor actively inhibiting tubular transport;
3. Hyperglucagonaemia, evident in hyperparathyroidism, possibly mediates in vivo changes of renal function in all metabolic stones;
4. at the level of the pancreatic α-cell factors responsible for hyperinsulinism might well stimulate glucagon secretion resulting in increased urinary uric acid and electrolytes.

Acknowledgment

We are greatly indebted to A. Wellmann and K. Schwille for skillful technical assistance.

References

1. Lemann, J., Piering, W.F. and E.J. Lennon: Possible role of carbohydrate induced calciuria in calcium oxalate kidney stone formation; N.Engl. J.Med. 280, 232, 1969.

2. Kim, H., Kalkhoff, R.K., Costrini, N.V., Cerletty, J.M. and M. Jacobson: Plasma insulin disturbances in primary hyperparathyroidism; J. Clin. Invest. 50, 2596, 1971.

3. Ching, K.N., Karam, J.H., Choy, F.B., Kolb, O.F., Grodsky, G.M. and P.H. Forsham: Hyperinsulinuria in patients with renal calcium stones; Clin. Chim. Acta 40, 383, 1972.

4. Pullman, T.N., Lavender, A.R. and H. Aho: Direct effects of glucagon on renal haemodynamics and excretion of inorganic ions; Metabolism 16, 358, 1967.

5. Schwille, P.O. and B. Barth: Clearance and stop flow studies in glucagon mediated hyperelectrolyturia; Acta endocr. Suppl. 159, 207, 1972.

6. Aguilar-Parada, E., Eisentraut, A.M. and R.H. Unger: Pancreatic glucagon secretion in normal and diabetic subjects; Am. J. Med. Sci. 257, 415, 1969.

7. Schwille, P.O.: Urinary acid base parameters in stone formers; Renal Stone Symposium, Bonn, 1972 (in press).

8. Gutman, R.A., Fink, G., Voyles, N., Selawry, H. Penkos, J. C., Lepp, A. and L. Recant: Specific biologic effects of intestinal glucagon-like materials; J. Clin. Invest. 52, 1165, 1973.

9. Schwille, P.O., Jügelt, U. and A. Sigel: Allopurinolabhängige Elektrolyt- und Stoffwechseländerungen im Urin von Nierensteinkranken; Urologe A 11, 185, 1972.

10. Coe, F.L. and L. Raisen:Allopurinol treatment of uric acid disorders in calcium stone formers; Lancet January 20, 129, 1973.

11. Hoffman, R.S., Martino, J.A., Wahl, G. and R.A. Arky: Fasting and refeeding. III. Antinatriuretic effect of oral and intravenous carbohydrate and its relationship to potassium excretion; Metabolism 20, 1065, 1971.

12. Frey, H.M.M., Falch, D., Forfang, K., Norman, N. and D. Fremstad: effects of long term infusion of glucagon on carbohydrate metabolism, insulin and growth hormone secretion in patients with congestive heart failure; Diabetes 21, 939, 1972.

OBSERVATIONS CONCERNING THE INCIDENCE OF DISTURBANCE OF LIPID AND CARBOHYDRATE METABOLISM IN GOUT

O. Frank

Special Hospital for Rheumatic Diseases

Baden, Austria

Disorders of lipid metabolism and impaired glucose tolerance in gout are well known. Their relation to gout could be verified by means of statistic evaluation but the reason for this association has remained obscure just as the frequent observation of coronary heart disease in gouty subjects.

In 4o unselected cases of primary gout and 45 males and 17 females with symptomless hyperuricemia cholesterol and triglyceride concentrations in serum, lipoprotein patterns and carbohydrate metabolism including an intravenous tolbutamide test were studied. The same investigations were done in 3 control groups consisting of normouricemic subjects suffering from osteoarthrosis and matched with gouty and hyperuricemic patients for sex, age and body weight expressed as percentage of normal weight. Table 1 shows a good conformity regarding the mean values and standard deviations of age and body weight between patients and controls.

Table 1

	n	SUA $\bar{x}$	Age	%Weight
GOUT	40	7.6	51.8 ± 9.2	126.5 ± 17.4
CONTROLS	40	5.6	52.4 ± 8.7	126.3 ± 14.0
HYPERURICEMIA	45	7.7	53.5 ± 10.1	113.7 ± 14.3
CONTROLS ♂	45	5.0	54.0 ± 10.3	113.4 ± 12.9
HYPERURICEMIA	17	6.3	56.9 ± 11.1	124.4 ± 12.7
CONTROLS ♀	17	4.7	58.8 ± 7.2	126.0 ± 12.8

Table 2

	n	HYPERLIPOPROT. type			DIABETES overt lat.		FHD	HYPER- TENSION	CHD
		IIa	IIb	IV					
GOUT	40	1	1	21	4	14	3	9	13
CONTROLS	40	1	1	4	3	14	9	10	6
HYPERURICEMIA♂	45	2	3	7	1	17	4	9	8
CONTROLS	45	1	1	6	-	17	5	12	5
HYPERURICEMIA♀	17	2	1	1	3	7	3	6	6
CONTROLS	17	1	1	2	-	8	3	8	4

FHD = FAMILY HISTORY OF DIABETES

CHD = CORONARY HEART DISEASE

As shown in table 2 in primary gout type IV hyperlipoproteinemia was found in 21 cases or 52.5 % but in the control
group this abnormality was identified in 3 patients or 7.5 %
only. The difference is highly significant with a chi square
of 17.6. In 1 patient each hyperlipoproteinemia type IIa and
IIb was found both in gout and in controls. In males and
females with symptomless hyperuricemia the frequency of type
IV hyperlipoproteinemia was essentially smaller with 15.6
and 5.9 % respectively. There was not found a clear difference against the control groups. Against that in hyperuricemic patients type II hyperlipoproteinemia was recorded
somewhat more frequently than in gout.

The mean values of cholesterol and triglycerides in
serum were significantly higher in gout than in controls
but only the triglycerides exceeded the normal range. In
hyperuricemic patients there was no significant difference
of cholesterol and triglyceride concentrations against the
control groups (table 3 below).

Table 3

	n	CHOLESTEROL	P	TRIGLYCERIDES	P
GOUT	40	277 ± 66	<0.001	240 ± 110	<0.001
CONTROLS	40	227 ± 42		151 ± 65	
HYPERURICEMIA	45	235 ± 54	>0.05	193 ± 155	>0.05
CONTROLS ♂	45	225 ± 45		168 ± 107	
HYPERURICEMIA	17	240 ± 59	>0.05	141 ± 87	>0.05
CONTROLS ♀	17	230 ± 45		164 ± 81	

We could not find a significant difference regarding the frequency of overt or subclinical diabetes in gout and symptomless hyperuricemia compared with the control groups. A family history of diabetes was found more rarely in gout than in controls, the difference proved to be not significant.

13 patients with gout and 6 control persons showed symptoms of coronary heart disease evidenced from typical complaints of angina of effort and ischemic ECG changes. Likewise in hyperuricemia coronary heart disease was found more often than in the respective controls, but just as in gout the difference was not significant. The incidence of hypertension in gout and hyperuricemia was almost the same as in controls.

In gouty patients with type IV hyperlipoproteinemia a subclinical diabetes was found in 47.5 %, in normolipemic cases in 23.5 %. Urolithiasis, tophi and a duration of disease of more than 10 years were more common in hyperlipemic patients, the differences were not significant.

The pathogenetic differentiation of disorders of metabolism in gout is difficult because of the complexity of the conditions with partly mutual interrelations between the particular disorders. It must be considered that gout presents largely a clinical entity but is completely inhomogeneous in pathogenesis. Primary gout can arise from a variety of partly hitherto unknown mechanisms. In symptomless hyperuricemia the difficulties are augmented because of lacking clinical symptoms. Because of the rarity of primary gout or hyperuricemia in females the raised serum uric acid in that group is to explain by age and obesity in the majority.

The most striking finding is the high incidence of type IV hyperlipoproteinemia in gout. The highly significant difference against the controls points at a close relation between these metabolic disorders. Undoubdedly there are biochemical interrelations between type IV hyperlipoproteinemia and hyperuricemia but they explain more the incidence of hyperuricemia in primary disorders of lipid metabolism than vice versa. Because of genetic origin and familiar incidence of both type IV hyperlipoproteinemia and gout a genetic relation between these disorders appears to be plausible. This assumption could not be proved but is corroborated by the results of J e n s e n et al. who could find a significantly smaller interpair variance of lipids and uric acid in monozygotic than in dizygotic twins suggesting a genetic control of lipids and uric acid.

The incidence of overt or subclinical diabetes did not show a significant difference between patients and controls. On ground of this observation it can be assumed that diabetes has no direct relation to gout. With regard to the common incidence of glucose intolerance associated with hyperinsulinemia and decreased insulin sensitivity in gout obesity is to be regarded as cause of glucose intolerance, a relationship which is well documented.

The association of gout with obesity and disorders of lipid and carbohydrate metabolism results in an accumulation of risk factors for development of atherosclerosis. The high incidence of coronary heart disease and hypertension in gout and on the other hand of hyperuricemia in atherosclerotic patients brought about the assumption that gout and hyperuricemia per se are to be regarded as risk factors. Concerning the increasing incidence of both gout and coronary heart disease and the efforts for preventive measures even at present time this question is of great actuality and has been widely discussed. In many studies a relationship was suggested between hyperuricemia and gout and atherosclerosis but this assumption could not be confirmed by epidemiologic studies. So in Tecumseh Community Health Study serum uric acid in patients with coronary heart disease did not show a significant difference against the mean of the population studied. Therefore it must be considered that coronary heart disease has no direct relation to gout or hyperuricemia but presumable the incidence of coronary heart disease in gout is to be explained by obesity and partly by type IV hyperlipoproteinemia which are common attributes in gout. This explanation is in accordance with our results.

GOUT AND HYPERLIPIDAEMIA

T. GIBSON   and   R. GRAHAME

Guy's Arthritis Research Unit, London, S.E.1

INTRODUCTION

A number of previous reports have demonstrated an association between gout and hyperlipidaemia (Barlow 1968; Rondier et al 1970).

The mechanism linking the two disorders has not been defined and although obesity and excessive alcohol consumption are factors common to both diseases, attempts to implicate them in this context have not been successful (Wiederman 1972).

Another suggestion which may provide an alternative mechanism for the association has been made by Hennecke and Sudhof (1970). This postulates that hypertriglyceridaemia accompanying gout is associated with fatty infiltration of the liver. The same hepatic changes are a plausible explanation for the high incidence of abnormal bromosulphthalein retention in gout subjects reported by Grahame et al (1968).

This study is a reappraisal of the relationship between gout, obesity, excess alcohol consumption and liver dysfunction.

METHOD

Fasting serum cholesterol, triglyceride and uric acid levels were measured in 40 male gout patients and an equal number of controls matched for age, sex and ponderal index (height divided by cubed root of weight). A third comparative group was studied and comprised healthy controls matched for age and sex only. None

499

of the controls had a history of gout or of diseases associated with hyperlipidaemia.

All the subjects were questioned about their drinking habits. Those taking blood uric acid lowering agents were given Colchicine as a substitute in the week preceding investigations. Samples of blood were taken after an overnight fast.

The serum cholesterol and triglyceride estimations were performed by the methods of Cramp and Robertson (1968) and Sevine and Zak (1964). Blood uric acid measurements of the gout subjects was by the method of Caraway (1963) and that of the controls was performed in a different laboratory by the method of Lofland and Crouse (1965).

Gout patients also underwent investigation of liver function by estimation of serum bilirubin, alkaline phosphatase, aspartate amino transferase and bromosulphthalein retention. The last mentioned investigation was performed by the intravenous injection of 5mg/1kg. body weight B.S.P. and estimation of the percentage remaining in the circulation 45 minutes later.

Twenty-four hour urine uric acid estimations were also performed on the gout patients after they had undergone a 3 day low purine diet (less than 300 mg. purine daily). Uric acid assay was by the method of Caraway (1963).

RESULTS

Details of the subjects and their lipid and blood uric acid results are illustrated in Table 1. The controls not matched for obesity had a significantly higher ponderal index than the other two groups ($p < 0.0001$) i.e. they were much less obese. The mean blood uric acid of the gout patients was significantly higher than that of the matched controls ($p < 0.0001$). There was no correlation between blood uric acid values and serum triglyceride ($r = 0.09$) of the gout patients. The mean cholesterol values of all three groups was similar and no significant difference existed between them.

The highest mean serum triglyceride level was that of the gout subjects (202.6 mg.%) but this was not significantly different from the mean value of the controls matched for obesity (170.0 mg.%) ($p < 0.1$). A significant difference existed only between the mean triglyceride of the gout patients and the controls not matched for obesity (154.2 mg.%) ($p < 0.05$). There was a significant inverse correlation between ponderal index and the serum triglyceride of the gout patients ($r = -0.355 \quad p < 0.05$) i.e. the more obese had

TABLE 1

Details of Subjects

	Gout	Matched Controls	Unmatched Controls
Number (all male)	40	40	40
Mean age years (range)	52.4 (26-68)	52.0 (28-65)	46.5 (28-64)
Mean ponderal index ($\pm$ S.D.)	12.11 ($\pm$ 0.55)	12.1 ($\pm$ 0.53)	12.63 ($\pm$ 0.45)
Mean blood uric acid ($\pm$S.D.) mg./100 ml.	8.97 ($\pm$ 1.97)	6.09 ($\pm$ 0.68)	5.66 ($\pm$ 0.95)
Mean serum triglyceride ($\pm$ S.D.) mg./100ml.	202.6 * ($\pm$ 122)	170.0 ($\pm$ 70.6)	154.2 * ($\pm$ 60.1)
Mean serum cholesterol ($\pm$ S.D.) mg./100 ml.	276.0 ($\pm$ 57.9)	280.9 ($\pm$ 59.7)	254.0 ($\pm$ 52.1)

* $t = 2.031$ $p < 0.05$

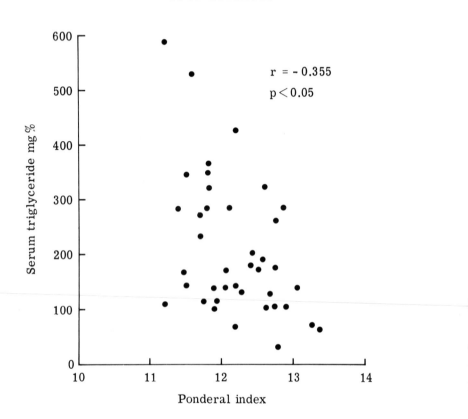

Fig. 1.   Correlation of ponderal index with serum triglyceride

higher levels (Fig. 1.)

Lean patients with gout were arbitrarily chosen as those with a ponderal index of more than 12.5. When their mean triglyceride level was compared with their matched controls, again, no significant difference was apparent (Table 2).

TABLE 2

Details of Non-Obese Subjects (P.I. > 12.5)

	Gout	Matched Controls
Number	12	12
Mean ponderal index ($\pm$ S.D.)	12.79 ($\pm$ 0.23)	12.8 ($\pm$ 0.34)
Mean serum triglyceride ($\pm$ S.D.) mg/100ml.	151 ($\pm$ 90.2)	147 ($\pm$ 64.02)

In this series, 17 (42%) of the gout patients drank more than 3 pints of beer (or the equivalent) daily and these were considered heavy drinkers. None of the control subjects was a heavy drinker. The mean serum triglyceride of the heavy drinkers (271mg%) was significantly higher than that of the non- drinking gout patients (150.0 mg.%) (p < 0.01) (Fig.2.) and than that of the obesity matched controls (178 mg.%) (p < 0.025).

Standard liver function test abnormalities were found in three gout patients who had elevation of serum aspartate amino transferase (Table 3). These also had abnormal B.S.P. retention (more than 5% at 45 minutes after injection) and a total of 14 (35%) of the gout subjects had retention exceeding the normal. There was no obvious relationship between heavy drinking and abnormal B.S.P. Only 6 of those with evidence of liver dysfunction were heavy drinkers. Abnormal B.S.P. retention was not predominent amongst those with higher triglyceride levels (Fig. 3) and there was no correlation between serum triglyceride and the percentage of B.S.P. retained (r = 0.129 N.S.).

The mean 24 hour uric acid excreted by the gouty subjects was 664 mg. There was no correlation between urine uric acid and

GOUT SUBJECTS

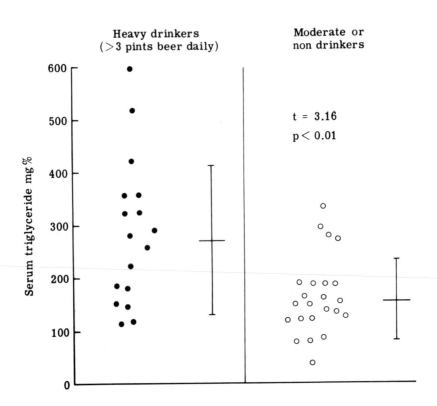

Fig. 2. Serum Triglyceride of heavy and non drinking gout subjects

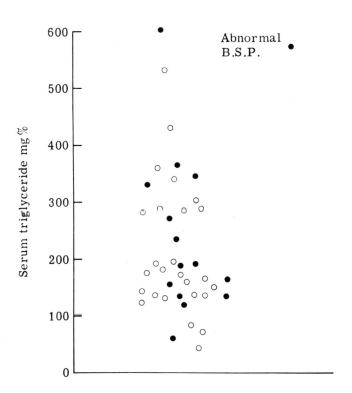

Fig. 3. Distribution of abnormal B.S.P. retention against
serum triglyceride levels.

either serum triglyceride (r = 0.04 N.S.) or serum cholesterol
(r = 0.16 N.S.).

TABLE 3

Enzyme Abnormalities in Gout Patients

Aspartate amino transferase (Normal<16 I.V.)	% B.S.P. retained at 45 minutes	Alcohol consumption (pints beer/daily)	Serum Triglyceride mg./100 ml.
40 units	7	0	156
56 "	19	3	132
33 "	6	2	122

DISCUSSION

The results confirm that gouty patients tend to have higher
fasting serum triglyceride levels than healthy non-obese subjects
but no difference exists with respect to cholesterol values.
These observations are in accord with those of Feldman and
Wallace (1964) and Darlington and Scott (1972).

It has been possible to show that obesity has a marked
effect on the levels and may be the major   factor responsible for
this observation. Previous accounts have concluded that that
difference cannot be explained on this basis (Emmerson and Knowles
1971; Wiederman et al 1972).

Alcoholic excess is well known to be associated with
hypertriglyceridaemia (Chaite et al 1972). It is therefore not
surprising that amongst gout patients, renowned to include a
high proportion of heavy drinkers (Grahame and Scott 1970), there
exists a number whose high serum triglyceride levels are associated
with the consumption of alcohol. The high triglyceride levels of
the heavy drinkers in this study could not be attributed to their
obesity because their mean level was significantly higher than
even that of their obesity matched controls.

Abnormal B.S.P. retention was apparent in a smaller percentage
of the gout patients than that reported elsewhere (Grahame et al
1968) and since this investigation was not performed in the control
subjects the incidence of abnormal results cannot be given
emphasis. It was quite clear however that evidence of liver
dysfunction was not related to the serum triglyceride level nor
invariably to abuse of alcohol.

A possible relationship between hyper-excretion of uric acid and hypertriglyceridaemia was entertained. Darlington and Scott (1972) have noted a low order correlation between serum triglyceride and urine uric acid but the results reported here do not confirm this observation.

The high serum triglyceride levels apparent in a large number of gout patients may offer some explanation for the high incidence of ischaemic heart disease in gout (Hall 1965). The obvious influence of obesity and alcohol excess on the triglyceride levels of gout patients suggested by this study not only provides some explanation for the observed high incidence of hypertriglyceridaemia but also suggests the means by which it may be treated.

SUMMARY AND CONCLUSIONS

A comparative study between the fasting lipid levels found in 40 gout patients, 40 equally obese healthy subjects and 40 individuals who were less obese, suggests that hypertriglyceridaemia found in association with gout can be explained on the basis of obesity and drinking habits. No relationship was found between high levels of fasting serum triglyceride and liver dysfunction nor was it possible to correlate uric acid excretion with serum triglyceride.

REFERENCES

Barlow K.A. (1968) Metabolism 17. 289 (Hyperlipidaemia in primary gout).

Caraway W.T. (1963) Standard Methods of Clinical Chemistry Edited by Seligson D. Academic Press N.Y. and London 4. 239.

Chait A. Mancini M. February A.W. and Lewis B. (1972) Lancet 2.62 (Clinical and metabolic study of alcoholic hyperlipidaemia).

Cramp D.G. and Robertson G. (1968) Analytical Biochemistry 25. 246 (The fluorometric assay of triglyceride by a semiautomated method).

Darlington L.G. and Scott J.T. (1972) Annals of Rheumatic Diseases 31.487 (Plasma lipid levels in gout).

Emmerson B.T. and Knowles B.R. (1971) Metabolism 20. 721 (Triglyceride concentrations in primary gout and gout of chronic lead nephropathy).

Feldman E.B. and Wallace S.L. (1964) Circulation 29. 508 (Hypertriglycerideamia in gout).

Grahame R. Haslam R.M. and Scott J.T. (1968) Annals of Rheumatic Diseases 27.19. (Sulphobromophthalein retention in gout and asymptomatic hyperuricaemia).

Grahame R. and Scott J.T. (1970) Annals of Rheumatic Diseases 29. 461 (Clinical survey of 354 patients with gout).

Hall A.P. (1965) Arthritis and Rheumatism 8. 846 (Correlations among hyperuricaemia, hypercholesterolaemia, coronary artery disease and hypertension).

Hennecke A. and Sudhof H. (1970). Deutsche Medizinische Wochenschrift 95. 59   (Liver involvement in gout).

Lofland H.B. and Crouse L. (1965) Automation in Analytical Chemistry (Technicon Symposia) (Uric acid determination using neocuproine copper reagent).

Rondier J. Truffert J. Le Go A.   Brouilhet H. Saporta L. de Gennes J.L. and Delbarre F. (1970). Revue Europeene d'Etudes Cliniques et Biologiques (Paris) 15. 959 (Gout and hyperlipidaemia).

Sevine J. and Zak B. (1964) Clinica Chimica Acta 10.381 (Total Cholesterol in serum).

Wiederman E. Rose H.G. and Schwartz E. (1972) Americal Journal of Medicine 53. 299 (Plasma lipoproteins, glucose tolerance and insulin response in primary gout).

LIPID AND PURINE METABOLISM IN BENIGN SYMMETRIC LIPOMATOSIS

M. M. Müller and O. Frank

Dept. of Medical Chemistry, University of

Vienna and Special Hospital for Rheumatic

Diseases, Baden, Austria

The benign symmetric lipomatosis (Adenolipomatosis of Launois-Bensaude (1), or Madelung's disease (2)) is characterized by multiple, symmetric deposites of subcutaneous fat on the neck, the nape, the back and upper part of the trunk. The etiology of this disease is not known: in almost every published case, however, excessive alcohol consumption is common (3). In some cases this disease in accompanied by an additional diabetes (4), by hyperlipoproteinemia of type IV with elevated triglycerides and by hyperuricemia (5).

Examinations of lipid and purine metabolism were only rarely undertaken. For our studies we used two patients with benign symmetric lipomatosis, one showing a symptomeless hyperuricemia.

MATERIALS AND METHODS

Case reports: Patient E.: The 41 year old patient suffered since 1967 from lipomatous deposits on his nape, neck and both upper arms. The lipomas of the nape were surgically removed three times. The family health record revealed no abnormalities. This patient was devoted to alcoholism.

Patient W.: The 53 year old patient had started suddenly a year before to deposit fat around his

shoulders, back, nape, and upper arms. Already in 1970
hyperuricemia and liver damage was diagnosed. In the
health records of the family the brother turned out
to have a primary gout and the mother diabetes. The
health history of the patient revealed heavy alcohol
consumption.

Studies on lipid metabolism: The sera of the
patients were examined for total lipids, cholesterol,
cholesterol esters, triglycerides, total phosphatides
and FFA using Boehringer kits. Lipoproteins were
separated on cellulose acetate and stained with Poisson
red. In addition 1 ml serum was extracted according to
Folch (6), and the phosphatides were separated by means
of thinlayer chromatography on Kieselgel G (Merck)
according to Zöllner (7). The spots were visualized
by ammonium molybdate/perchloric acid and evaluated
densitometrically.

Normal adipose tissue and lipomas from each
patient were excised, washed several times with saline
and homogenized. From this homogenate the lipids were
extracted. Total lipids were measured in this extracts
and then the composition of the neutral lipids was
determined by means of thinlayer chromatography (7).

Studies on purine metabolism: For the measurement
of the de novo purine synthesis 10 µCi glycine-C14 (U)
were administered orally in the morning to the patients
who were on a low purine diet. Following glycine-C14
administration the urines were collected in 3 portions
on the first day and in one portion on the remaining
days. 5 ml urine were applied to a Dowex cation-
exchange column (Dowex 50 WX 4, 200 - 400 mesh;
25 x 1,7 cm). The uric acid was eluted with water. The
effluent was collected in 3 ml fractions. The tubes
corresponding to uric acid were pooled, evaporated to
dryness, dissolved in Bray's solution and the radioactivity measured in the liquid scintillation counter.

Washed and hemolyzed erythrocytes of the patients
were used for the determination of phosphoribosylpyrophosphate synthetase (PRPP-synthetase) according to
Hershko (8), and for purinephosphoribosyltransferases
according to Kelley. Uric acid was determined enzymatically
using Boehringer kit.

The fasting patients were infused with fructose:
in the frist hour they received a saline infusion
followed by a two hour infusion (with the same drop

rate) of 1.0 g fructose / kg body weight / hour.

Various techniques: Glucose was assayed by
glucose oxidase method according to Boehringer.
Carbohydrate metabolism was examined by means of the
iv-tolbutamide test; calculation was performed
according to Lange and Quick (10). Protein was
determined by the biuret method according to Sols (11).

RESULTS AND DISCUSSION

The results of our laboratory data are summerized
in table 1.

Lipid Metabolism

Hyperlipidemia or hypertriglyceridemia was absent
in both patients. The FFA in serum of the patient W.

Parameter	Patient E.	Patient W.
Total lipids (mg/100 ml)	655	712
Total cholesterol (mg/100 ml)	191	242
Cholesterol esters (mg/100 ml)	147	206
Triglycerides (mg/100 ml)	70	116
Free fatty acids (mVal/l)	0.20	0.54
Total phospholipids (mg/100 ml)	270	232
Lysolecithin (% of PL)	4.5	3.4
Sphingomyelin (% of PL)	11.7	22.4
Lecithin (% of PL)	77.9	67.4
Colamine-Cephalin (% of PL)	6.8	6.8
Electrophoresis of lipoproteins	normal	normal
Blood sugar (mg/100 ml)	92	80
Tolbutamide test (T_3)	-5.9	-7.4
Serum creatinine (mg/100 ml)	1.43	1.20
Urinary creatinin (g/d)	1.62	1.60
Serum uric acid (mg/100 ml)	5.8	8.9
Urinary uric acid (mg/d)	640	850

Table 1. Laboratory data.

were slightly elevated. The composition of serum
phosphatides and the distribution of serum lipoproteins
were normal too.

Patient E.'s total lipids from lipomatous adipose
tissue was actually twice the normal level (see table 2).
However, the TLC separation of neutral lipids showed
no significant differences. The lipomatous just as the
normal adipose tissue was made up of 89 % triglycerides,
3 % free cholesterol, 4 % cholesterol esters,
4 % phosphatides (lecithin). Only traces of FFA could
be detected. These results are in accordance with the
finding of Gellhorn (12). This author also did not
find any differences in the composition of lipomas and
normal adipose tissue.

Carbohydrate Metabolism

Although the benign symmetric lipomatosis may be
accompanied by diabetes (4), our patients did not
show elevated blood glucose levels. The intravenous
tolbutamide test was normal in both patients.

Purine Metabolism

Both patients showed a somewhat elevated excretion
of uric acid (table 1). The hyperuricemic patient W_o
and also patient E. incorporated an increased amount
of glycine-C14 into urinary uric acid (figure 1):
Looking at the diagram one notices that within the
first 6 hours the maximum uptake of radioactivity is
reached by both patients. This is in contrast to
normal persons whose maximum reaches a lower level and

Parameter mg/mg wet weight	Patient E. normal	Patient E. lipoma	Patient W. normal	Patient W. lipoma
Total lipids	0.39	0.60	0.27	0.28
Triglycerides	0.35	0.56	0.24	0.24
Free cholesterol	0.01	0.02	0.01	0.01
Cholesterol esters	0.02	0.01	0.01	0.01
Phospholipids	0.01	0.01	0.01	0.02

Table 2. Lipid composition of normal adipose tissue and
lipomas from patients with benign symmetric lipomatosis.

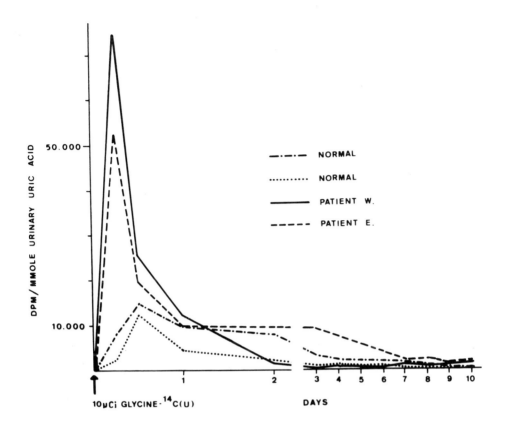

Figure 1. Radioactivity of urinary uric acid following oral administration of glycine-C-14(U).

is attained 6 to 12 hours following the glycine administration. During the first 7 days the incorporation of glycine-C14 into urinary uric acid for both control persons amounted to 0.13 % and 0.24 % respectively, whereas, in both patients with benign symmetric lipomatosis the incorporated glycine reached 0.54 % (patient W.) and 0.55 % (patient E.). These investigations show that patient W. with hyperuricemia as well as patient E. have an elevated purine synthesis de novo.

    In order to demonstrate whether this elevated purine synthesis is caused by an enzyme defect we examined a few key enzymes of purine metabolism in the patients' red cells. Both purinephosphoribosyltrans- ferase activities were within the normal range (tab. 3)

Patient	A-PRT	HG-PRT	PRPP-S
Normal (8)	15.36 ± 2.44	78.60 ± 11.51	24.00 ± 5.21
Patient E.	14.49	76.20	17.00
Patient W.	14.69	65.30	52.10

Table 3. Adenine phosphoribosyltransferase, hypoxanthine guanine phosphoribosyltransferase, and phosphoribosyl-pyrophosphate synthetase activities in hemolysates from patients with benign symmetric lipomatosis. Enzyme activities: nM/mg protein/hr

On the other hand, it is noteworthy that the hyperuricemic patient W. with 52 nM/mg protein/hr showed twice the normal activity of PRPP-synthetase. An elevated activity of this enzyme causes an increased production of PRPP (phosphoribosylpyrophosphate), a precursor in purine synthesis. This increased PRPP-synthetase activity producing an elevated uric acid production could be one of the reasons for patient W.'s elevated purine synthesis.

Gouty patients or people with primary asymptomatic hyperuricemia show a sustained increase in serum uric acid following fructose infusion. Both patients were infused (1.0 g/kg body weight/hr) for two hours. It was striking that the patient W., immediately after beginning the infusion, showed a significant rise in serum uric acid (fig. 2). During the entire infusion time the concentration of uric acid continued to increase. The uric acid level remained constant after completion of the infusion. This type of serum uric acid curve is usually representative for gouty patients (13). During fructose infusion in patient E. the serum uric acid did not change significantly, a type of reaction characteristic for metabolically healthy persons.

On the basis of our studies, a disorder of lipid and carbohydrate metabolism as origin of this disease might be excluded. It must be noted that both patients had an elevated purine synthesis de novo. The pathological mechanism of the disturbance in purine metabolism and its connection to the metabolism of adipose tissue remains obscure. As a clinical consequence of our investigations we would like to

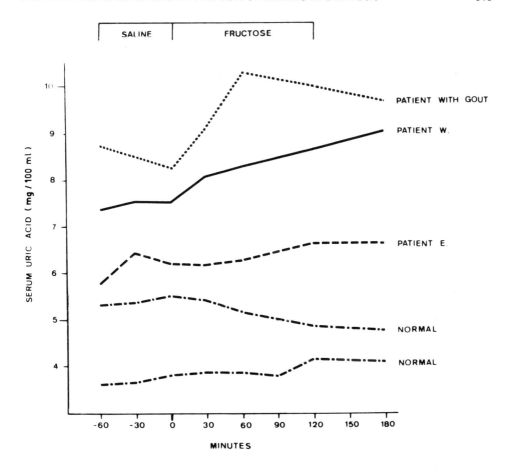

Figure 2. Changes of serum uric acid during fructose
infusion (1.0 g/kg body weight/hr).

suggest examinations of purine metabolism in every
case of benign symmetric lipomatosis and particular
attention should be attributed to alcohol
consumption in the case history.

SUMMARY

In two patients with benign symmetric lipomatosis
(one with hyperuricemia: serum uric acid 8.9 mg/100 ml)
the following investigations were performed:
(1)   Total lipids, total cholesterol, cholesterol
       esters, phospholipids, triglycerides and FFA
       in serum and lipomatous and normal adipose tissue.
       All parameters were within the normal range. The

distribution of serum lipoproteins was normal too.
(2)   Glycine-C14 was overincorporated into urinary
      uric acid in both patients.
(3)   Adenine- and hypoxanthine-guanine phosphoribosyl-
      transferase in hemolysates were within the normal
      range. Phosphoribosylpyrophosphate synthetase
      activity in the hemolysate of the hyperuricemic
      patient was increased.
(4)   Following infusion of fructose a sustained
      increase in serum uric acid was found in the
      hyperuricemic patient.

REFERENCES

(1)   Launois P.E., Bensaude R.: Soc. Med. Hop. Paris
          Bull. Memorois April 1, 298 (1898)
(2)   Madelung O.W.: Arch. Klin. Chir. $\underline{37}$, 106 (1888)
(3)   Brunner W.: Dtsch. Z. Chir. $\underline{244}$, $\overline{335}$ (1935)
(4)   Steinberg T.: Diabetes $\underline{16}$, 715 (1967)
(5)   Greene M.L., Glueck C.J., Fujimoto W.Y.,
      Seegmiller J.E.: Amer. J. Med. $\underline{48}$, 239 (1970)
(6)   Folch J., Lees M., Sloane-Stanley G.H.:
          J. Biol. Chem. $\underline{226}$, 497 (1957)
(7)   Zöllner N., Wolfram G.: Klin. Wschr. $\underline{40}$, 1098(1962)
(8)   Hershko A., Razina A., Mager J.: Biochim. Biophys.
          Acta $\underline{184}$, 64 (1969)
(9)   Kelley W.N., Rosenbloom F.M., Henderson J.F.,
      Seegmiller J.E.: Proc. Nat. Acad. Sci, USA
          $\underline{57}$, 1735 (1967)
(10) Lange H.J., Knick B.: Klin. Wschr. $\underline{43}$, 215 (1965)
(11) Sols A.: Nature (London) $\underline{160}$, 89 (1947)
(12) Gellhorn A., Marks P.: J. Clin. Invest.
          $\underline{40}$, 925 (1961)
(13) Al-Hujaj M., Schönthal H.: Med. Welt $\underline{22}$, 1887 (1971)

Renal Disease and Uric Acid Lithiasis in Urate Overproduction

AN INTEGRATIVE HYPOTHESIS FOR THE RENAL DISEASE OF URATE OVERPRODUCTION

B.T. EMMERSON

University of Queensland Department of Medicine

Princess Alexandra Hospital, Brisbane, Australia

In patients with both gouty arthritis and renal disease, it may be difficult to determine whether the gout followed the renal disease or the renal disease was secondary to the hyperuricaemia and gout. However, in genetic overproducers of urate, the abnormality of urate metabolism has been present from birth and is the primary abnormality. Renal disease in such subjects is most likely to be secondary to this metabolic abnormality. Thus, a study of kidney disease in patients with HGPRTase deficiency should give useful information concerning renal damage due to uric acid, especially as the study of Talbott and Terplan (1) concerning renal disease in gout showed uric acid crystals to be an almost constant finding.

Before describing observations on HGPRTase deficient patients, it is necessary to consider the effect of progressive glomerular insufficiency upon the urinary excretion of urate. Our observations of the 24 hour urinary excretion of urate on a purine free diet in patients with a wide range of glomerular filtration rates has shown that the urinary urate falls steadily with progressive reduction in the creatinine clearance (Observation 1).

In the case histories of patients with HGPRTase deficiency, one frequently notes that renal colic and uric acid calculi are often the first evidence of any abnormality and these manifestations frequently antedate the development of gouty arthritis by a number of years (Observation 2).

We have also studied a heterozygote for HGPRTase deficiency, with a reduced level of HGPRTase activity in erythrocyte lysates, who had a serum urate of 4.9 mg/100 ml and was asymptomatic.

Detailed study revealed that she excreted 760 mg of urate per 24 hours in her urine and that her urate production amounted to 853 mg/ 24 hours. She also incorporated excessive glycine into urate and had a urate clearance of twice normal. This woman therefore demonstrates that the serum urate concentration may be normal in the presence of urate overproduction, provided urate excretion is able to keep pace with the urate production (Observation 3).

Another patient with HGPRTase deficiency and minimal neurological signs presented with intractable gout at the age of 35 years and developed an episode of acute renal failure when his fluid intake was reduced below 3 litres/24 hours. This was attributed to the formation of uric acid crystals within his renal tubules so that an intense alkaline diuresis was instituted, resulting in a urine volume which, over one 24 hour period, exceeded 15 litres. This succeeded, however, in completely reversing his acute renal insufficiency and, when a regular urine volume of 5 litres/day was maintained, his renal function returned to normal. His usual 24 hour urinary urate excretion exceeded 2.5 g. Thus, an acute deterioration of renal function may occur in urate overproduction, which may be completely reversed by an intense diuresis (Observation 4).

We have also studied one very large pedigree with HGPRTase deficiency, details of which we have reported to this meeting in relation to the genetic linkage between HGPRTase deficiency and colour-blindness, and it has been apparent in this family that some members develop renal colic in adolescence, gouty arthritis in their thirties and die of renal failure in their forties, whereas other members of the family, with a similar degree of enzyme deficiency, live to their eighth decade without evidence of crystalluria, renal disease or gout. While it is clearly difficult to assess the causative factors in this situation, we have been struck by the fact that those who develop symptoms early, usually have urine volumes of less than one litre/24 hours, whereas those who do not develop the symptoms until an advanced age, habitually have urine volumes considerably higher than this, often close to 3 litres/24 hours (Observation 5).

We have also noticed that hypertension, when it occurs in HGPRTase deficient subjects before middle age, is not an early feature, but is usually seen in people with severe gouty arthritis who already have evidence of renal disease and gouty arthritis (Observation 6). We interpret this as suggesting that hypertension may be a sequel to the formation of microtophi within the kidney.

When the kidneys of urate overproducers are examined at autopsy, it is apparent that urate is found in two forms. The

first of these is as acicular crystals within microtophi, usually
being found within the interstitial tissue of the kidney and the
second is as amorphous accumulations of urates, usually being found
first within the lumen of a tubule and being potentially able to
cause obstruction of that tubule. We may thus look upon microtophi as being related chiefly to the urate concentration within
the kidney tissues and thus as being chiefly dependent upon the
serum urate concentration, although deposition within the renal
interstitium may well be determined by additional local factors.
On the other hand, the amorphous urate accumulations within the
tubules are clearly dependent upon the concentration of uric acid
in tubular urine. This is dependent upon both the total urate
excretion, the urine volume and also the urine pH. Now both of
these types of deposit are well recognized as occurring in gouty
kidneys (2), as is the fact that their presence predisposes to
hypertension, arterial disease, renal calculus formation, urinary
tract infection and moderate degrees of renal insufficiency (3).

But what are the factors which cause the deposition of urate?
The following sequence, based on these observations of HGPRTase
deficient patients, is proposed. While the renal excretory
capacity for urate is adequate, an overproducer of urate will have
a normal serum urate and will thus not be troubled by gout. At
that time, however, his urinary uric acid excretion will be high
and he will be subject to urate crystalluria and calculus formation
at any time when conditions for uric acid crystal formation are
favourable. When such urate deposition occurs within tubules
causing obstruction, it will result in impairment of renal excretory
capacity, firstly, of urate, but also of other substances. In
time, this will lead to a reduced renal excretion of urate, so
that hyperuricaemia results, and this can, in time, lead to the
development of gouty arthritis. This will not completely prevent
however, further urate crystal deposition in tubules when
conditions in the tubule favour this, and this will lead to
progressive impairment of renal excretory capacity for urate, until
a stage is reached when urinary uric acid has fallen to such a
level that the urinary uric acid concentration is no longer high
enough to promote crystal and calculus formation. However, by this
time, severe hyperuricaemia will have supervened, leading to
severe gout. This will then predispose to urate deposition as
acicular crystals within the renal parenchyma with microtophus
formation, which will thereby lead to further renal insufficiency.

While this proposed mechanism may not necessarily explain all
the observed facets of uric acid nephropathy, it is suggested that
it does contribute further to our understanding of the pathogenetic
mechanisms involved in this condition.

REFERENCES

1. TALBOTT, J.H. and TERPLAN, K.L. (1960). The kidney in gout.
    Medicine (Baltimore) 39:405.

2. SEEGMILLER, J.E. and FRAZIER, P.D. (1966). Biochemical
    considerations of the renal damage of gout. Ann. Rheum. Dis.
    25:668-672.

3. FINEBERG, S.K. (1958).   Gout nephropathy. J. Am. Geriatr.
    Soc. 6:10-16.

ACUTE INTRATUBULAR CRYSTAL DEPOSITION CAUSING PERMANENT RENAL

DAMAGE IN THE PIG

D.A.Farebrother[1], P.J.Hatfield[2], H.A.Simonds[2], J.S.

Cameron[2], A.S.Jones[3], A.Cadenhead[3]

[1]Wellcome Research Laboratories, Beckenham, Kent, U.K.
[2]Guys Hospital Medical School, London SE 1, U.K.
[3]Rowett Research Institute, Aberdeen, Scotland

The description of guanine gout in pigs by Virchow (1,2) and the recent reported recurrent leg weakness in pigs (3) were believed to have a common aetiology and stimulated re-investigation of purine metabolism in the pig.

Pigs were given a large purine load (guanine 150mg/kg/day) or a xanthine oxidase inhibitor (allopurinol 300mg/kg/day). When given separately no histological changes were seen but when given together, the two pigs receiving them became polyuric, polydypsic and wont off their food. One pig became so ill that it was taken off the mixture after a week, and both pigs were killed three weeks after going onto the mixture. The finding of a crystal nephropathy led to the second experiment, in which the effects of the crystals on the kidney were studied in more detail (4,5).

12 large white/landrace pigs were fed on a purine-free diet of barley, skimmed milk and water. 8 pigs were given allopurinol (300mg/kg) and guanine (150mg/kg) for a 3 week period. 4 pigs were kept as controls. To establish the genesis of change, 2 treated and 1 control pig were killed after 10 days and 21 days of dosing. The other 4 treated pigs were returned to a normal diet for 4 weeks, before being killed.

Food and water intake and urinary excretion were measured daily, and blood urea, creatinine clearance, and blood and urinary osmolarity measured at weekly intervals.

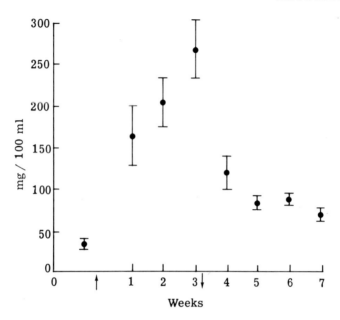

FIGURE 1 Blood urea.
Effect of dosing pigs with allopurinol (300mg/kg) and guanine
(150mg/kg) for 3 weeks, on blood urea levels.
There was an increase thoughout the period of dosing.   When
dosing finished, there was a rapid fall toward normal values;
the predosing level was not reached.
Each point is the mean of 4 pigs, and the bar is the standard
error of the mean.
↑ = dosing started.   ↓ = dosing stopped.

        At autopsy, the heart, lung, liver, kidney and spleen were
weighed, and thin slices of these organs, together with tissue
from the rectus abdominis, pancreas, salivary gland and duodenum
were fixed in absolute alcohol or 10% formalin buffered to pH
6.8.

RESULTS:
        Within a few days of going onto the mixture the average
urinary volume increased from 3 to 6 litres/day.   The urinary
osmolality rapidly fell below that of the slightly elevated
plasma osmolality.   Renal function deteriorated rapidly; blood
urea levels rose from 30 to 300mg/100ml., and creatinine
clearances fell from 100 to 20mls/min.

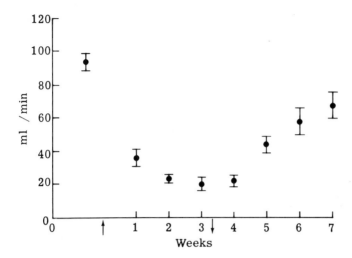

FIGURE 2 Creatinine clearance.
Effect on endogenous creatinine clearance of dosing pigs
allopurinol (300mg/kg) and allopurinol (150mg/kg) for 3 weeks.
During the first week of dosing there was a rapid fall, this fall
continued at a slower rate for the remainder of the dosing period.
There was a rise in clearance values during the 4 week recovery
period, but the predosing rate was not obtained.
Each point is the mean of 4 pigs and the bar is the standard
error of the mean.
↑ = dosing started. ↓ = dosing stopped.

At autopsy of pigs killed at 10 and 21 days, the kidneys
were enlarged and oedematous; they were twice the control
weight as g/kg of body weight. The cut surfaces showed orange
streaks radiating from the medulla to the capsule.

Microscopically, the alcohol fixed sections contained large
numbers of birefringent crystals within tubules; some were
spherical with a maltese cross birefringence pattern, whilst
others were amorphous. The crystals were shown by chemical
analysis to consist of a 2:1 mixture of xanthine and oxypurinol
(6).

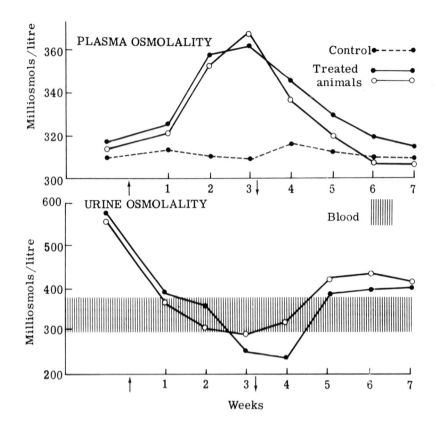

FIGURE 3
Effect on osmolarity of the plasma and urine (milliosmals/1) of
dosing pigs with allopurinol (300mg/kg) and guanine (150mg/kg)
together for 3 weeks.
(a) During the dosing period there was a small rise in plasma
osmolarity, which returned to normal during the 4 week recovery
period.
(b) In the first week of dosing there was a rapid fall in urine
osmolarity, which continued to fall to below the osmolarity of
plasma during the 2nd and 3rd weeks of dosing. Within 2 weeks
of stopping dosing, the urine osmolarity rose to just above
that of the plasma (400 millismals/1), but failed to rise any
more.
Each point is the mean of 2 pigs. The shaded area on the
lower graph (b) is the limits of plasma osmolarity.
↑ = dosing started. ↓ = dosing stopped.

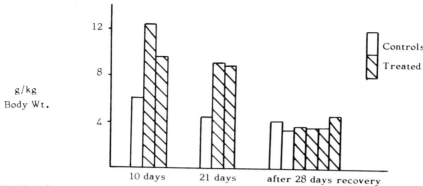

FIGURE 4

Effect on kidney weight of dosing pigs with allopurinol (300mg/kg)
and guanine (150mg/kg) together for 3 weeks. (g/kg body weight).
The kidneys of pigs killed at both 10 and 21 days weighed twice
that of the control pig killed at the same time. At the end of
the recovery period the kidneys of treated and controls were
similar.

At 10 days:

In animals killed at 10 days of dosing, the crystals were
always found in either the distal tubules or collecting ducts,
where a variety of pathological changes were seen. Many tubules
were dilated; in some the epithelium had become flattened and
degenerated giving it a scalloped appearance; in other tubules
the epithelium had regenerated and had grown over the crystals,
The basement membranes of some tubules had ruptured, and at
these sites there were surrounding foci of acute inflammation.
The interstitium was oedematous, and contained small localised
areas of acute inflammatory cells surrounding damaged tubules.

At 21 days:

In animals killed at 21 days the, end of the dosing period,
the lesions associated with the crystals were equally frequent,
but the majority were more advanced. In these, the deposits
were surrounded by chronic inflammation, consisting mainly of
mononuclears but with some macrophages and multinucleate giant
cells. There was often little or no evidence of the original
tubule,so that the inflammatory focus appeared to be in the
interstitium. The interstitium showed widespread inflammatory
infiltration. Tissue not involved by inflammation appeared
normal.

After recovery:
        In the pigs which had had no drug for four weeks, the blood
urea fell from 300 to 60mg/100ml, and creatinine clearance rose
from 20 to 60mls/min., but there was little improvement in
urinary concentrating power.

        At autopsy, the kidneys were a similar size and weight to
those of the controls, but the cut surfaces and capsules showed
scarring.

        Microscopically, there was a great reduction in the number
of crystals which were surrounded by macrophages and fibrous
tissue. The inflammation had regressed leaving widespread
fibrous tissue mostly along the medullary rays. Occasionally
several rays were linked together by fibres which weaved between
and around tubules and glomeruli, trapping them and producing
glomerulosclerosis. The glomeruli and tubules away from the
scar tissue appeared normal.

Other tissue examined:
        No pathological changes or crystals were seen in the other
tissues examined.

DISCUSSION:
        Virchow (1,2) suggested that pigs suffered from guanine
gout because they lacked the enzyme guanase. In a recent re-
investigation of purine metabolism in the pig, Simmonds et al.
(5), showed that the pig did not lack the enzyme. Similar to
most animals, the end product of purine metabolism in the pig is
the highly soluble allantoin, but after dosing with allopurinol,
the almost insoluble xanthine is excreted in the urine.

        Feeding large quantities of guanine or allopurinol separa-
tely did not produce any histological changes, but when administered together, crystalline deposits were found in the kidneys,
which were shown to consist of xanthine and oxypurinol. The
position of the crystals in the distal tubules suggests that
their formation at that site was due to concentration of urine.

        Large numbers of crystals were present in the distal tubules
and collecting ducts of animals killed during or at the end of
the 3 week dosing period, but crystals were not seen in any other
tissue examined. This finding was unexpected as xanthine
crystals have been reported in the muscle of xanthinuric patients
(7), and in patients with primary gout treated with allopurinol
over a long period (8).

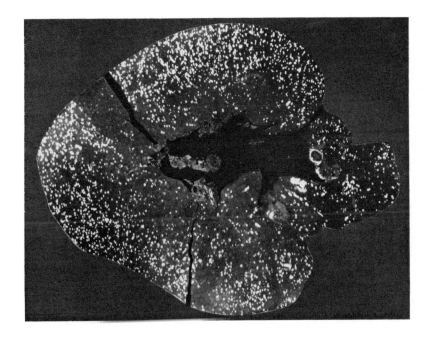

FIGURE 5
Crystal deposits in the kidney of a pig killed after dosing for
10 days with allopurinol (300mg/kg) and guanine (150mg/kg).
                        5µ section. Polarised light.
                        magnification X.6

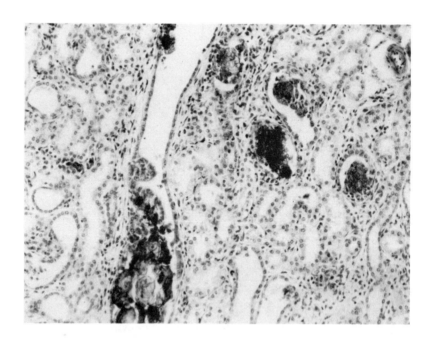

FIGURE 6
Crystal deposits in the renal distal tubules and collecting ducts
of a pig killed after 10 days dosing with allopurinol (300mg/kg)
and guanine (150mg/kg).
In the tubule cut L.S., crystals are compressing the epithelium,
causing cellular degeneration (top), and there is epithelial
proliferation with cells growing over the crystals as if attempting to keep the tubule complete (centre). The basement membranes
appear to be complete.
The crystals in the tubules (left) are also overgrown by
epithelium. The basement membranes can be seen around the normal
parts of the tubules, but are lost in the region of the crystals
where there is inflammatory infiltration.
                     5µ H.E.   magnification X.180

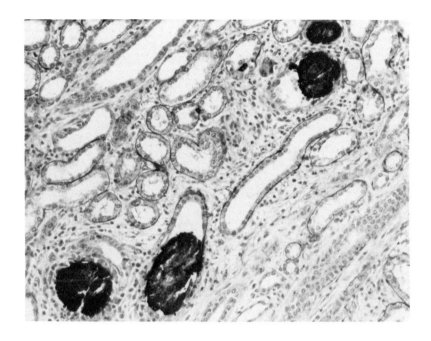

FIGURE 7
Kidney section from a pig killed after 21 days treatment with
allopurinol (300mg/kg) and guanine (150mg/kg).
The basement membranes are complete in the normal tubules. In
the tubule (top centre) part of the epithelium is normal, with
a basement membrane, but in the area of the crystal, both
disappear in an inflammatory focus. The basement membrane of
the tubules containing crystals (bottom left) have gone, although
there are still epithelial cells. The tubules are surrounded
by inflammatory cells. The crystals (top right) surrounded by
inflammatory cells appears outside the tubule.
                    5μ section. Hexamine Silver, H.E.
                    magnification X.150

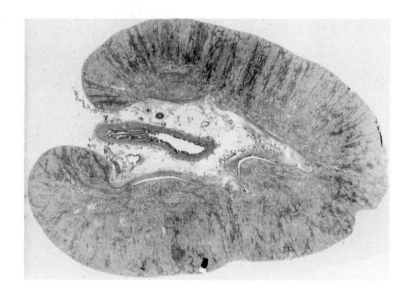

FIGURE 8
Cross section of whole kidney showing widespread fibrous tissue,
principally involving the medullary rays. Kidney from pig dosed
for 21 days, allopurinol, 300mg/kg and guanine (150mg/kg, then
allowed 28 days recovery period.

5µ section. Alcian Blue P.A.S.
magnification X.6

The pathogenesis of the tubular damage appeared to be the deposition of crystals, followed by tubular epithelial damage, then proliferation of the epithelium with the crystals being overgrown, and lastly degeneration of the basement membrane, which could not be demonstrated in the absence of an inflammatory response. The associated finding of oedema of the adjacent interstitium, could be due either to: an inflammatory substance released from the degenerating cells, or to urine leaking through the damaged cells into the interstitial tissue and the releasing histamine or other inflammatory factors (9,10).

The crystals were probably deposited in the first 2 or 3 days of treatment. This is supported by the early onset of polyuria and the great increase in kidney weight at 10 days; much of the increased renal mass being due to oedema. Renal function deteriorated throughout the entire dosing period, which is illustrated by the continued change in the blood urea and creatinine clearance. During the recovery period, renal function improved, blood urea levels fell and creatinine clearances rose, but the improvement in concentrating ability was less marked.

The initial increase in urine volume was considered to be due to a crystal induced osmotic diuresis, and then, as tubules became blocked, a solute diuresis would occur in the remaining patent nephrons. In the chronic stage when the intratubular crystals had disappeared, tubular damage and the consequent loss of concentrating ability, plus the loss of nephrons, due to scarring would be responsible for the continued diuresis.

This study has shown that deposition of crystals in the renal tubules even for short periods, can produce irreversible damage, not only locally, but also affecting other parts of the nephron. It is hoped that this experiment, together with a long term study now underway, will provide a model for crystal nephropathy in man.

References:
(1) R. Verchow; Verchow's Arch. Path. Anat. Physiol. 35, 358, 1866.
(2) R. Verchow; Verchow's Arch. Path. Anat. Phsyiol. 36, 147, 1866.
(3) T. Walker, B.F. Fell, A.S. Jones, R. Boyne and M. Elliot; Vet. Rec. 79, 472, 1966.
(4) H.A. Simmonds, P.J. Hatfield, J.S. Cameron, A.S. Jones, and A. Cadenhead; Biochem-Pharmac, 1973. In press.
(5) D.A. Farebrother, P.J. Hatfield, H.A. Simmonds, J.S. Cameron, A. Cadenhead and A.S. Jones. Manuscript in preparation.

(6) R.B. Parker. Manuscript in preparation.
(7) R.W.E. Watts, J.T. Scott, R.A. Chalmers, L. Bitensky and
    J. Chayen; Q.J. Med. 40, 1, 1971.
(8) R.W.E. Watts, W. Sneddon, and R.A. Parker; Clin.Sci. 41, 153,
    1971.
(9) J. Brod, The Kidney; 1973. Published by Butterworth & Co.,
    London.
(10)A. Golden, J.F. Maher, The Kidney: Structure and Function
    in Disease; 1971. Published by Williams & Williams Co.,
    Baltimore.

ANALYSIS OF TREATMENT RESULTS IN URIC ACID LITHIASIS WITH AND WITHOUT HYPER-URICEMIA

C. E. Alken, P. May and J.S.Braun

University Urologic Clinic
Homburg/Saar

Numerous literature assertations and our own observations confirm an increase in uric acid stone patients in the past 20 years.

Owing to successful conservative medicinal therapy with Uralyt-U and Allopurinol, an entirely ambulatory treatment of most patients with uric acid stones is possible. Prerequisite for this, however, is a consequential specialized urological control of the therapy course, in particular during the dissolution phase. The uric acid stone prophylaxis can be overseen by the family physician.

With consequential treatment of otherwise uncomplicated cases, over 9o% successful chemolitholysis with Uralyt-U and Allopurinol could be attained.

A more problematic and strictly selective body of patients were those stationarily treated for uric acid stones. Stationary treatment is only carried out in

cases of anuria or urinary obstruction, or in radio-
logically functionless or severely functionally
damaged kidneys as well as by partially calcified or
large concrements with therapy-resistant urinary
tract infection and naturally in cases of lodged ureter stones.

Of a total of 203 uric acid stone patients whose
diagnosis, owing to one of the aforementioned findings,
necessitated stationary treatment between 1965 and
1971, 153 were able to undergo routine control
reexaminations.

In 14 of the patients, familiar histories of gout or
uric acid stones were known. Over one-half of the
patients were overweight. In only 65 patients could a
primary stone be diagnosed, based on the case history,
while 21% of the cases were primary receditives and
in 35% of the cases frequent stone formation or
passage of a stone were known in the previous history.

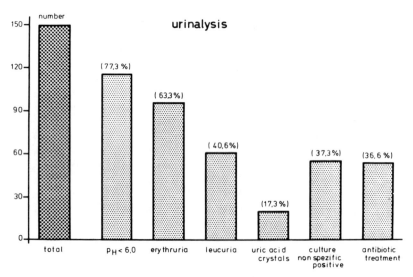

Fig. 1   Urine examination of uric acid stone patients.

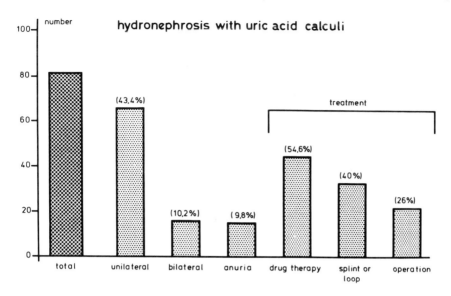

Fig. 2   Frequency of urine flow obstruc-
                  tion or anuria in uric acid
                  lithiasis.

As expected, in 116 patients, an acidic urinary pH
was observed and in over onethird, an antibiotic
therapy was necessary owing to an ascertainable bacterial urinary tract infection. This relatively high
percentage of infected uric acid stones can be
correlated to the similar high percentage of serious
obstructive disorders.

In 81 of the total 153 patients, the concrements lead
to obstruction of the preceeding canalicular system,
in 16 instances in both sides. In almost 10% of the
cases, an anuria was observed at the time of admission with serum creatinine values over 10 mg%, which
in six cases normalized by the time of discharge.

A rapid retrogression of the obstruction indicators
with reduction in size or passage of the concrement
were seen in 44 cases solely under conservative medicinal treatment. In 32 cases with total double-sided
or single-side ureter obstruction it was also possible
to avoid operative procedures. A ureter catheter could
be passed over the uric acid stone obstruction. Upon
simultaneous initiation of medicinal treatment with
Uralyt-U[R1] and Zyloric [R2] with antiphlogistic
therapy and antibiotic protection, the concrement
was either dissolved or so reduced in size that the
ureter catheter could be removed.

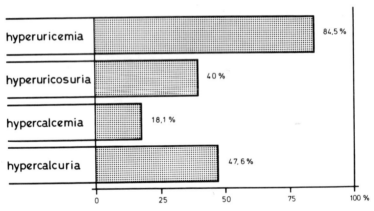

laboratory investigation

Fig. 3   Laboratory examinations during
                         uric acid lithiasis.

As long as no acute therapeutic measures were necessitated, in the rule, the serum uric acid levels were

R_1 - Dr. Madaus Co.
R_2 - Wellcome Co.

measured three times and in 84,5% of the cases, a
hyperuricemia was found; in 40% hyperuricuria was
observed. These figures lie well above comparable
results in the literature which generally state that
a maximum of 50% of the uric acid stone patients demonstrate a hyperuricemia.

Of note is the relatively high percentage of patients
with hypercalcemia and in particular, hypercalcuria
with simultaneous hyperuricuria. These observations
are, at least partially supported by SCHWILLE's uric
acid lithiasis examinations. As opposed to these
findings, recent observations of an increased
occurence of hyperuricemia and hyperuricuria in calcium oxalate stone patients have been reported.

In uric acid stone patients with hyperuricemia, we
principally combined a conservative therapie with
Uralyt-U and Allopurinol.

In a total of 83 of 117 patients, a purely medicinal
treatment was successful in an average of 48 days. In
34 further instances, additional instrumental allevia-
tion of acute obstructions was necessary. In 23% of
the cases it was necessary to operate owing to acute
urinary obstruction or anuria or due to unsuccessful
medicinal therapy.

Impressed by the good results of purely conservative
treatment of uric acid lithiasis, one might easily be
lead to believe that every uric acid concrement could
be thus treated and medicinally dissolved. However,
our clinical observations indicate that frequently,
only the urologist can determine whether a conservative

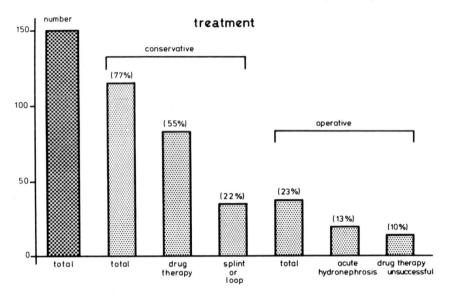

Fig. 4   Treatment results.

therapy alone is sufficient, or whether initially
or during the course of treatment, instrumental or
operative measures are necessary. Such is the case,
for example, if a reduced concrement leaves the renal
pelvis and upon entrance into the ureter, causes an
obstruction.

ISOLATION OF A RED PIGMENT FROM THE URIC ACID CALCULI

Bernardo Pinto

Laboratorio de Exploraciones Metabo-

licas,Aragon 420,Barcelona-13,Spain

Common observations indicate that Uric Acid renal stones
have a pink-redish color. This is on strong contradiction
with the white color, usually observed, on the shelf
uric acid. Most authors have suggested that it is due to
the special orientation of the rombic crystalls within
the calculi macles. However, ealier observations at this
laboratory shown that a pink-yellow color appeared on
the alkaline extracts from Uric acid calculi. Such a
finding can not be explained by crystallografic means.
The purpose of this paper is to show the preliminary
results regarding a red pigment of possible inorganic
structure existing in the Uric Acid calculi and responsible for the special type of color of these calculi,
however we still ignore its place of origin, exact
structure and role played on uric acid deposition and
crystallization.

MATERIAL AND METHODS

General Materials and Apparatuses. Renal stones were a
gift from Dr. Crespí, Ciudad Sanitaria, Barcelona (Spain)
Spectroscopy grade Potassium Bromide was bougth from the
Merck & Co., (Darmstadt, W.G.). Dowex-50 resin, 200-400
mesh was purchased from Sigma St. Louis, Missouri U.S.A.
Rest of reagents were the highest purity grade commercial-
ly available.
Infrared spectra   were obtained on a 700 and 377 Perkin

541

Elmer infrared spectrophotometers. Spectrophotometric
readings and spectra   on the U.V.-Visible range were
taken on a U.V.-Vis Hitachi-Perkin Elmer type 139, spec-
trophotometer. Centrifugations were performed on a Sor-
vall-superspeed RC-2B refrigerated centrifuge. Water
evaporations were done on a Büchi flash evaporator.Frac-
tions after chromatography columns were collected with
a Shandon-MBI fractions cutter.
Isolation of the Pigment. All the operations are performed at room temperature unless otherwise is indicated.
The Isolation process involves four different stages:
alkaline extraction, acidification, concentration and
ion exchange chromatography.
1- Alkaline Extraction. The uric acid renal stones weighing 31.719 g. were grounded to powder by using an agate
mortar and pestle. The stone powder is then extracted
with 1150 ml.of 2 M. OHNa. The non extractable stone
residue was dried at 100º C for 48 hours and after cooling,at room temperature, it weighed 5.061 g.
2- Acidification. The above mentioned alkaline stone
solution (pink yellowish color) was brought to pH 1 by
using concentrated HCl. After extract acidification a
color shift from pink-yellowish to green was commonly
observed. The acidified stone extract was kept at 4º C
for 48 hours. After centrifugation at 3,000 x g. for 20
minutes and heating the sediment at 100º C. for 24 hours.
A rather small precipitate, weighing 2.20 g., was obtained.
3- Concentration. The green supernatant from previous
step was concentrated to 50 ml. at 35º C by using a
flash evaporator. A tremendous white precipitate of 7.860
g. dried weight was obtained after centrifugation a 3,000
x g for 20 minutes.
4- Ion Exchange Chromatography. A Dowex 50, H^+ form,
column of 40 x 3 cm., equilibrated by thorough washing
with 0.01 M HCl, was prepared. On the top of the column,
the 50 ml. supernatant, from previous step, was poured.
After the column was thoroughly washed with 3 l. of 0.01
M HCl, elution began with 2 M HCl by collecting 20 ml.
fractions. The 290 mμ absorption was measured on each
fraction. When no absorption at 290 mμ was obtained,the
eluent was changed for 0.01 M OHNa, the alkaline elution
continued until the pigment came off. The pigment elution

can be observed directly, nevertheless the 520 mμ absorp
tion was taken. Those fractions containing the highest
520 mμ absorption were pooled together and concentrated
to dryness at 35º C on a flash evaporator. The red dry
powder residue weighed 18.26 mg.

Infrared and U.V.-Vis. Spectra . The dried sample –
potassium bromide disck was mixed on a 1/200 proportion
at 10 Tm. pressure and then the spectra   were taken on
the Infrared spectrophotometer.The U.V.-Vis range spectra   were obtained by dissolving 1 mg. of the respective
samples in 3 ml. of 0.1 M OHNa and 0.1 M HCl. The same
solutions were used as blanks.

Miscelaneous Determinations. The stone examination was
performed by using standard crystallografic, under pol-
larize light, and infrared spectrum combined procedures.
Attempts to get the melting point were carried out on a
Kofler device.

                          RESULTS

Infrared Spectrum. Two bending bands at 3,400 and 1,430
cm^{-1} and two stretching bands at 860 and 670 cm^{-1} are
the main findings to have into account.

U.V. and Visible range spectrum. No peak of absorption
on the U.V. region was noted, however two small peaks at
610 and 520 were observed depending upon the pH of the
solution. A shift from 610 to 520 was obtained when the
pH changed from acid to alkaline.

Miscelanoous properties. The melting point was higher
than 400º C and the compound is dialysed out under normal
dialysis conditions.

                        DISCUSSION

Data shown herein appear to indicate the presence in the
Uric Acid renal stones of a small molecule of inorganic
type. Though, no definitive structure is jet known and
the elementary chemical analysis it is now being investigated, however the infrared spectrum seems to suggest
the existence of a XO_3 group as the main component of its
structure. The isolation procedure does not indicate the
possibility of a carbonate group and additionally the
shift on the visible range associated to the pH changes
are neither explained by a carbonate structure.
The limited number of bending and stretching bands in

the infrared region together with the high melting point
are enough data suggesting its inorganic composition.
If not pinpoint structure is already available,obviously,
the mechanism of the color shift in the visible range
spectrum is unknown, however changes from free ion to
salt may be the cause for such changes.
If very little is known regarding the structure and
physical-chemical characteristics, its place of origin,
the presence in the uric acid stones and the physiological
significance is even less known.

Urate Deposition in Tissue

CONCENTRATION OF URATE BY DIFFERENTIAL DIFFUSION: A HYPOTHESIS FOR INITIAL URATE DEPOSITION

Peter A. Simkin

Division of Rheumatology, Department of Medicine

University of Washington, Seattle, Washington

For the sake of argument, I've decided to build this talk around the base of the great toe. We now accept the idea that acute gouty arthritis has a physical cause: the sodium urate monohydrate crystal. We can logically extend that concept, then, to say that there must be a physical reason or reasons why one joint receives such a disproportionate share of these crystals and of the miseries they inflict. Ideally, that same reason should explain why attacks are provoked by exertion, why they occur in the middle of the night and why gout usually strikes patients in their forties when hyperuricemia may have been present since puberty.

None of our present concepts really explains podagra. We have no histologic or chemical evidence of a basic difference in the synovium, cartilage or synovial fluid of the big toe. Nor do we have evidence that the crystals found there differ from those in other gouty joints.

I would like to suggest that involvement of this joint results from unusual degrees of physical stress, that this stress leads to degenerative joint disease, that further use of the damaged joint leads to synovial effusion, and that the nocturnal resolution of these effusions transiently concentrates uric acid. When this sequence of events occurs in a hyperuricemic patient, the increment in urate concentration may surpass its solubility limits and lead to crystal deposition.

Let's start by considering the biomechanics of the foot in normal walking. As the center of gravity moves forward in the stance phase of gait, the foot pronates throwing half of the body weight onto the base of the great toe. As the center of gravity crosses the ball of the foot and moves ahead the force to the floor is largely transmitted by the big toe. This is called the push-off phase of normal gait. In the first slide I have shown diagrammatically

547

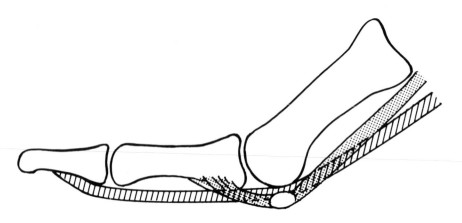

Figure 1. Lateral view of the first toe.

the muscles acting across this digit. The first metatarsalphalangeal
joint itself is flexed directly by the flexor hallucis brevis one
head of which has been shown in gray with its included sesamoid bone.
Passing between the two heads of the flexor brevis and inserting on
the distal phalanx is the flexor hallucis longus which is striped.

What I want you to notice about this slide is the poor mechanical advantage that both of these muscles have for the major downward force required in normal walking. In effect, both muscles act
as slings beneath the first MTP joint and both must transmit a force
of longitudinal compression to this small joint which is substantially
greater than the downward force they impart to the floor. The recent studies of Radin and his collaborators indicate that longitudinal compression plays a major role in the pathogenesis of degenerative
disease. We need a precise vector analysis of the forces across this
joint but that requires precise measurements and force determination
which have not, to my knowledge, been done.

The high prevalence of hallux valgus testifies to the longitudinal forces and the resultant joint disease found at this site.
In view of the severe stress it is not surprising that Kellgren and
Lawrence found much more degenerative disease here than in any other
weight bearing joint. They found radiographic changes in 43% of men
between the ages of 55 and 64. A subsequent series of dissections
of the big toe indicates that radiographs miss a lot of disease and
that 43% is likely to be conservative.

These mechanical considerations are presented at some length
in an attempt to persuade you that the most significant differences
we know of between the first MTP and other small joints are the
excessive stress and the frequent degenerative changes found there.
In view of these facts it seems reasonable to suggest that synovial

effusions must often occur at this small joint, that such effusions may be provoked by excessive walking, that they are most likely to occur in middle-aged subjects who have degenerative disease, and that they may resolve during the night when the foot is elevated and put to rest.

Were such an effusion to resolve by diffusion, the process could lead to an increase in the intrasynovial concentration of urate. This would occur because diffusion is dependent on molecular size. Since water is much smaller than urate, it diffuses faster. As water leaves the joint, urate will lag behind and its concentration will transiently rise.

A simple experimental model was developed which shows that diffusion gradients can, indeed, lead to concentration of urate in normal human connective tissues. In normal volunteer subjects, fenestrated plastic catheters were placed subcutaneously in the flexor surface of the forearm. An injection of ten ml of Ringer's solution was then made which was equal to plasma in its concentration of urate, and which also contained trace amounts of tritiated water and ^{14}C labelled urate or inulin and small amounts of epinephrine. The catheter was then cut near its insertion into the skin and from three to ten serial samples of subcutaneous fluid were collected in attached capillary tubes. Aliquots of each sample were assayed for uric acid and for both isotopes. The results of one such study are shown in the next slide. The concentrations of the labelled molecules, and of unlabelled urate are expressed as a percentage of their concentration in the injectate and are plotted against time. You will see that the concentration of tritiated water (open circles) falls rapidly but there is a rise in concentration of labelled urate (in triangles) and total urate(in squares).

The concentration of total urate rose in nine of the ten experiments with an average increment of 0.5 mg%. In the study shown here the urate level of the injectate was 7.6 mg% and of the last sample was 8.9 mg%. Remember, of course, that this is in the forearm. The increment at the big toe might be less, but on the other hand, it might also be more.

The absolute rise in ^{14}C shown here occurred in only three of the ten studies but in each case there was a marked divergence between the concentration of ^{14}C and tritium. The mean rates of change for each isotope could be determined from regression lines through the experimental points. Summary data are shown on the next slide. The rate for water (-0.72 %/min) is strikingly faster than that for urate (-0.18 %/min) or inulin (-0.08 %/min). They could only occur as a result of diffusion away from the injection site. It is this process of differential diffusion which concentrates total urate in this experimental model and which I would like to implicate in the pathogenesis of podagra.

In summary, there should be a physical reason for the urate crystal deposition which leads to podagra. A model has been developed which shows that urate may be concentrated by diffusion gradients in normal human connective tissues. It may be that hyperuricemic

P. A. SIMKIN

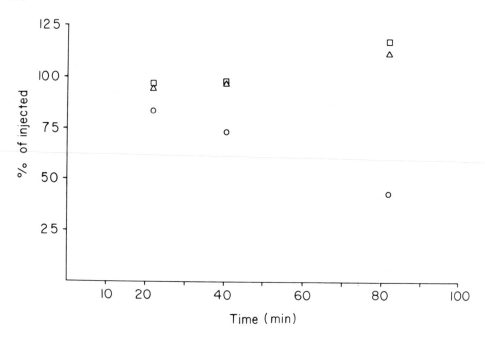

Figure 2. Diffusion experiment in the human forearm. Concentrations
of tritiated water (open circles) labelled urate (triangles) and
total urate (squares) are relative to the amount injected.

Synovial fluid is similarly concentrated beyond its solubility
limits during the nocturnal resolution of small effusions at the
base of the great toe.

EXPERIMENTAL HYPERURICEMIA IN RATS

R. Bluestone, J.R. Klinenberg, J. Waisman and
L.S. Goldberg Medical Service, VA-Wadsworth Hospital
Center, Depts of Medicine, Cedars-Sinai Medical Center,
and UCLA School of Medicine, L.A., California, U.S.A.

Renal Disease is a well recognized complication of gout, and the typical histologic features of chronic gouty nephropathy have been carefully studied, however, early evolution of the nephropathy and its precise relationship to hyperuricemia and uricosuria have been difficult to study in the human subject. Unfortunately, animal models of sustained hyperuricemia are difficult to produce and maintain. Recently, Stavric and colleagues described the production of hyperuricemia and urate nephropathy in rats fed orally with oxonic acid and uric acid. oxonic acid is a potent, hepatic uricase inhibitor; and this chemical, combined with uric acid or RNA dietry supplements, consistently leads to hyperuricemia in this species.

We have performed experiments to investigate the optimal conditions for producing sustained hyperuricemia in the rats. Further, we have tried to define objective methods to assay both the histologic changes in the kidney and the renal tissue deposition of uric acid.

Male or female wistar rats (200-300G) fed a 20G/day laboratory chow diet containing 2% by weight of oxonic acid and 3% of uric acid immediately develop sustained hyperuricemia and markedly elevated urinary urate concentrations. By housing animals in individual metabolic cages it is possible to ensure a constant intake of food and adequate 24 hour collections of urine. Under these conditions, the animals apparently thrive and continue to gain weight over a 3 week period. Sacrifice after 3 weeks of this diet reveals slightly swollen kidneys invariably showing an

551

irregular cortical surface with fine granules, pits or obvious
scars. Their cut surface reveals yellow deposits and streaks in
the pyramid, often extending to the cortex. Microscopical
examination shows completely normal glomeruli and vessels confirmed
by normal ultrastructural features on electron microscopy. However,
gross lesions are produced in the tubular portion of the nephron,
the collecting tubules and the interstium. These lesions consist
of dilated, atrophic tubules containing collections of polymorphs
and deposits of amorphous and acicular material. These amorphous
and acicular deposits stained positively using a De Galantha stain,
and polarising microscopy on alcohol-fixed sections show that the
acicular deposits are biregringent crystals. Thus it appears that
the tubular deposits probably contain amorphous uric acid and
crystalline sodium urate. The interstitium contains areas of acute
inflammation, associated with clusters of mononuclear cells,
fibrosis and occasional tophi characterized by large multinucleated
giant cells containing crystals. By grading the histologic
features on a 0-3 scale, a semi-quantitative measure of the
nephropathy is obtained for each animal.

It was highly desirable to have an objective measurement of total
tissue urate. Such a measurement would reflect intra-vascular and
interstitial uric acid in that tissue, together with deposited
crystalline sodium urate. An elution technique is therefore used
which permits extraction of uric acid from a whole kidney and gives
reproducible results. In this technique the dissected, stripped,
weighed kidney is homogenized in lithium carbonate thereby freeing
solubilised tissue, interstitial fluid and intravascular uric acid.
The homogenate is then dialysed against glycine buffer overnight at
4oC. The freed uric acid diffuses freely into the dialysate. The
uric acid concentration in the dialysate then reflects the amount
of uric acid eluted from that kidney and can be expressed as MG
urate per G wet weight of kidney tissue. It is now our practice
to use one kidney for the described histologic examinations and
one for the elution procedure from each animal studied.

After three weeks on the diet containing oxonic acid and uric acid
the following biochemical features were detected. There is clearly
a statistically significant hyperuricemia, increased urinary
urate concentration and increased tissue urate deposition.

This experimental model provides opportunities for many novel
experiments. Studies on the early evolution of the inflammatory
response to renal urate deposition are possible. Conversely, the
long-term sequelae of hyperuricemia and tissue urate deposition
can be observed and correlated with renal function. Moreover, the
modifying effect of the anti-inflammatory agents on the renal med-

TABLE I

HYPERURICEMIC NEPHROPATHY
BIOCHEMICAL FEATURES

Group	Diet	Plasma Urate (mean mg% ± SEM)	Urinary Urate (mg/24 hrs)	Eluted Urate (mg/G kidney)
I (n=18)	CHOW 20 G/DAY	2.7 (± 0.3)	22 (± 1)	0.5 (± 0.02)
II (n=21)	CHOW 3% URIC ACID 2% OXONIC ACID	4.8 (± 0.5)	184 (± 17)	1.5 (± 0.2)
p for difference		<0.001	<0.001	<0.001

These abnormalities are associated with uniform histologic changes in the renal tubules and interstitium

TABLE II

HYPERURICEMIC NEPHROPATHY
HISTOLOGIC FEATURES
(mean scores ± SEM)

Group	Diet	Tubules Atrophy	Exudate	Mitotic Figs.	Deposits	Interstitium Exudate	Fibrosis
I (n=18)	CHOW 20G/DAY	0	0	0	0	0	0
II (n=21)	CHOW 3% URIC ACID 2% OXONIC ACID	1.8(±0.2)	2.0(±0.2)	0.5(±0.2)	1.9(±0.2)	1.3(±0.2)	0.9(±0.2)

p for difference in every feature <0.001

ullary inflammatory response can now be observed. Indeed, preliminary experiments reveal that salicylate has a remarkable prevent-
ative effect on this type of nephropathy. Lastly, such studies
will hopefully provide valuable information pertinent to the whole
problem of the practical therapeutics of asymptomatic hyperuricemia.

Urate Binding to Plasma Proteins

THE BINDING OF URATE TO PLASMA PROTEINS

James R. Klinenberg*, Rodney Bluestone**
David Campion***, and Michael Whitehouse***
*Depts. Med., Cedars-Sinai Med. Ctr. & UCLA Sch. of Med.
**Med. Serv., VA-Wadsworth Hosp. Ctr.
***UCLA Sch. of Med., Dept. of Med., Los Angeles, Ca., USA

Although the etiologic role of microcrystalline monosodium urate monohydrate in the pathogenesis of acute gouty arthritis has been well established, the mechanism by which urate precipitates in joints, soft tissues and the kidneys remains unclear. Some studies have suggested that the solubility of urate is the key factor and that when the solubility of urate in blood synovial fluid and urine is exceeded, urate then precipitates. The question is more complex than this, however, since only approximately 25 per cent of patients with hyperuricemia develop acute gouty arthritis. Furthermore, there is poor correlation between uric acid concentration and the frequency of acute gouty attacks. This has led to speculation that there are factors in serum which are partially responsible for solubilization of urate and that alterations in these factors could be responsible for urate deposition.

The possible binding of uric acid by plasma macro-molecules has been extensively studied in the past using methods of ultra-filtration, equilibrium dialysis and electrophoresis. Such investigations have yielded conflicting but generally negative results, and it was concluded that there was no significant irreversible binding of urate to plasma proteins. However, in 1965 Alvsaker (1), using the techniques of gel filtration and immuno-electrophoresis followed by autoradiography, demonstrated the reversible interaction between certain macro-molecules and urate. Specifically, he demonstrated the interaction between an alpha$_1$-alpha$_2$-globulin and urate, and suggested that this was a specific urate binding protein. He also was able to demonstrate a deficiency of this protein in many patients with primary gout.

We reported in 1970 (2) a reproducible method for measuring the binding of urate to plasma proteins using the technique of equilibrium dialysis, in which the protein is enclosed within a semi-permeable membrane and dialyzed against a large volume of buffer containing free urate. After equilibration, the difference between the uric acid concentration in the membrane and that in the outside buffer represents the amount of urate bound and can be expressed as micrograms urate bound per ml of plasma, or per mg of protein. Using this technique it was found that although there was some binding of urate to many plasma proteins, the most quantitatively significant binding was to albumin. We were unable to demonstrate significant binding to a specific $alpha_1$-$alpha_2$-globulin. The factors which significantly influenced the binding of urate to albumin in this system included: pH, ionic strength, temperature, free urate concentration and protein concentration.

There was a linear relationship between the amount of urate bound and the free urate concentration in the buffer up to a free urate concentration of 20 mg/100 ml (Fig. 1). By using a free urate concentration of 15 mg/100 ml in our assay system, we were able to demonstrate binding in more physiological urate concentrations and maximize the chances of detecting differences between samples. It was found that most reproducible results could be obtained with a dialysis period of 16 hours. During this long period of time bacterial contamination in the dialysis bag was frequently a problem at higher temperatures despite meticulous sterilization techniques and, therefore, the majority of our studies have been performed at 4°C. Binding was greatest at low ionic strength and, therefore, the studies were generally done in 0.05 molar buffer.

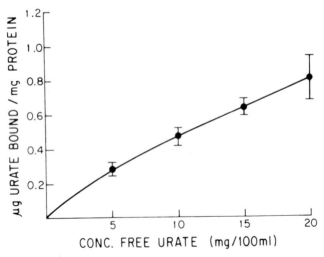

Fig. 1.   Binding of urate to albumin.

	Number of Subjects	mcg. urate bound/ml. plasma (at free urate concentration of 15 mg./100 ml.)	
		Mean	Std. Deviation (±)
Normal Volunteers	12	29	4
Tophaceous Gout - severe	3	15	1
Tophaceous gout	3	27	4
Non-tophaceous gout	14	26	4
Asymptomatic hyperuricemia	5	27	6
Renal disease	10	23	3
Patients taking 4 gm aspirin/day	5	5	4

Table I.   Urate binding capacity of plasma.

Although we have screened a variety of patients with gout, asymptomatic hyperuricemia, renal disease, etc., we were able to demonstrate a significant decrease in urate binding capacity of plasma in only a small group of patients with severe tophaceous gout and in uremic patients. However, during these studies (Table I) it was noted that all control subjects with rheumatoid arthritis, taking aspirin on a regular basis, had reduced plasma urate binding capacities. Since this was thought to be a drug effect, it seemed likely that other uricosuric agents would affect plasma urate in a similar way and we, therefore, studied the effects of commonly used agents in the treatment of gout on urate binding capacity and plasma urate in six normal volunteer subjects (3).

Table II demonstrates that allopurinol, aspirin, phenylbutazone, probenecid and sulfinpyrazone all significantly reduced the plasma urate concentration, while colchicine and indomethacin had no significant effect.

Drug	No. of Subjects	Mean plasma urate mg per 100 ml			P for difference between days 0 and 8
		Day 0	Day 8	Day 12	
Allopurinol	6	4.8	2.3	4.2	< 0.01*
Aspirin	6	5.1	2.2	4.7	< 0.01*
Colchicine	4	4.8	4.5	4.6	> 0.5
Indomethacin	4	6.1	5.9	6.2	> 0.5
Phenylbutazone	6	5.1	3.0	3.9	< 0.02*
Probenecid	6	4.2	2.1	5.0	< 0.001*
Sulphinpyrozone	6	5.6	2.2	4.8	< 0.001*

* Significant

Table II.   Effect of drugs on plasma urate.

Table III demonstrates that of these agents, aspirin, phenylbutazone and probenecid also significantly decreased urate binding, while the other agents had little effect.   This drug-induced
reduction of plasma urate binding detected in vitro might be because of competition between urate and drug for common protein
binding sites.

Drug	No. of Subjects	Mean plasma urate bound μg per ml			P for difference between days 0 and 8
		Day 0	Day 8	Day 12	
Allopurinol	6	29.2	27.5	28.7	> 0.1
Aspirin	6	24.8	6.8	30.7	< 0.001*
Colchicine	4	32.8	33.0	29.5	> 0.5
Indomethacin	4	29.3	25.5	28.8	> 0.02
Phenylbutazone	6	24.3	13.8	19.0	< 0.001*
Probenecid	6	33.7	15.5	21.2	< 0.001*
Sulphinpyrozone	6	30.3	26.0	29.3	> 0.1

* Significant

Table III.   Effect of drugs on plasma urate binding.

Drug	Concentration (mM)
Salicylate	0. 10
Phenylbutazone	0. 15
Octanoate	0. 20
Sulfinpyrazone	0. 20
Aspirin	0. 25
Indomethacin	0. 30
Probenecid	0. 45

Free urate concentration - 15 mg % (0. 9mM)

Table IV.   Concentrations of drugs displacing 50%
of bound urate from human albumin.

By expressing drug concentrations as that required to dis-
place 50 per cent of the baseline bound urate, the relative urate
displacing capacities can be compared (Table IV).   Aspirin is
again the most potent urate displacer, probenecid the least.   We
have subsequently demonstrated a variety of agents which inhibit
the binding of urate to plasma proteins and to date all such agents
have been uricosuric.   Other substances, such as bilirubin, have
also been shown to interfere with the urate-plasma protein interaction.

A prospective study of 5 jaundiced patients (Table V) has
shown a significant decrease in urate binding during the period
of jaundice and a statistically significant inverse correlation
between the serum urate and serum bilirubin concentrations (5).
Studies of the binding of urate to plasma from rats with surgi-
cally-induced jaundice have also shown significant interference
of the binding of urate to plasma proteins by bilirubin (Table VI).

A major question regarding the interpretation of the physiological significance of these studies has generally evolved around
the question as to whether the low temperature and low ionic
strength used in these in vitro studies would have any relevance
to the in vivo situation.   Therefore, it was felt necessary to
find a technique to examine the binding of urate to proteins under
the physiological conditions of pH 7.4, ionic strength of 0.16,
and at a temperature of $37^{\circ}C$.

Since previous studies had indicated that most of the urate
binding capacity of whole plasma resided in the albumin fraction,
we studied the possible competition between urate and drugs for
common binding sites by dialyzing purified human serum in albumin
against uric acid in vitro and adding therapeutically realistic
doses of drugs directly to the dialysis medium (4).

The amount of urate bound in the absence of drugs was regarded
as the baseline and the amounts bound in the presence of various
drugs were expressed as percentages of the control values (Fig. 2).
It can be seen that probenecid, sulfinpyrazone and sodium salicylate all significantly displaced urate from human albumin binding
sites, while colchicine had no effect.

Such studies have now been performed using an Amicon Ultrafiltration Cell and a continuous ultrafiltration technique of
Blatt, Robinson and Hixler (6). A 50 ml ultrafiltration cell with

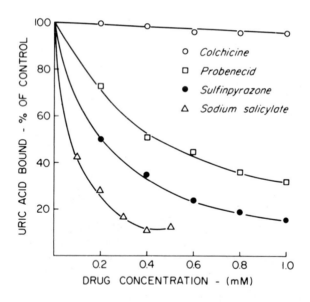

Fig. 2.   Displacement of urate from
           human serum albumin by drugs

Patient	Bilirubin (mg%)	Serum Urate (mg%)	Urinary Urate (mg/24 hrs)	Urate Clearance (ml/min)	Urate Binding (μg/ml)
1	21.5	5.3	932	12.2	27
	2.5	6.7	632	6.5	40
2	11.5	4.9	970	13.7	20
	6.7	6.0	683	7.9	30
3	3.5	3.6	315	6.1	20
	0.8	5.0	556	7.7	32
4	13.2	3.8	860	15.7	14
	1.0	3.7	575	10.8	41
5	6.0	5.6	484	6.0	25
	0.7	7.6	514	4.6	35
MEAN CHANGE	− 8.8	+ 1.2	− 120	− 3.2	+ 14.4
P VALUE	< 0.05	< 0.05		< 0.1	< 0.01

Table V.   Jaundice and urate handling.

DILUTION OF PLASMA (normal : jaundiced)	BOUND URATE (υg urate/ml plasma)
Normal Plasma	38
4 : 1	24
2 : 1	15
1 : 1	6
Jaundiced Plasma	0

Table VI.   Rat plasma urate binding:
effect of jaundice.

PM-10 Amicon membrane was connected via a one liter reservoir to a
nitrogen gas cylinder. A pH 7.4, 0.16 M phosphate buffer was used
in all experiments. The ultrafiltration chamber during the first
run contained 50 ml of buffer; during the second run it contained
50 ml of buffer and 5 mg/100 ml of de-fatted human serum albumin,
or pooled plasma. During both runs the reservoir contained a solution of sodium urate and phosphate buffer at exactly 15 mg/100 ml.
Using this technique, binding of urate to albumin has also been
confirmed at higher temperatures (Fig. 3). An analysis of the
binding kinetics at 23°C and 37°C revealed that at free urate concentration between 1 and 12 mg/100 ml, a constant fraction of
sodium urate is bound to serum human albumin. Per cent bound at
37°C was 19.6 ± 0.3; at 22.5°C was 22.6 ± 0.3, and at 4°C was
30.3 ± .6.

Figure 4 demonstrates that using this technique the effects
of salicylate and other drugs in displacing urate from the albumin binding sites has also been confirmed.

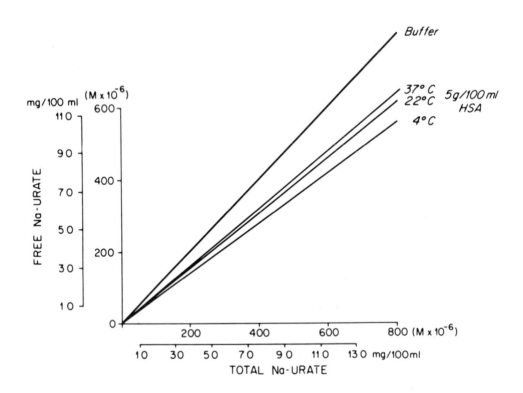

Fig. 3.   Binding of urate to albumin.

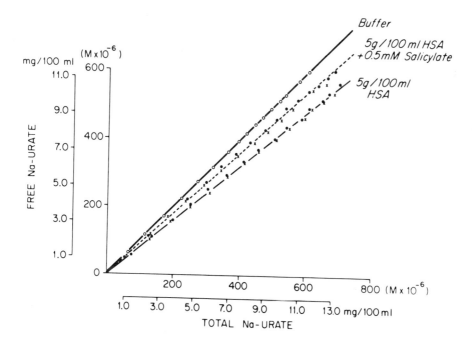

Fig. 4.   Effect of salicylate on urate binding
            as measured by ultrafiltration.

We have thus shown by two different techniques that urate is
bound to human serum albumin and that there is significant binding under physiological conditions using the in vitro ultrafiltration system. Nevertheless, the physiological in vivo significance of plasma urate binding remains unknown. If in vivo urate
binding does occur, it surely would raise several important speculative considerations. First, impaired binding might result in
reduced urate solubility, therefore contributing to deposition of
uric acid from hyperuricemic plasma into joints, tophi and kidneys. Attempts to gain insight into this problem are currently
underway in our laboratories. Secondly, an increase of free
urate induced by drugs might increase the total glomerular urate
clearance, thereby augmenting the renal tubular effects of these

drugs. Thus, albumin-bound phenylbutazone and salicylate may well
be inactive as analgesics, as well as many anti-inflammatory
agents, but at the same time they could have a uricosuric effect
even distinct from other direct tubular effects they might have in
their unbound state. Finally, since it seems most likely that
this binding of urate to human plasma proteins would prevail at
the glomerulus, it is probably necessary for us to reconsider the
hypothesis that essentially all urate is freely filtered by the
kidney and return to an older concept that at least a portion of
urate which is bound is not filterable.

REFERENCES

1) Alvsaker, J.O.: Uric acid in human plasma. III. Investiga-
    tions on the interaction between the urate ion and human
    albumin. Scand. J. Clin. Lab. Invest. 17: 467-475, 1965.

2) Klinenberg, J.R., and I. Kippen: The binding of urate to
    plasma proteins determined by means of equilibrium dialysis.
    J. Lab. Clin. Med. 75: 503-510, 1970.

3) Bluestone, R., I. Kippen, and J. R. Klinenberg: Effect of
    drugs on urate binding to plasma proteins. Brit. Med. J. 4:
    590-593, 1969.

4) Bluestone, R., I. Kippen, J. R. Klinenberg, and M. Whitehouse:
    Effect of some uricosuric and anti-inflammatory drugs on the
    binding of uric acid to human serum albumin in vitro.
    J. Lab. & Clin. Med. 76: 85, 1970.

5) Schlosstein, L., I. Kippen, R. Bluestone, M.W. Whitehouse,
    and J. Klinenberg: The association between hypouricemia
    and jaundice. Clin. Chim. Acta (In Press).

6) Campion, D., R. Bluestone, and J. R. Klinenberg: Uric Acid:
    Characterization of its interaction with human serum
    albumin. JCI (In Press).

PHARMACOLOGY

Allopurinol and Thiopurinol in the Treatment of Gout

EFFICACY OF SINGLE DAILY DOSE ALLOPURINOL IN GOUTY HYPERURICEMIA

Gerald P. Rodnan, M.D. James A. Robin, M.D.
Sanford F. Tolchin, M.D.

Division of Rheumatology and Clinical Immunology

Department of Medicine, University of Pittsburgh

Pittsburgh, Pennsylvania

During the past ten years, since its introduction by Rundles and his colleagues, allopurinol has gained widespread favor in the treatment of gout and other hyperuricemic states (1). Although it is conventional to prescribe allopurinol in multiple divided doses each day, the drug is metabolized in man in such a way as to suggest that a single daily dose may prove equally effective.

Allopurinol is quickly cleared from the plasma, the half-clearance time being on the order of two hours of less (2). In the main this is due to rapid conversion to the oxidation product - alloxanthine or oxipurinol. Like its parent compound, although to a lesser degree, oxipurinol also suppresses xanthine oxidase and thus leads to a reduction in serum and urinary uric acid levels. Like uric acid, but unlike allopurinol, oxipurinol appears to be reabsorbed by the tubules, since its renal clearance is much less than the glomerular filtration rate in man and it is not bound to plasma proteins (3). The half-clearance time of oxipurinol from the plasma is much longer than that of allopurinol, being 18 to 30 hours. It thus appears that much of the effectiveness of treatment with allopurinol may be due to the prolonged activity of oxipurinol. This being the case it seems reasonable to predict that a single daily dose of allopurinol would prove equally effective as multiple divided doses in the reduction of hyperuricemia.

In order to test this proposition, we have measured serum urate levels in 20 patients with gout, in an open labeled cross-over trial involving the study of two regimens of treatment with allopurinol, namely, that of taking the drug in a dose of 100 mg. three times a day versus taking a single tablet containing 300 mg. each day. These tablets were prepared and supplied to us by Burroughs Wellcome & Company.

All twenty of the patients were men, all presented histories of recurrent paroxysms of mono-articular inflammation characteristic of acute gouty arthritis – extending over periods of from one to 25 years – and all had hyperuricemia. The diagnosis of gout was confirmed in five individuals from whom synovial fluid was available by the demonstration of crystals of monosodium urate.

After obtaining their informed consent, the patients were entered into a 7-week long study. The trial for each patient consisted of four periods, the first of which was a 2-week "wash-out" period during which time no allopurinol was given. In the second period – Weeks 3 and 4 – half of the patients were given 100 mg. of allopurinol three times a day (Group A), and the other half (Group B) were given tablets containing 300mg. of allopurionl and instructed to take one such tablet each morning. The third period (Week 5) was a second "wash-out" period, during which, again, no allopurinol was taken. In the final period, of two weeks duration, the alternate regimen of allopurinol was administered. Each patient received colchicine and/or indomethacin throughout the trial in order to prevent acute gouty inflammation.

The patients were scheduled to return at the end of each week during the trial. Samples were obtained for enzymatic spectrophotometric measurement of serum urate, and the patients were questioned concerning possible adverse effects. In addition, blood and urine specimens were obtained from a number of patients and forwarded to Dr. Gertrude Elion for oxypurine analyses.

RESULTS : The 10 patients in Group A first received 100 mg. tablets of allopurinol three times a day, and subsequently (Weeks 6 and 7) were given a single tablet containing 300 mg. of the drug each morning. There was a substantial reduction in serum urate levels during both treatment periods, with average values of 6.1 and 6.4 mg. per 100 ml. respectively . A relatively close similarity in the degree of reduction in serum urate was observed in all but one of the individual patients, all of whom appeared to respond to the drug.

In the case of the ten patients in Group B the order in which allopurinol was administered was reversed, treatment with the single

Table 1 : Serum urate levels (mg. % ± standard deviation) at end of each week of study. The patients in Group A first received allopurinol, 100 mg. three times a day, and then were given (weeks 6 and 7) a single daily tablet of 300 mg. of the drug. In the case of the patients in Group B the order of treatment was reversed.

	Week of Study						
	1	2	3	4	5	6	7
	No treatment		Treatment		No Treatment	Alternating Treatment	
Group A (10)	9.0 ±1.6	9.7 ±1.5	6.4 ±1.5	6.1 ±1.2	8.6 ±1.2	6.2 ±1.0	6.4 ±1.5
Group B (10)	8.6 ±2.1	9.4 ±1.7	5.5 ±0.7	4.9 ±0.6	8.1 ±0.8	5.5 ±0.6	5.1 ±0.7

daily dose of 300 mg. preceding the use of a divided dosage schedule.
The reductions in serum urate levels which were observed again proved
comparable, with value of 4.9 mg. and 5.1 mg. per 100 ml. at the
conclusion of the two 2-week periods of treatment. Once again, there
was a notably close uniformity of response to the drug in all ten of
the patients.

These findings are summarized in Table 1, in which we have depicted serum urate levels at the conclusion of each week of the trial.
It can be seen that the mean serum urate concentration at the end of
the second control, or no allopurinol, period (Week 5) was slightly
lower than that found at the end of the first control period (Week 2).
We attribute this difference to the relatively brief duration of the
second control period, which was apparently inadequate to permit the
complete "wash-out" of the effects of the allopurinol taken during
Week 3 and 4 .

An analysis of variances was performed on the measurements of
serum urate at the conclusion of each of the two "wash-out" periods,
that is to say at the end of Week 2 and Week 5. These serum urate
levels were found to differ significantly. The mean serum urate at the
end of the shorter (one-week) "wash-out" period was approximately one
mg.% lower than the mean value at the end of the longer (two-week)
initial control period. As noted, we believe that this is due to the
difference in the duration of the periods and represents residual
allopurinol effect in the shorter period.

As expected, the variance in serum urate levels at the ends of
the two treatment periods indicated a significant difference between
individual patients : that is to say, not all patients responded to
the fixed daily dose of allopurinol to the same degree. However,
there was no significant difference in serum urate levels between the
two regimens or periods of treatment .

Examining serum from 10 patients, Dr. Elion found no significant
defference in the minimum levels of oxipurinol or in the conversion
of allopurinol to oxipurinol in patients taking one 300 mg. tablet
of allopurinol per day as compared to those taking 100 mg. of the
drug three times a day.

Side effects were minimal in all patients in this study. Seven
experienced transient diarrhea during control periods. This was attributed to colchicine and ceased when indomethacin was substituted.
One man developed a mild erythematous rash during the secon week of
treatment with 100 mg. of allopurinol taken 3 times a day: this faded
in a few days and did not recur when he again took allopurinol in

the alternate dosage schedule. None of the patients experienced attacks of gouty arthritis during the trial, attesting , we believe, to
the efficacy of colchicine and/or indomethacin prophylaxis.

CONCLUSION : On the basis of this short-term study, the administration of allopurinol in a single daily dose of 300 mg. appears to be
an effective means of lowering the elevated serum urate levels of
patients with gouty hyperuricemia and compares favorably with the
results obtained with the use of allopurinol in multiple daily divided
doses.

REFERENCES

1. Rundles, R.W., Wyngaarden, J.B., Hitchings, G.H., Elion, G.B.,
    and Silberman, H.R. Trans. Assoc. Amer. Physicians. 76: 126-140
    1963.

2. Elion, G.B., Kovensky, A., Hitchings, G.H., Metz, E., and
    Rundles, R.W. Biochem Pharmacol. 15: 863-880, 1966.

3. Elion, G.B., Yü, T.F., Gutman, A.B., and Hitchings, G.H. Amer.
    J. Med. 45: 69-77, 1968.

WITHDRAWAL OF ALLOPURINOL IN PATIENTS WITH GOUT

J.T. Scott and W.Y. Loebl

Charing Cross Hospital and Kennedy Institute of

Rheumatology, London

A few years ago the possibility was raised that allopurinol therapy might lead to the incorporation of abnormal genetic material into cell nuclei. Another potential hazard was suggested by the finding of crystals of allopurinol and oxypurinol in muscle biopsy specimens from patients taking allopurinol. It subsequently appears that the first eventuality does not in fact occur and that the second is of no clinical significance. At the time, however, it was thought to be reasonable to determine the effect of discontinuing allopurinol therapy in gout patients who were not over-producers of uric acid and whose renal function was not impaired.

Medication was discontinued in 33 patients with primary gout (mean age 58 years). In 5 of them the drug would have been discontinued in any case because of rash or malaise; in 28 the drug was stopped in the absence of side effects, after full discussion with individual patients. None of these 28 patients had impairment of renal function and none were over-producers of uric acid. Mean length of history of gout before starting treatment was 15 years (range 4.5 to 42 years), with an average of 4.5 attacks per year. Mean duration of allopurinol treatment before the drug was discontinued was 93 weeks (range 19-236 weeks), and the mean duration off allopurinol at the time this analysis was made was 86 weeks (mean 14-210 weeks).

Mean plasma uric acid before starting therapy was 8.4 mg/100 ml (standard deviation 1.1). Mean plasma uric acid after allopurinol therapy was established was 5.5 mg/100 ml (S.D. 1.2). After withdrawal of allopurinol uric acid levels rose to

pre-treatment figures (mean 8.8 mg/100 ml, S.D. 1.2). This rise
in plasma uric acid took place rapidly and the level usually
achieved its pre-treatment value within one week regardless of
the duration of therapy. There was some variation here, however,
and in a few instances pre-treatment values were not regained
until two or three weeks after stopping the drug. Renal
function, as estimated by blood urea, appeared to show no change,
during or after therapy. Mean blood urea before allopurinol was
41 mg/100 ml (S.D. 10), during allopurinol treatment 39 mg/100 ml
(S.D. 12), and after allopurinol was discontinued 37 mg/100 ml
(S.D. 11).

        In 12 patients there has been a recurrence of gouty
arthritis and 21 patients have remained well and free of
symptoms. 13 patients have re-started the drug (after a mean
duration off treatment of 55 weeks). The usual reason for this
has been recurrence of gouty arthritis, but in 4 patients the
drug was started relatively early after discontinuance, because
either the patient or his physician preferred to do so; one
patient was restarted because of crystalluria; and one patient
was restarted on treatment after having remained symptom-free
for 3 years because he was due to have a prostatectomy, and it
was thought wise to avoid the risk of post-operative gout. 20
patients are still not taking allopurinol after a mean duration
off treatment of 107 weeks. 16 of these patients remain free of
gout although 4 have had mild recurrences. Details are shown in
the Figure.

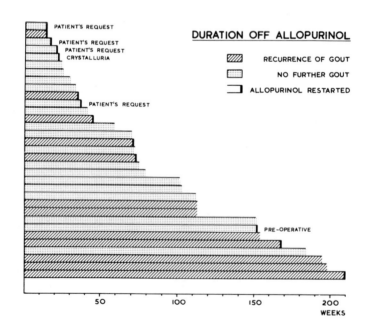

Comparison of patients suffering a recurrence of gout after stopping treatment with those who have as yet had no recurrence is indicated in the Table:-

Status after stopping allopurinol

	Recurrence gout	No recurrence yet
Mean age stopping allopurinol	58.2 years	57.3 years
Mean duration clinical gout	16 years	15 years
Mean duration allopurinol treatment	82 weeks ($\pm$ S.D. 54)	99 weeks ($\pm$ S.D. 59)
Mean daily dose allopurinol	388 mg	368 mg
Mean initial P.U.A.	8.6 mg/100 ml	8.4 mg/100 ml
Mean P.U.A. during treatment	6.2 mg/100 ml ($\pm$ S.D. 1.4)	5.1 mg/100 ml ($\pm$ S.D. 0.9)

There is no significant difference between the two groups with regard to the mean age at stopping allopurinol, mean duration of clinical gout, mean duration of allopurinol treatment, mean daily dose of allopurinol, and the mean plasma uric acid. Those patients having no recurrence of gout, however, are seen to have had a better control of plasma urate level while taking the drug (mean 5.1 mg/100 ml) than those who have had a recurrence of gouty arthritis (mean 6.2 mg/100 ml).

In summary, discontinuance of allopurinol is followed by a rapid rise of plasma urate to pre-treatment levels. This is not necessarily accompanied by a return of clinical gouty arthritis during the period encompassed by the present study, particularly if previous control of the plasma urate has been efficient. Nevertheless, gouty arthritis does recur in some patients, especially in those who have been off treatment the longest (4 out of 5 patients who have discontinued treatment for over 3 years) and no doubt gout will recur in many of the others if treatment is not resumed. Although allopurinol can therefore be discontinued in patients with uncomplicated gouty arthritis for relatively long periods without apparent harm, the results of this study do not alter our present concept of therapy, namely that once a decision has been made to lower the plasma uric acid level then treatment should be continued indefinitely.

EFFECT OF LONG-TERM ALLOPURINOL ADMINISTRATION ON SERIAL GFR IN NORMOTENSIVE AND HYPERTENSIVE HYPERURICEMIC SUBJECTS

JOSEPH B. ROSENFELD

DEPT. OF MEDICINE AND RENAL UNIT, BEILINSON
MEDICAL CENTER, PETAH TIKVAH, ISRAEL
TEL-AVIV UNIVERSITY MEDICAL SCHOOL

Acute experiments of uric acid infusion in animals and the clinical observation that in patients with gout there is an increased incidence of renal lesions, suggested that high levels of plasma uric acid may be detrimental to the kidney and cause a decrease in renal function. It is not known, however, whether hyper-uricemia, not clinically associated with gout, causes kidney damage, nor is it known whether therapeutic red-uction of serum uric acid levels slows the rate of the deterioration of renal function in such patients. To establish whether maintenance of normal plasma uric acid concentration in patients with hyperuricemia assoc-iated with various diseases has any effect on serial determination of GFR or slows the rate of kidney deter-ioration, this study was planned.

SELECTION AND ALLOCATION TO TREATMENTS

117 patients in whom plasma uric acid concentration was greater than 7 mg% in men and 6.5 mg% in women were included in this study. Factors, other than hyper-uricemia, known to be associated with progressive deter-ioration of renal function, are presence of intrinsic renal disease and hypertension. Therefore, patients in whom one, both or neither of these factors are operative have been included in the study. As patients in whom renal function is reduced may respond differently from

581

those in whom renal function is still good, patients
with normal and with reduced GFR were included.

Some of the patients were receiving diuretic drugs
of the thiazide class, which are known to cause hyperuricemia and hypokalemia.   Either or both of these may
influence renal function.   Therefore in all cases
receiving thiazides, special attention was paid to
correcting possible hypokalemia and to keeping serum
uric acid levels below 5 mg%, by increasing accordingly
the dose of Allopurinol.

By keeping the independent variables like blood
pressure and plasma uric concentration constant, it is
possible to analyse the dependent variables, the creatinine clearance or the plasma creatinine and to follow
the rate of change of the renal function in patients
treated with Allopurinol as compared to the untreated
controls.

The 117 patients were divided into the following
five groups:

Group I   -   Normotensive patients with glomerular
              filtration rates above 80 ml/min - 11 pts.
Group II -   Hypertensive patients with glomerular filt-
              ration rates above 80 ml/min - 26 pts.
Group III -   Hypertensive patients with glomerular filt-
              ration rates between 40 and 80 ml/min -
              28 pts.
Group IV -   Renal cases, normotensive with glomerular
              filtration rates between 40 and 80 ml/min -
              24 pts. and
Group V   -   Renal cases with hypertension, with glomer-
              ular filtration rates between 40 and 80 ml/
              min - 28 pts.

No case with glomerular filtration rate lower than
40 ml/min was included in the study.

These patients were randomly allocated in every
group to a treated and non-treated control group.   In
the treated group, the objectives of the treatment were
to lower the uric acid to 5 mg% in men and to 4 mg% in
women by adjusting the daily dose of Allopurinol in
order to achieve these concentrations.

MEAN URIC ACID

Group	Allopurinol	Non-Allopurinol	
1	6.90	6.86	N.S.
2	7.40	8.47	N.S.
3	7.75	7.74	N.S.
4	7.44	7.85	N.S.
5	8.32	7.56	N.S.

MEAN CREATININE CLEARANCE

Group	Allopurinol	Non-Allopurinol	
1	91.7	92.0	N.S.
2	99.9	93.6	N.S.
3	54.2	52.7	N.S.
4	51.2	58.5	N.S.
5	54.6	69.7	N.S.

SYSTOLIC BLOOD PRESSURE (SITTING)

Group	Allopurinol	Non-Allopurinol	
1	141.4	130.0	N.S.
2	171.0	158.1	N.S.
3	193.3	181.4	N.S.
4	148.0	137.5	N.S.
5	177.5	182.5	N.S.

DIASTOLIC BLOOD PRESSURE (SITTING)

Group	Allopurinol	Non-Allopurinol	
1	82.1	83.0	N.S.
2	103.0	99.4	N.S.
3	108.3	109.3	N.S.
4	80.5	80.6	N.S.
5	105.0	103.1	N.S.

MEAN AGE

Group	Allopurinol	Non-Allopurinol	
1	51.6	37.8	N.S.
2	48.7	43.8	N.S.
3	54.9	59.7	N.S.
4	49.4	51.1	N.S.
5	57.4	41.4	Significant

The groups were well-matched.   No significant difference was present between the Allopurinol and the non-Allopurinol groups for either the mean uric acid, the mean creatinine clearance, the systolic or the diastolic blood pressure.   There was only one exception. In the mean age for the fifth group, there was a difference with a significance at the 1% level.

Of the original 117 patients, 98 were available for final analysis and participated more than $2\frac{1}{2}$ years in the study.   Six patients died, 3 in the Allopurinol group and 3 in the placebo group.   Seven of the placebo group and 6 of the Allopurinol group dropped out during the course of the study.   All patients who participated more than $2\frac{1}{2}$ years in the study were included in the final analysis.

The patients, whether on Allopurinol or placebo, received identical-looking tablets.   The study was not a double-blind one, as the doctors who were following the patients knew the group to which every patient belonged.   No special dietary restrictions were imposed on the patients.   The patients were seen at least once a month.   Each time, in addition to body weight and blood pressure, blood was drawn for plasma uric acid, electrolytes, and plasma urea.   The patients collected, for every monthly visit, a 24-hour urine collection for creatinine clearance.   Once every three months, maximal urine osmolality was determined, and blood drawn for determination of plasma cholesterol, glucose, Hb., serum iron, total iron binding capacity, in addition to the other monthly determined parameters.

RESULTS

The initial and final serum creatinine, as well as the creatinine clearance, for the Allopurinol and the placebo patients, are presented.   In each column, the standard mean of the error is depicted, as well as the statistical significance for the observed changes. See Figures I to V, as follows:

FIGURE I

Initial and final serum creatinine and creatinine
clearance in Group I - Normotensives with GFR above
80 ml/min

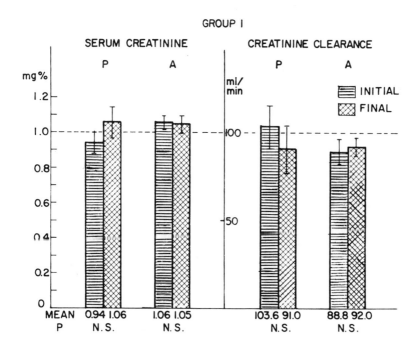

P = placebo

A = Allopurinol

FIGURE II

Initial and final serum creatinine and creatinine
clearance in Group II – Hypertensive patients with
GFR above 80 ml/min

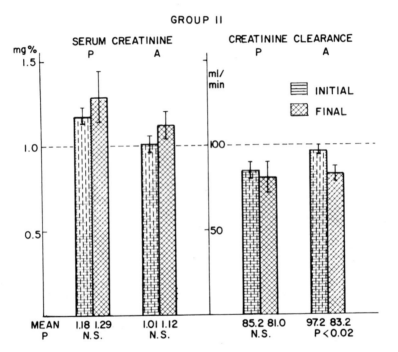

P = placebo

A = Allopurinol

FIGURE III

Initial and final serum creatinine and creatinine clearance in Group III - Hypertensive patients with GFR between 40 and 80 ml/min

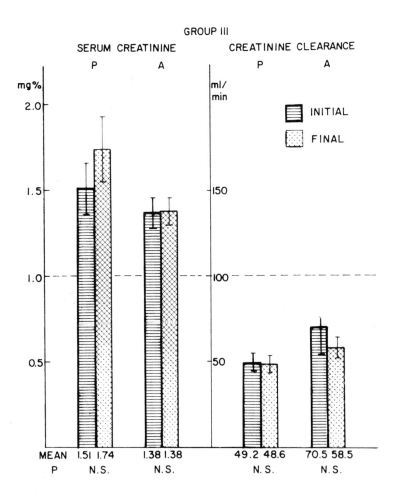

P = placebo

A = Allopurinol

FIGURE IV

Initial and final serum creatinine and creatinine
clearance in Group IV - Renal cases, normotensive, with
GFR between 40 and 80 ml/min

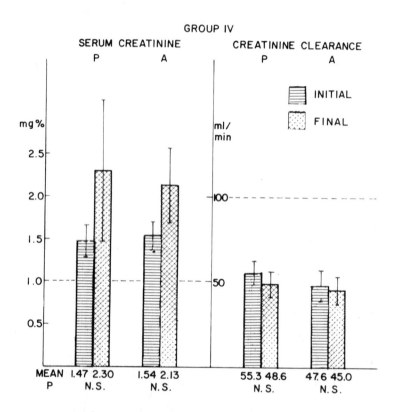

P = placebo

A = Allopurinol

FIGURE V

Initial and final serum creatinine and creatinine clearance in Group V - Renal cases with hypertension, with GFR between 40 and 80 ml/min.

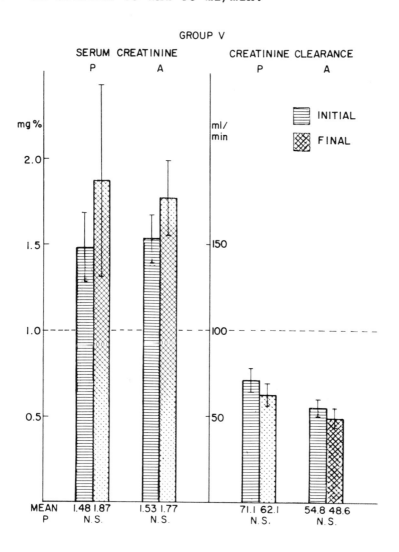

P = placebo

A = Allopurinol

FIGURE VI

Correlation coefficient between plasma creatinine and creatinine clearance – Group I

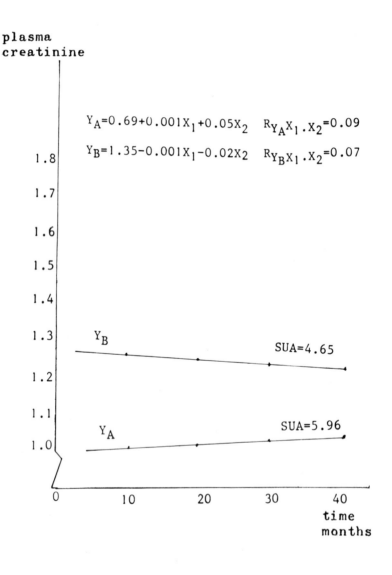

FIGURE VII

Correlation coefficient between plasma creatinine and creatinine clearance – Group II

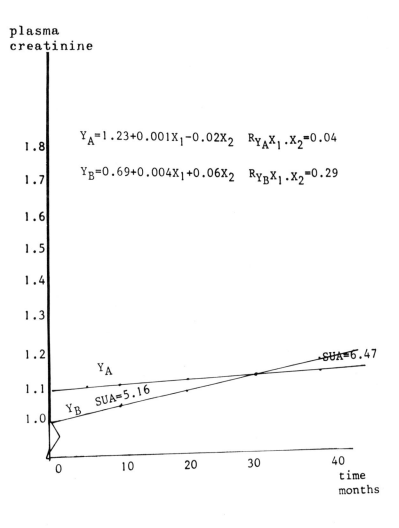

FIGURE VIII

Correlation coefficient between plasma creatinine and
creatinine clearance – Group III

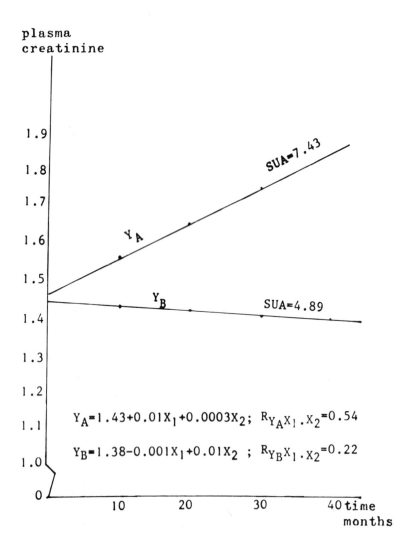

FIGURE IX

Correlation coefficient between plasma creatinine and creatinine clearance – Group IV

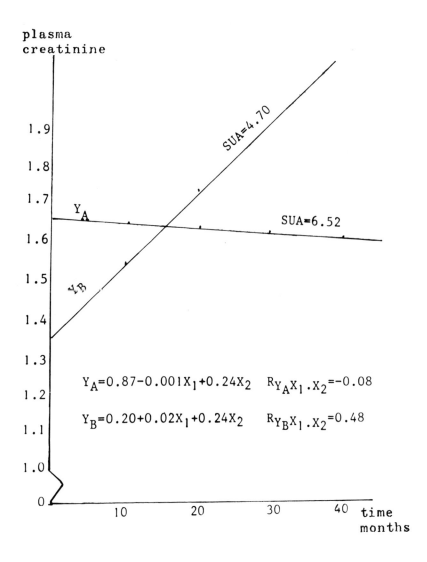

$Y_A = 0.87 - 0.001X_1 + 0.24X_2$ $R_{Y_A X_1 . X_2} = -0.08$

$Y_B = 0.20 + 0.02X_1 + 0.24X_2$ $R_{Y_B X_1 . X_2} = 0.48$

FIGURE X

Correlation coefficient between plasma creatinine and
creatinine clearance – Group V

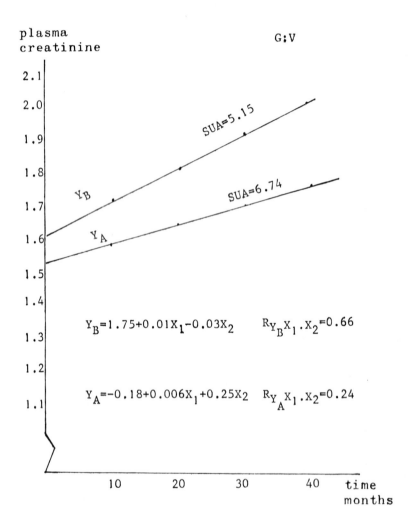

In the normotensive hyperuricemic group, which was not affected by any renal disease – one can see that the serum creatinine levels in both groups, treated and non-treated, did not change significantly.   The serum creatinine in the placebo group increased from 0.94 to 1.06 and in the Allopurinol decreased from 1.06 to 1.05 mg%. A 19% decrease was noted in the placebo group and a 5% increase in the creatinine clearance of the Allopurinol group.   All these changes are non-significant.

In the second group, the serum creatinine levels in the treated group are slightly worse as compared to the non-treated, but this difference is too small to be really significant.   The creatinine clearance in the Allopurinol group decreased from a mean 97.2 ml/min to a mean 83.2 ml/min, with a significance at the 2% level. In the group of hypertensive patients with clearances below 80 ml/min, the treated group showed a non-significant reduction in the creatinine clearance.   It is to be emphasized, however, that this greater decrease in creatinine clearance in the treated group is possibly caused by inaccurate collection, as the plasma creatinine levels in this group compared to the non-treated group are almost the same.   The serum creatinine level is a more accurate index of the deterioration of renal function as time elapses than creatinine clearance. This is true for both renal groups, the hypertensive and the normotensive ones.   The change for both renal groups, placebo and allopurinol, as expressed by serum creatinine or creatinine clearance, showed a decrease of about 10% in the creatinine clearance.   This, however, was statistically not significant.   The results are not different when the correlation coefficient between plasma creatinine and creatinine clearance for each treated and non-treated group are plotted against time. (See Figures VI to X).

After a 30-month follow-up, which extended to 40 months for some patients, one may conclude that in asymptomatic hyperuricemic patients with different etiologies, renal disease, or hypertensive patients, the normalization of the plasma uric acid concentration does not change substantially the rate of the change in the renal function in these patients.   It is possible that this was due to the fact that most of our patients had their serum uric acid between 6.5 and 8.0 and only 29

had a serum uric acid above 8.0 mg%.   It would be of
interest, therefore, to carry out a similar study in
patients with a higher range of serum uric acid - between 8 and 11 mg% - and G.F.R. above 40 ml/min.   A
similar study is also warranted for patients with reduced creatinine clearance in the range of 15 to 20 ml/
min.

        It is common usage in many nephrological and
general medical services to administer Allopurinol
indiscriminately to any patient with reduced G.F.R. and
uric acid level above 7.0 mg%.   No conclusive evidence
has ever been published of the real possible beneficial
effect of this treatment.   These results suggest that
only symptomatic patients with hyperuricemia should be
treated.

METABOLIC STUDIES OF THIOPURINOL IN MAN AND THE PIG

R. Grahame, H. A. Simmonds, A. Cadenhead and B. M. Dean

Departments of Rheumatology and Medicine, Guy's Hospital

and Professorial Medical Unit, St. Bartholomew's Hospital

London; Rowett Research Institute, Aberdeen, U.K.

Thiopurinol (mercapto-4-pyrazolo (3,4-d) pyrimidine) differs in structure from Allopurinol only by the substitution of a SH (thiol) group for an OH group on the 4-carbon atom. Delbarre and co-workers in 1968 drew attention to the fact that Thiopurinol is a potent inhibitor of uric acid synthesis, comparable in efficiency to Allopurinol. In their series of 56 patients suffering from gout, Thiopurinol was found to be clinically effective and generally well tolerated. A striking finding was that the fall in plasma and urine uric acid levels was not accompanied by a rise in the corresponding levels of xanthine and hypo-xanthine, suggesting that xanthine oxidase inhibition was not the predominating mechanism. These findings were confirmed by Serre and co-workers who treated 50 patients with gout satisfactorily.

In this paper we present the results of a more detailed investigation of the effects of Thiopurinol on total purine excretion in three patients suffering from gout, one of whom showed a partial deficiency of HGPRTase. We further report metabolic studies of Thiopurinol in the pig designed to determine both the effect of Thiopurinol on purine metabolism in this animal and the fate of 6-14 -C Thiopurinol administered orally during the period of investigation. Finally comparative in vitro studies were undertaken using intact human and pig erythrocytes.

Cases Studied

Case 1 This 46 year old pub proprietor gave an eight year history of recurrent attacks of gouty arthritis affecting the great toes and the knees. He was a heavy beer drinker and

obese (90 kilograms). There was no family history of gout.
Although his pre-treatment plasma uric acid level was only modestly
elevated (7.3 mg/100 ml.) his urinary excretion was 1,140 mg. in
24 hours. His creatinine clearance was 135 mg/min. He had been
satisfactorily treated with Allopurinol which was stopped one
month prior to the commencement of the present study. With 400 mg.
of thiopurinol per day the plasma uric acid fell in one week from
7.2 to 5.0 mg/100 ml., a drop of 31%. At the same time the urinary
uric acid excretion fell by 52%, with no change in the urinary
excretion of xanthine or hypo-xanthine or in their ratio.

TABLE 1

EFFECT OF THIOPURINOL ON PURINE

METABOLISM IN GOUT* (CASE 1)

Day	1	2	3	4	5	6	7
Thiopurinol mg./day	Nil	50	100	150	400	400	400
Plasma Uric Acid mg./100 ml.	7.2	6.5	6.2	6.2	5.3	-	5.0

URINARY OXYPURINE EXCRETION (mg./24 hours)

Uric Acid	660	-	485	445	340	406	315
Xanthine	5.6	-	7.7	6.8	6.9	6.0	5.0
Hypoxanthine	7.5	-	8.9	8.8	7.4	6.5	6.0
Xanthine/ Hypoxanthine ratio	0.75	-	0.85	0.8	0.9	0.9	0.85

* Patient previously treated with Allopurinol - off therapy 1 month

Case 2   A 35 year old man with a history or recurrent gouty arthritis commencing in 1957. He weighed 99 kilograms and was mildly hypertensive (B.P. 160/100). His pre-treatment plasma uric acid level had been 10.2 mg/100 ml. with a uric excretion of 1.100 mg. per day. The creatinine clearance was 149 ml/min. In 1969 he had been placed on Allopurinol 400 mg. a day and this had adequately corrected his hyperuricaemia and controlled his gouty attacks. In 1971, he volunteered to take part in a study of Thiopurinol which took place after Allopurinol had been left off for one week. This may have been an inadequate period of time as shown in Table 2. The plasma uric acid at the commencement of the study was only 5.7 mg/100 ml. Nevertheless, there was a 37% fall in plasma uric acid level following the introduction of Thiopurinol in a dose of 400 mg. per day. In this case however, there was no comparable fall in the urine uric acid level, and no change in the oxypurine excretion.

TABLE 2

EFFECT OF THIOPURINOL ON PURINE

METABOLISM IN GOUT* (CASE 2)

Day	1	2	3	4	5	6	7
Thiopurinol mg./day	Nil	Nil	50	100	200	300	400
Plasma Uric Acid mg./100 ml.	5.7	5.7	5.9	5.5	4.0	3.9	3.6

URINARY OXPURINE EXCRETION (mg./24 hours)

Uric Acid	416	511	555	530	512	458	473
Xanthine	8.9	7.8	11.4	8.9	8.7	8.6	9.0
Hypoxanthine	9.5	9.7	12.4	11.7	12.9	12.6	11.7
Xanthine/ Hypoxanthine ratio	0.94	0.80	0.92	0.76	0.68	0.68	0.77

* Patient previously treated with allopurinol – off therapy 1 week

Case 3 A 20 year old student had presented at the age of 15
with attacks of gout, and had been found to be suffering from a
partial deficiency of HGPRTase. He was a member of Dr. Sperling's
famous family of patients suffering from this heriditary enzymatic
defect. (Sperling et al, 1970). This patient had the added
misfortune of being strongly allergic to Allopurinol and had to be
maintained on uricosuric drugs and alkalis, a somewhat hazardous
undertaking in this situation. In view of this, he was admitted
for a trial of Thiopurinol and the results are shown in Table 3.
In the dose used (200 mg. per day), Thiopurinol appeared to have
no demonstrable effect on uric acid metabolism in this patient.
The study was abandoned after 4 days when the patient appeared to
be developing a rash. This failure of Thiopurinol to act in the
absence (or partial absence) of the enzyme HGPRTase has been
previously noted by Delbarre, and has led to speculation that the
node of action of Thiopurinol is that of a "synthetic" nucleotide
which acts by inhibition 5-phosphoribiosyl-pyrophosphate-amido-
transferase and requires that the drug be ribo-phosphorylated by
the enzyme HGPRTase.

In order to investigate this latter possibility (which suggests
that tissue incorporation might well occur, as has been suggested
but never proven for Allopurinol) we have first established the
effect of Thiopurinol on purine metabolism in the pig.

METABOLIC STUDIES IN THE PIG

In the pig allantoin represents the principal end product of
purine metabolism case and uric acid excretion amounts to less
than 100 mg/day. On 400 mg of Thiopurinol per day (Table 4)
the only urinary constituent to show any effect from the therapy
was the urinary uric acid level which fell by about 30%. Total
purine excretion was relatively unaffected and fell at the most
by 6%. As in the three human studies no alteration in Xanthine
or hypoxanthine excretion was noted. Thiopurinol also had no
effect on the urinary excretion of the pyrimidines orotidine and
orotic acid in the pig. Similar results for pyrimidine excretion
were also recorded in our first three cases illustrating an
important difference between Allopurinol and Thiopurinol therapy
in this respect.

Radioisotope Studies

After the pig had been on the above (maximal human) dose of
Thiopurinol for one week (400 mg/day) on day 7, 1 millicurie of
(6-14C) Thiopurinol* was administered orally with the unlabelled
drug and its metabolic fate, in faeces and urine, followed for one
week. The animal was then sacrificed and the tissues examined for

TABLE 3

EFFECT OF THIOPURINOL ON PURINE METABOLISM

IN PARTIAL HGPRTase DEFICIENCY*

Day	1	2	3	4	5	6
Sulphympyrozone mg./day	400	Nil	Nil	Nil	Nil	Nil
Thiopurinol mg./day	Nil	Nil	50	100	200	100
Plasma Uric Acid mg./ml.	10.0	11.6	13.4	13.8	13.5	14.0

URINARY OXIPURINE EXCRETION (mg./24 hours)

Uric Acid	1431	705	687	548	732	712
Xanthine	27.8	21.8	20.3	14.0	16.2	17.3
hypoxanthine	75.0	57.0	50.0	38.2	43.4	44.2
Xanthine/ Hypoxanthine ratio	0.37	0.38	0.40	0.37	0.37	0.38

* Patient previously treated with Sulphympyrozone - off therapy 1 day

---

any residual radioactivity. The corresponding results for an animal fed 1 millicurie of 6- 14C Allopurinol are presented for comparison, the only difference being that unlabelled Allopurinol was fed thirty times maximal human dosage (300 mg/kg/day). The results with these two drugs were comparable in both cases in that almost total recovery of radioactivity was recorded; radioactivity not accounted for in the urine (for Allopurinol 92% Thiopurinol 66%) being recovered in the faeces (Allopurinol 6%, Thiopurinol 36%). The slower excretion in the urine coupled with the greater

TABLE 4

THE EFFECT OF THIOPURINOL ON URINARY PURINES

AND PYRIMIDINES IN THE PIG*

Purine	Control		Week 1		Week 2	
	mg.	mmoles	mg.	mmoles	mg.	mmoles
Allantoin	617	3.9	579	3.66	588	3.72
Uric Acid	69	0.41	50	0.30	48	0.29
Xanthine	22	0.15	19	0.13	27	0.18
Hypoxanthine	10	0.07	9.7	0.07	9.8	0.07
Total		4.53		4.26		4.26
Orotidine + orotic Acid		5.5		6.7		5.7

*   Mean results of two animals

percentage of radioactivity found in the faeces, suggests, that
compared with Allopurinol, Thiopurinol was not as effectively
absorbed from the gut. In both cases less than 0.02% was found
in any given tissue as is shown in Table 5, mostly in the kidney
and not in the form of tissue nucleotides indicating that neither
drug is measurably incorporated into body tissues.

Drug Metabolites

It has not yet been possible to identify the urinary
metabolites of Thiopurinol. By direct electrophoresis of the urine
samples on thin layer plates followed by autoradiography. It
appears that there is one major and four relatively minor radioactive metabolites indicating that some degradative process must
have gone on. One of these minor constituents is most certainly
oxipurinol, but oxipurinol could not be detected in the urine of
patients or the pig by conventional very sensitive chemical techniques,
indicating that oxipurinol excretion must represent less than 1%
of the dose administered.

TABLE 5

(14C) RADIOACTIVITY IN PIG TISSUES AFTER ORAL ADMINISTRATION

OF (6-14C) ALLOPURINOL OR (6-14C THIOPURINOL)*

Organ	% administered (6-14C) Allopurinol	% administered (6-14C) Thiopurinol
Heart	< 0.01	< 0.01
Liver	< 0.01	0.02
Kidney	0.01	0.02
Spleen	< 0.01	< 0.01
Lung	< 0.01	< 0.01
Duodenum	< 0.01	< 0.01

* Calculated for whole organ

Mode of Action

The mode of action of Thiopurinol is still unsolved but from these results it would appear unlikely that it acts as a pseudo feedback inhibitor of de novo purine synthesis. We can also say with certainty that the effect can not be due to in vivo conversion of Thiopurinol to oxipurinol. Elion et al (1968) have also reported that in the mouse little conversion to oxipurinol is found.

The lack of effect of Thiopurinol on urinary uric acid excretion in one patient and on total purine excretion in the pig is similar to that reported by Serre et al (1970) in man. These workers found that while Allopurinol reduced urinary uric

acid excretion in both hyperuricosuric and normouricosuric man,
Thiopurinol only effectively reduced urinary uric acid levels in
hyperuricosuric man. No metabolites of Thiopurinol have been found
within the intact red cell which may well be due to disulphide
peptide bonding on the cell membrane. Further studies in vitro
are at present being undertaken in erythrocytes to help elucidate
the mode of action of Thiopurinol.

Comment

No-one would challenge the fact that the introduction of
Allopurinol has been an important landmark in the therapeutics of
purine disorders. However, over the past few years the following,
relatively minor, complication of Allopurinol therapy have been
reported, the clinical significance of which has not been fully
evaluated.

1. The finding of crystals of xanthine, hypoxanthine and
    oxipurinol in the muscles of patients taking Allopurinol,
    by Watts and his co-workers in 1971.

2. Xanthine nephropathy - as reviewed by Aplin and co-workers
    in 1972.

3. The disturbance of pyrimidine metabolism as evidenced by
    the production of orotic aciduria and orotidinuria as
    reported originally by Fox and his co-worker (1970a).

4. Suggested tissue incorporation of Allopurinol ribotide
    (Fox et al 1970b).

The present comparative studies were undertaken to explore the
possibility that the use of Thiopurinol a drug which does not induce
xanthinuria, might have certain advantages over Allopurinol in the
treatment of gout. Though we do not claim to have proven this, the
absence of xanthinuria or any effect on pyrimidine metabolism in the
form of orotidine or orotic aciduria, coupled with the finding that
in the pig, tissue incorporation does not occur, are three points
which we believe favour the possible increased use of this drug.

Acknowledgements

We are greatly indebted to Dr. P. Thorogood who kindly
synthesised the 6-14C Thiopurinol and to Dr. T. Rising for the
14C Thiopurinol Radioisotope results.

References

1. Ablin et al Metabolism (1972) 21/8 771

2. Delbarre et al Presse Med. (1968) 49 2329

3. Elion et al (1968) Israel J. Chem. 6 787

4. Fox et al Science (1970a) 168 861

5. Fox et al (1970b) New. Eng. J. Med. 283 1177

6. Serre et al Sem. Hop. Paris (1970) 46 3295

7. Sperling et al (1970) Rev. Europ. Etudes, Clin et Biol. 15 942

8. Watts et al Quart J. Med. (1971) 15 1-14 7

Effects of Allopurinol
and Its Metabolities on Purine
and Pyrimidine Metabolism

EFFECTS OF ALLOPURINOL AND OXIPURINOL ON PYRIMIDINE BIOSYNTHESIS IN MAN

T. D. Beardmore and W. N. Kelley

Hahnemann Medical College, Philadelphia, Pennsylvania
19102 and Duke University Medical Center, Durham
North Carolina 27710

The final steps of pyrimidine biosynthesis de novo which are catalyzed by two sequential enzymes, orotate phosphoribosyltransferase (OPRT) and orotidylic decarboxylase (ODC), involve the PP-ribose-P dependent conversion of orotic acid to orotidine-5'-monophosphate (OMP) followed by decarboxylation at the 7 position to form uridine 5'-monophosphate (UMP) (Fig. 1). UMP is then utilized further in the synthesis of nucleic acids and co-enzymes. Defects at this site in this metabolic pathway are important for they can result in "pyrimidine starvation" from depletion of the intracellular pool of pyrimidine nucleotides. In man the rare genetic disease, orotic aciduria, involves a deficiency of both OPRT and ODC (Type 1) (Smith, Sullivan and Huguley, 1961) or, less commonly, only ODC (Type II) (Fox, O'Sullivan and Firken, 1969). The resultant clinical state is characterized by failure of growth and development, megaloblastic anemia, and the increased urinary excretion of orotic acid (Types I and II) and orotidine (Type II).

Certain pharmacologic agents when administered to man produce orotic aciduria and orotidinuria. Two of these agents are allopurinol (4-hydroxypyrazolo (3,4-d) pyrimidine) and its chief metabolite oxipurinol (4,6-dihydroxypyrazolo- (3,4-d) pyrimidine) (Fox, Royse-Smith and O'Sullivan, 1970; Kelley and Beardmore, 1970).

Allopurinol is a structural analogue of hypoxanthine and as such inhibits xanthine oxidase and the oxidation of hypoxanthine to xanthine and xanthine to uric acid (Feigelson, Davidson and Robins, 1957). It is this action that impairs synthesis of uric acid and results in the observed hypouricemic effect of the drug. This effect accounts for the therapeutic usefulness of allopurinol and oxipurinol

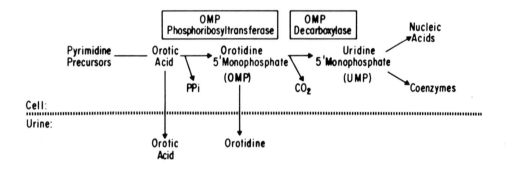

Fig. 1.   The biconversion of orotic acid to UMP in the pathway of
pyrimidine biosynthesis de novo.

in the treatment of gout and other hyperuricemic states (Rundles,
Metz and Silberman, 1966). Allopurinol also acts as a substrate for
xanthine oxidase and is itself converted to oxipurinol, a structural
analogue of xanthine. Oxipurinol can also act as a competitive
inhibitor of xanthine oxidase.

These two compounds, allopurinol and oxipurinol, have a
number of effects in addition to the one (xanthine oxidase inhibition) for which they are administered to man. The effects of these
drugs on pyrimidine metabolism will be discussed in this report.

CLINICAL EFFECTS

Therapy with allopurinol in man is followed by the prompt and
persistant increase in the excretion of orotic acid and orotidine
in the urine. During allopurinol therapy the mean excretion of orotic
acid and orotidine in seven gouty patients was 14.1 mg/24 hr (range
8.2 - 29.8) and 65.2 mg/24 hr (range 31.7 - 97.5) respectively

(Kelley and Beardmore, 1970; Beardmore, Cashman and Kelley, 1972). Similar values are seen with oxipurinol therapy. This represents a 7-10 fold increase over pretreatment values for orotic acid (<2.0 mg/24 hr) and orotidine (6.7 mg/24 hr) but is much less than the quantity excreted by patients with hereditary orotic aciduria who may excrete up to 1500 mg/24 hr. This pattern of excretion of pyrimidines is analogous to that observed during therapy with the pyrimidine analogue, 6-azauridine (Fallon, et al., 1961).

The excretion of orotidine is also increased when allopurinol is administered to patients with the Lesch-Nyhan syndrome and hereditary orotic aciduria. Two children with the Lesch-Nyhan syndrome had base line orotidine levels of less than 2.0 mg/24 hr which increased to 18.1 and 24.3 mg/24 hr with treatment. Similar findings were seen in two patients with hereditary orotic aciduria who exhibited an increased excretion of orotidine from base line values of 34.8 and 21.3 mg/gm creatinine to 150 and 134 mg/gm creatinine. However, in neither genetic disorder was there a significant change in orotic acid excretion with allopurinol administration.

The increased excretion of orotic acid and orotidine suggests a drug related metabolic block involving the conversion of orotic acid to UMP. Further evidence for this was provided by an investigator of the metabolism of isotopically labeled orotic acid in vivo and in cell culture (Beardmore and Kelley, 1971; Kelley, et al., 1971). When 7-C^{14}-orotic acid is administered to patients during allopurinol and oxipurinol therapy there is an increase in labeled compounds, C^{14}-orotic acid and C^{14}-orotidine in the urine and a decrease in $C^{14}O_2$ in expired gas when compared to the control pretreatment period. After eight days of therapy up to a 43% decrease in $C^{14}O_2$ and a 580% increase in excretion of labeled compounds in the urine is seen (Fig. 2). These effects are seen as early as four hours after drug ingestion and are readily reversible several days after the drugs are stopped. The addition of allopurinol or oxipurinol to human skin fibroblasts grown in culture leads to a 20% and 40% decrease respectively in the incorporation of orotic acid-6-^{14}C into nucleic acids (Fig. 3). Essentially the same results were observed in cultured cells lacking hypoxanthine-guanine phosphoribosyltransferase (HGPRT). These drugs did not alter the incorporation of ^3H-uridine into nucleic acids. The oxidation of allopurinol to oxipurinol was precluded in these experiments since fibroblasts lack xanthine oxidase.

From these studies it was possible to conclude that 1) Oxipurinol and to a lesser extent, allopurinol, inhibit the conversion of orotic acid to UMP in vivo and in cell culture. This effect appears to be due to inhibition of ODC and/or OPRT. 2) The inhibitory effect of these compounds occurs in patients who lack

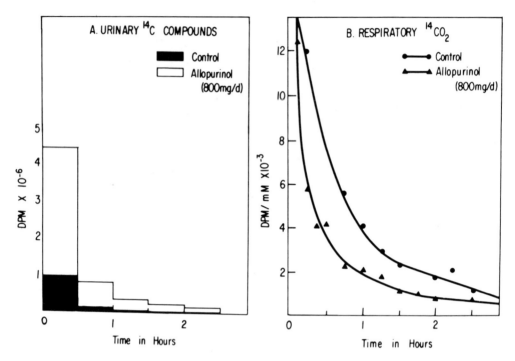

Fig. 2. The effect of allopurinol therapy, 800 mg per day for 8 days, on the metabolism of orotic acid-7-^{14}C in vivo; A, urinary excretion of ^{14}C-metabolites of orotic acid; B, excretion of $^{14}CO_2$ in expired air. (From Beardmore and Kelley, 1971).

HGPRT or OPRT and therefore this effect is not totally dependent on the presence of either enzyme. 3) The inhibitory effect of these drugs is not a consequence of xanthine oxidase inhibition.

INHIBITION STUDIES IN VITRO

What is the mechanism of this drug induced interference with pyrimidine metabolism? The finding of orotic acid and orotidine in the urine is similar to that seen after therapy with the drug 6-azauridine. This compound is converted by a pyrimidine kinase to 6-azauridylic acid which competitively inhibits ODC (Handschumacher, 1960). By analogy a nucleotide derivative of allopurinol, oxipurinol or a metabolite may be a competitive inhibitor of ODC. In addition, a purine or pyrazolopyrimidine base could act as an inhibitor of OPRT or both mechanisms could work in concert. Finally, depletion of an essential substrate such as PP-ribose-P could also contribute.

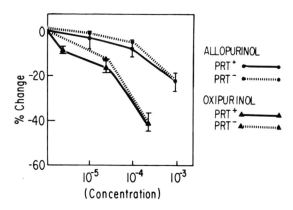

Fig. 3. Effect of allopurinol and oxipurinol on the incorporation
of orotic acid-^{14}C into nucleic acids. Values represent mean ± S.E.
expressed as per cent change from control values:●——●, allopurinol,
normal hypoxanthine-guanine phosphoribosyltransferase activity;
●---●, allopurinol, hypoxanthine-guanine phosphoribosyltransferase
deficiency; ▲—— ▲, oxipurinol, normal hypoxanthine-guanine phosphoribosyltransferase activity; ▲ — — — ▲, oxipurinol, hypoxanthine-
guanino phosphoribosyltransferase deficiency (From Kelley, et al.,
1971).

Allopurinol therapy results in the accumulation of the purine
bases, hypoxanthine and xanthine and probably also leads to an
increased formation of their ribonucleotide derivatives. The
effect of these compounds along with the 1-N ribonucleotide derivative of allopurinol (supplied by Dr. Gertrude B. Elion, Burroughs
Wellcome) on OPRT and ODC was examined in vitro. Xanthosine-5-
monophosphate (XMP) and allopurinol ribonucleotide were found to
be competitive inhibitors of ODC (Table 1) with Ki's of 7 X 10^{-7}M
and 8 X 10^{-7}M respectively (Kelley and Beardmore, 1970). This
suggested that the inhibition of ODC in vivo might be due to an
accumulation of allopurinol and xanthine with a consequent increase
in the intracellular concentration of their ribonucleotide derivatives. However, the findings that oxipurinol was a more effective
inhibitor than allopurinol and that the inhibition in vivo was not

TABLE 1

EFFECTS OF ALLOPURINOL AND ITS MAJOR METABOLITES AND PRECURSORS OF
URIC ACID SYNTHESIS ON INHIBITION OF OPRT AND ODC IN VITRO

Compound (1 X 10^{-3}M)	% Inhibition ODC	OPRT
XMP	99	0
Allopurinol Ribonucleotide	98	2
Allopurinol	0	4
Oxipurinol	0	0
Hypoxanthine	2	12
Xanthine	2	0
Allopurinol Ribonucleoside	-	13
AMP	19	-
GMP	0	-
IMP	0	-

(From Kelley and Beardmore, 1970)

totally dependent on HGPRT, the enzyme necessary to catalyze the
formation of the nucleotides studied, were inconsistant with this
hypothesis. Further clarification was provided by utilizing red
cell lysates from patients lacking HGPRT or OPRT activity (Beardmore
and Kelley, 1971; Fox, Wood and O'Sullivan, 1971). When oxipurinol
is preincubated with phosphoribosylpyrophosphate (PP-ribose-P) in
cell lysates with normal enzyme content, the products formed produce 95% inhibition of ODC. In lysates lacking HGPRT 95% inhibition
of ODC occurs. The formation of this inhibitor can be prevented
in this preparation if orotic acid is included in the preincubation
mixture. Similarly if oxipurinol and PP-ribose-P are preincubated
with cell lysates lacking OPRT, 80% inhibition of ODC is observed.
The formation of this inhibitor can be prevented in this preparation
if hypoxanthine is included in the preincubation mixture. These
studies provide strong indirect evidence that metabolite(s) of oxipurinol ,which are dependent on the presence of PP-ribose-P for their
formation, inhibit ODC (Beardmore and Kelley, 1971; Fox, Wood and
O'Sullivan, 1971) and that their formation in cell extracts can be
catalyzed by either HGPRT or OPRT. From this data we have suggested
that oxipurinol is converted to a 7-ribonucleotide derivative, which
is catalyzed by OPRT, and to a 1-ribonucleotide derivative, which is
catalyzed by HGPRT (Fig. 4). Thus in patients with normal HGPRT and
OPRT activity it is possible that at least 4 compounds are formed
which are capable of inhibiting ODC activity; 1-allopurinol ribonucleotide, xanthosine-5'-monophosphate, 1-oxipurinol ribonucleotide
and 7-oxipurinol ribonucleotide. The oxipurinol derivatives appear
to be the most important. To further test this hypothesis the
metabolites formed from oxipurinol must be positively identified and

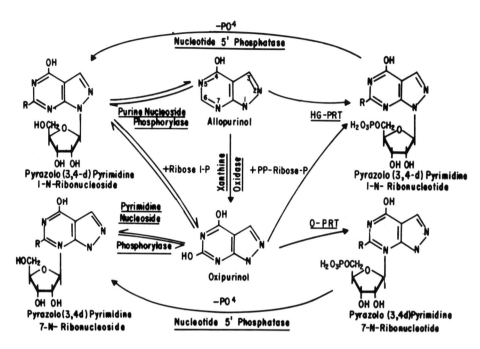

Fig. 4. Metabolism of allopurinol and oxipurinol to ribonucleotides that act as inhibitors of orotidylic decarboxylase. OPRT = orotate phosphoribosyltransferase; HGPRT = hypoxanthine-guanine phosphoribosyltransferase.

shown to be inhibitors of ODC activity.

EFFECTS ON ENZYME ACTIVITY

Fox and O'Sullivan were the first to observe that allopurinol therapy produces an 8 to 10 fold increase in the specific activity of OPRT and ODC in circulating erythrocytes (1970; 1971). Others have demonstrated an increase in OPRT and ODC activity in rat spleen and liver following allopurinol therapy (Krooth, 1971; Brown, Fox, and O'Sullivan, 1972). We have recently attempted to further define the nature of this effect in man. The increase in activity of OPRT and ODC is apparent within a few days, becomes maximum in several weeks, and does not effect other phosphoribosyltransferase enzymes (Beardmore, Cashman and Kelley, 1972) (Fig. 5). The increased activity persists as long as therapy with allopurinol or oxipurinol is continued. It is not confined to red cells since ODC activity increases about two fold in circulating leukocytes following

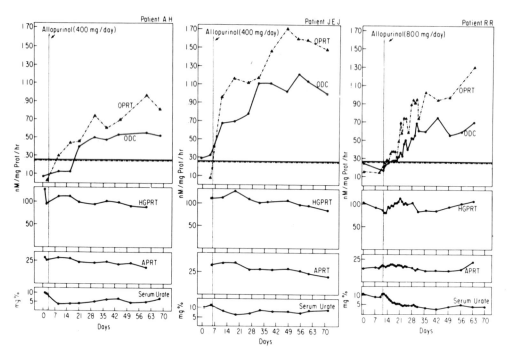

Fig. 5. The effect of allopurinol therapy on erythrocyte orotidylic decarboxylase (ODC), orotate phosphoribosyltransferase (OPRT), hypoxanthine-guanine phosphoribosyltransferase (HGPRT), and adenine phosphoribosyltransferase (APRT), and serum urate in three patients with gout. The upper limits of normal OPRT and ODC activity in erythrocytes (mean ± 2 S.D.) are indicated by the dotted and solid horizontal lines, respectively. (From Beardmore, Cashman and Kelley, 1972).

initiation of therapy (Beardmore, Cashman and Kelley, 1972). However, the increased activity of OPRT and ODC observed in circulating erythrocytes and leukocytes is not associated with a detectable reduction in the inhibition of ODC in vivo (Beardmore, Cashman and Kelley, 1972).

Fox, et al. (1971) had reported that the apparent half-life of OPRT and ODC was longer in erythrocytes from patients treated with allopurinol as compared to untreated subjects. Our experimental

approach differed from that of Fox, et al. (1971 in that we followed
the apparent half life of OPRT and ODC serially before and during
therapy with allopurinol and oxipurinol.   As noted in Fig. 6.,
initiation of therapy with allopurinol was followed by increased
activity in both young and old cells.   In addition, the curves
depicting apparent half-life of enzyme activity before and after
treatment are parellel. These findings are not consistant with
induction of enzyme synthesis as a cause for the increased activity
since one would expect a greater increase in activity in younger cell
fractions.   These data are also inconsistant with enzyme stabiliza-

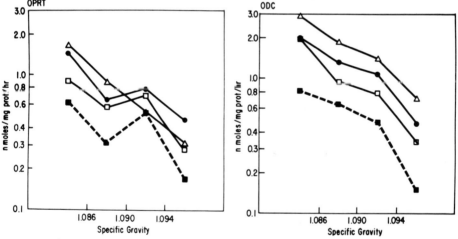

Fig. 6. Orotate phosphoribosyltransferase (OPRT) and orotidylic
decarboxylase (ODC) in circulating erythrocytes of different density
after initiation of therapy with allopurinol (800 mg/day) in patient
C.R. Mean activity is plotted against specific gravity (increasing
values correspond to increasing cell age in vivo).   Control, ■-■;
allopurinol therapy (solid lines); day 6, □-□; day 9 and 13, ●-●;
day 16 and 20 ▵-▵. (From Beardmore, Cashman and Kelley, 1972).

tion as the mechanism responsible since one would expect a greater
increase of activity in older cell fractions (i.e. the most dense).
Our initial interpretation of this phenomenon was that some type of
"activation" (used in its broadest sense) was occurring. However,
it is also apparent that stabilization of the two enzymes to the
extraction procedure could produce the same results. We are currently
investigating this phenomenon further.

SUMMARY

Allopurinol and oxipurinol inhibit pyrimidine biosynthesis
de novo by establishing a metabolic block which results in orotic
aciduria and orotidinuria. This block appears to be due to inhibition of ODC by one or more of at least four ribonucleotide derivatives; 1-allopurinol ribonucleotide, 1-oxipurinol ribonucleotide,
7-oxipurinol ribonucleotide and xanthosine-5'-monophosphate. However, to date the interference has not produced any recognized,
clinically significant abnormalities.

The appearance of increased OPRT and ODC enzyme activity is
also consistantly seen in erythrocytes as well as in nucleated
cells. Our data are most consistant with the possibility that the
enzymes are "activated" or stabilized to the extraction procedure.
Whatever the mechanism responsible for increased activity, this
phenomenon does not appear to be associated with a reduction in the
degree of inhibition of pyrimidine synthesis produced by allopurinol
in vivo.

REFERENCES

Beardmore, T. D., Fox, I. H. and Kelley, W. N. 1970. Effect of
    allopurinol on pyrimidine metabolism in the Lesch-Nyhan syndrome.
    Lancet 2: 830-831.

Beardmore, T. D. and Kelley, W. N. 1971. Mechanism of allopurinol
    mediated inhibition of pyrimidine biosynthesis. J. Lab. Clin.
    Med. 78: 696-704.

Beardmore, T. D., Cashman, J. and Kelley, W. N. 1972. Mechanism
    of allopurinol mediated increase in enzyme activity in man.
    J. Clin. Invest. 51: 1823-1832.

Brown, G. K., Fox, R. M., O'Sullivan, W. J. 1972. Alteration
    of quaternary structural behavior of an hepatic orotate
    phohsphoribosyltransferase-orotidine-5-phosphate decarboxylase
    complex in rats following allopurinol therapy. Biochem Pharma.
    21: 2469-2477.

Fallon, H. J., Frei, E., Block, J. and Seegmiller, J. E. 1961. The uricosuric and orotic aciduria induced by 6-azauridine. J. Clin. Invest. 40: 1906-1914.

Feigelson, P., Davidson, J. D. and Robins, R. K. 1957. Pyrazolo-pyrimidines as inhibitors and substrates of xanthine oxidase. J. Biol. Chem. 226: 993-1000.

Fox, R. M., O'Sullivan, W. J. and Firken, B. G. 1969. Orotic aciduria differing enzyme patterns. Amer. J. Med. 47: 332-336.

Fox, R. M., Royse-Smith, D. and O'Sullivan, W. J. 1970. Orotidin-uria induced by allopurinol. Science 168: 861-864.

Fox, R. M., Wood, M. H. and O'Sullivan, W. J. 1971. Studies on the coordinate activity and lability of orotidylate phosphoribosyl-transferase and decarboxylase in human erythrocytes and the effects of allopurinol administration. J. Clin. Invest. 50: 1050-1060.

Handschumacher, R. E. 1960. Orotidylic acid decarboxylase: Inhibition studies with azauridine 5'-phosphate. J. Biol. Chem. 235: 2917-2919.

Kelley, W. N., Rosenbloom, F. M., Henderson, J. F. and Seegmiller, J. E. 1967. Xanthine phosphoribosyltransferase in man: Relationship to hypoxanthine-guanine phosphoribosyltransferase. Biochem. Biophys. Res. Commun. 28: 340-345.

Kelley, W. N. and Beardmore, T. D. 1970. Allopurinol: Alteration in pyrimidine metabolism in man. Science 169: 388-390.

Kelley, W. N., Beardmore, T. D., Fox, I. H. and Meade, J. C. 1971. Effect of allopurinol and oxipurinol on pyrimidine synthesis in cultured human fibroblasts. Biochem. Pharmacol. 20: 1471-1478.

Krooth, R. S. 1971. Molecular models for pharmacological tolerance and addiction. Ann. N. Y. Acad. Sci. 179: 548-560.

Rundles, R. W., Metz, E. N. and Silberman, H. R. 1966. Allopurinol in the treatment of gout. Ann. Intern. Med. 64: 229-258.

Smith, L. H., Jr., Sullivan, M. and Huguley, C. M., Jr. 1961. Pyrimidine metabolism in man. The enzymatic defect of orotic aciduria. J. Clin. Invest. 40: 656-664.

THE MOLECULAR BASIS FOR THE EFFECTS OF ALLOPURINOL ON PYRIMIDINE METABOLISM

James A. Fyfe, Donald J. Nelson and George H. Hitchings

Wellcome Research Laboratories, Research Triangle Park

N.C. 27709

Investigations undertaken early in the course of clinical studies of allopurinol established the predominance of oxipurinol (alloxanthine) among its metabolites. However, one patient to whom ^{14}C-allopurinol had been given excreted a significant amount of a presumptive allopurinol ribonucleoside [1]. Somewhat later this ribonucleoside was synthesized chemically, and enzymatically by purine nucleoside phosphorylase, and was identified as 1-ribosyl-allopurinol (Fig. 1, I)[2]. With oxipurinol as substrate the corresponding 1-ribosyloxipurinol (Fig. 1, II) was obtained, and by using uridine phosphorylase a third ribonucleoside, oxipurinol-7-ribonucleoside was prepared. Meanwhile a third metabolite had been isolated from the urine of treated patients, and the availability of reference substances made it possible to identify this as 7-ribosyl-oxipurinol (Fig. 1, III).

ALLOPURINOL-I-R   OXIPURINOL-I-R   OXIPURINOL-7-R

Fig. 1

621

Two principal alternative routes of formation of these metabolites seemed possible, as typified by that for the 1-ribosyl nucleosides.

Equation 1

[structure] $+$ Ribose$-1-PO_3H_2 \longrightarrow$ [structure] $+ H_2PO_4^-$

(left structure bears N–H; product structure bears N–Ribose)

$+$

Pyrophosphorylribose$-5-PO_3H_2$

Equation 2

[structure, N–Ribose$-5-PO_3H_2$] $\longrightarrow$ [structure, N–Ribose] $+ H_2PO_4^-$

The demonstrated substrate activity of allopurinol and oxipurinol for nucleoside phosphorylases made this route plausible (Equation 1). However, conversion of the base to a ribonucleotide and subsequent dephosphorylation (Equation 2) could not be dismissed. A search for nucleotides in the acid soluble extracts of tissues and in nucleic acid fractions [1] gave negative results, but the methods permitted only the establishment of an upper limit of occurrence at about 10^{-6} M. Similarly, Kelley and Wyngaarden were unable to detect allopurinol derivatives among the cellular acid solubles [3]. Presently available, much more sensitive methods have now permitted the detection in tissues of ribonucleotides corresponding to all 3 of the ribonucleosides previously in hand. However, the precursors of the urinary ribonucleosides may not be predominantly the ribonucleotides, although the latter have been shown to be somewhat short-lived in tissues.

New information concerning the minor metabolic interactions of allopurinol was introduced by the finding of Fox et al. [4] confirmed by Kelley and Beardmore [5] that allopurinol treatment results in significant excretion of orotate and orotidine in both animals and man. The demonstration that unidentified inhibitors were formed by the incubation of red cell lysates with PRPP and allopurinol and oxipurinol [6,7] and that ^{14}C-orotate incorporation was inhibited

in the presence of oxipurinol [8] led both groups of authors to
postulate the formation of ribonucleotides with inhibitory action
on orotidylate decarboxylase.

In an accompanying paper Dr. Elion will present evidence for
the formation in vivo of allo- and oxipurinol ribonucleotides, their
characterization and their quantification in tissues after various
dosage regimens. It is the purpose of this paper to report the
kinetics of their inhibitory activities on a representative orotidylate decarboxylase in vitro, and the effects on pyrimidine
metabolism consequent to their formation in vivo.

Orotidylate decarboxylase is a key enzyme in the pathway of
pyrimidine nucleotide biosynthesis de novo (Fig. 2);

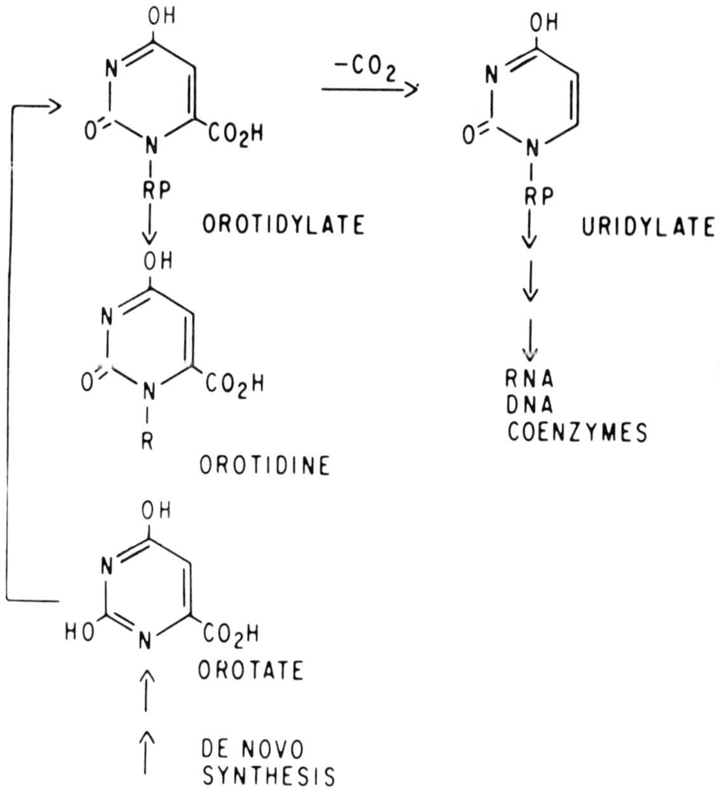

Fig. 2

the product of its action, uridylate, is the precursor of uridylate and cytidylate in RNA, deoxycytidylate and deoxythymidylate of DNA as well as uridine and cytidine-containing coenzymes. The enzyme used for the greater part of these studies was purified from a yeast concentrate to remove interfering substances. A rat liver preparation with a somewhat lower specific activity, was used for reference purposes with the more important inhibitors.

The studies with orotidylate decarboxylase presented some difficulties related to the instability of the purified free enzyme. It is notable that it was stabilized by both substrate and inhibitors.

Kinetic studies were complicated by a bimodal rate function. At concentrations of OMP above about 10 µM, high K_m (2 µM) high V_{max}, and for inhibitors high K_i values prevailed. At concentrations below about 2×10^{-6}M, lower K_m (0.5 µM) and lower V_{max} and K_i values were found. This difference seems to be a genuine property of the enzyme [9]. The difference in K_m values was similar with the rat liver enzyme, approximating 4 and 1 µM, respectively.

It is probable that it is the K_i values at low substrate concentration that are pertinent to the inhibition of the enzyme in vivo, since the upper limit for orotidylate in normal rat tissues is below 1 µM and the rise that follows high dose allopurinol treatment probably does not quite reach the high K_m range [10].

It may be instructive, therefore, to compare the tissue levels of nucleotides reported in Dr. Elion's paper, with the K_i values for the rat liver enzyme at the lower substrate concentrations. After a dose of 50 mg/kg i.v. to rats[1] the liver analog nucleotide values reached approximately the following levels (all $\times 10^{-8}$M): 1-Alo-5'-P, 200; 1-Oxi-5'-P, 70; and 7-Oxi-5'-P, 50. All 3 of these may exist, at least transiently, at levels above the respective K_i values, and in the case of 1-Oxi-5'-P so greatly in excess of the K_i value that one might have expected complete shut-down of the enzyme (Table 1)[11].

Nevertheless the effects on uracil-containing metabolite pools are small and transient (Fig. 3) [12] despite the fact that the turnover of uridylates appears to be rapid [13].

---

[1] The parameters of dosage and route of administration, and time after dosage were explored to find conditions for maximal inhibition of pyrimidine metabolism. Intravenous doses in the range 20 to 50 mg/kg produced inhibition of ^{14}C-orotate incorporation that was maximal 1 hour after dosing and was maintained through the 4th hour.

Table 1

ODC Inhibition and Analog Concentrations

Inhibitor	Low Ki	Max Conc Found	Ratio
1-Alo-P	100	200	2
1-Oxi-P	0.05	70	3500
7-Oxi-P	4	50	12.5

All concentrations X 10^8 M.
Concentrations found in rat liver 1 hr 50 mg/kg allopurinol i.v. [11].

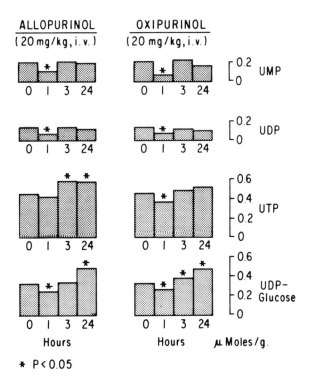

Fig. 3. Effect of acute administration of allopurinol and oxi-
         purinol on liver uridine-phosphate pools. Allopurinol or
         oxipurinol as a solution of the Na salt was injected in-
         travenously. At the time indicated the animals were sac-
         rificed, an aliquot of liver was extracted and the
         nucleotides in the extract were separated and quantified
         by high pressure liquid chromatography.

When one measures the incorporation of ^{14}C-orotate by infusion[2], in control and treated animals the inhibition is apparently greater (Table 2).

Table 2

Incorporation of Infused ^{14}C-Orotate

Substance	Time After Injection of Allopurinol Radioactivity Incorporated (dpm/g liver)		
	0	1 hr	8 hr
UMP	45,018	7,878	51,250
UDP	126,464	19,104	95,760
UTP	194,925	52,332	153,296
UDPS	271,068	53,009	216,200
UDPG	22,380	7,380	20,250
Total	659,855	139,703	536,756

Infusion 10 mμM = 660,000 dpm/g liver [10].

However, there are adequate reasons for the belief that these incorporations of the tracer do not accurately reflect the inhibition. On the one hand, one can argue that the inhibition is greater than apparent since the decreased incorporation in the treated animals is measured against a rate in the control animals which is certainly not maximal. (It is known, for example, that the same quantity of ^{14}C-orotate can be taken up and incorporated in a 5 min rather than 15 min infusion.) On the other hand, it is quite clear that the precursor pools have undergone much more dilution in the treated, than in the untreated animals. In the control livers, orotate, orotidine and orotidylate are undetectable by methods which would be capable of measuring tissue concentrations that were as high as 1 μM. In the inhibited animals the concentrations of these precursors rose to measurable levels. In a pool of extracts of the 1 hour livers, orotidylate was present at a concentration of 3 μM. Its specific activity relative to that of the orotate infused showed a dilution of 4.8-fold. Similarly, orotate pools rose to 4 μM with an RSA of 4.1 [10]. If one assumes that the dilution of orotate during the control infusion was much

[2] In these experiments 0.1 μ mole of ^{14}C-orotate was infused intravenously over the course of 15 min. The total radioactivity could be accounted for in the liver nucleotides at the end of the infusion, and, since the liver of a 200 g rat weighs about 10 g the administered orotate was approximately 0.01 μM/g. liver tissue, compared with a total uridine phosphate pool of 2.1 μM/g.

smaller, then the rate of incorporation of orotate in the allopurinol-treated animal becomes almost normal, after correction for
dilution, and is consistent with the observed effects on the
uridine-phosphate pool sizes.

This interpretation is further supported by observations on
animals chronically treated with allopurinol (Fig. 4). After 4-weeks
feeding of rats on allopurinol-containing diets there were no significant changes in uridine phosphate pool sizes, nor in the rate of
RNA synthesis. It is pertinent that blood levels of oxipurinol in
rats receiving the diet containing 0.1% allopurinol were about
6 μg/ml, a concentration similar to those of patients receiving
allopurinol therapy [14].

The most likely interpretation of the events following allopurinol administration is as follows: Ribonucleotides of allo-
and oxipurinol are formed in concentrations high enough to inhibit
orotidylate decarboxylase. However, there is rapid accommodation to
this inhibition by accumulations in the orotidylate and orotate
pools sufficient to insure a normal through-put. The elevation of
orotate diminishes the synthesis of Oxi-7-P. Simultaneously, the
elevations in tissue hypoxanthine and xanthine pools diminish the
synthesis of Alo-1-P and Oxi-1-P.

Moreover, it has been shown that the levels of ORPTase and
ODCase are elevated 2-5-fold in patients receiving allopurinol

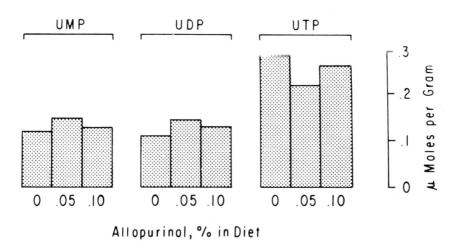

Fig. 4. Lack of effect of chronic allopurinol administration on
         uridine-phosphate pools of liver. Allopurinol was incor-
         porated in the diet as indicated, and the rats (200 g)
         were fed for 4 weeks, sacrificed, and the livers were
         extracted.

therapy [4,15]. The suggestion of Fox et al. [4] that this results from protection of the enzymes by substrates and inhibitors would be consistent with the observation [9] that such protection can be observed in vitro. In the steady state, orotate and orotidylate pools would be elevated to maintain a normal rate of uridylate biosynthesis, and normal pools of all metabolites further along on the biosynthetic pathway. The maintenance of elevated orotidylate levels would account for the increased excretion of orotate and orotidine.

REFERENCES

1   G.B. Elion, A. Kovensky, G.H. Hitchings, E. Metz & R.W. Rundles (1966) *Biochem. Pharmacol.*, 15: 863.

2   T.A. Krenitsky, G.B. Elion, R.A. Strelitz & G.H. Hitchings (1967) *J. Biol. Chem.*, 242: 2675.

3   W.N. Kelley & J.B. Wyngaarden (1970) *J. Clin. Invest.*, 49: 602.

4   R.M. Fox, D. Royse-Smith & W.J. O'Sullivan (1970) *Science*, 168: 861.

5   W.N. Kelley & T.D. Beardmore (1970) *Science*, 169: 388.

6   R.M. Fox, M.H. Wood & W.J. O'Sullivan (1971) *J. Clin. Invest.*, 50: 1050.

7   T.D. Beardmore & W.N. Kelley (1971) *J. Lab. clin. Med.*, 78: 696.

8   W.N. Kelley, T.D. Beardmore, I.H. Fox & J.C. Meade (1971) *Biochem. Pharmacol.*, 20: 1471

9   J.A. Fyfe, R.L. Miller & T.A. Krenitsky (1973) *J. Biol. Chem.*, in press.

10 D.J. Nelson, unpublished.

11 D.J. Nelson, C.J.L. Bugge, H.C. Krasny & G.B. Elion (1973) *Biochem. Pharmacol.*, in press.

12 D.J. Nelson, C.J.L. Bugge, H.C. Krasny & G.B. Elion (1973) *Fed. Proc.*, 32: 511.

13 N.L.R. Bucher & M.N. Swaffield (1966) *B.B.A.*, 129: 445.

14 G.B. Elion, Ts'ai-Fan Yü, A.B. Gutman & G.H. Hitchings (1968) *Am. J. Med.*, 45: 69.

15 T.D. Beardmore, J.S. Cashman & W.N. Kelley (1972) *J. Clin. Invest.*, 51: 1823.

PURINE AND PYRIMIDINE BIOSYNTHESIS IN NEUROSPORA CRASSA AND HUMAN SKIN FIBROBLASTS.ALTERATION BY RIBOSIDES AND RIBOTIDES OF ALLOPURINOL AND OXIPURINOL

W.Kaiser and K.Stocker

Medical Polyclinic, University of Munich

D-8 Munich 2, Pettenkoferstrasse 8a (Germany)

Allopurinol and oxipurinol inhibit according to Kelley et al. (1,2) both de novo purine and pyrimidine biosynthesis in human fibroblast cultures. This effect (at least at higher concentrations) was shown not to be dependent on the presence of xanthine oxidase or HGPRTase, so that it has to be attributed to some other mechanism than inhibition by allopurinol-ribonucleotide.

We studied the effects of allopurinol, allopurinol-1-ribonucleoside, allopurinol-1-ribonucleotide and oxipurinol, oxipurinol-7-ribonucleoside, oxipurinol-7-ribonucleotide on purine and pyrimidine biosynthesis in cultures of Neurospora crassa (wild strain 74 a) and human fibroblasts.

The substances were chemically synthesized according to known methods for preparation of purine nucleosides and by known phosphorylation reactions, they are chromatographically pure and highly stable.

METHODS

The first three steps of purine synthesis de novo were assessed by measuring the ^{14}C-FGAR-formation in the presence of azaserine, essentially as described by Rosenbloom et al. (3), Kelley and Wyngaarden (1). Pyri-

---

The substances were kindly provided to us by Dr. Steinmaus from Dr. Henning in Berlin.

midine biosynthesis was determined by absorption of
$^{14}CO_2$ from ^{14}C-carboxyl-labeled 7-orotic acid during
the incubation periods. Neurospora crassa cultures
(wild strain 74 a) were taken because of the relatively short doubling time of 4 - 5 hours.

RESULTS

I. Influence of allopurinol- or oxipurinolribonucleosides and -ribonucleotides on purine and
pyrimidine biosynthesis in Neurospora crassa.

Allopurinol shows significant inhibition in ^{14}C-FGAR
synthesis in the order of about 2o to 25 percent in
concentrations ranging from 10^{-5} to 10^{-3} M. Allopurinol-
1-ribonucleoside has no significant influences on this
synthesis during our incubation periods (3o minutes),
whereas allopurinol-1-ribonucleotide inhibits ^{14}C-
FGAR synthesis from 2o percent at 10^{-5} M to 45 percent
at 10^{-3} M. Oxipurinol inhibits ^{14}C-FGAR synthesis to
the same extent (25 percent) as does allopurinol. Oxi-
purinol-7-ribonucleoside inhibits in the range of 15
to 2o percent at 10^{-5} M and 10^{-4} M, to about 5o percent
at the concentration of 10^{-3} M. An inhibition by oxi-
purinol-7-ribonucleotide (of about 15 percent) can only
be demonstrated by high concentrations (10^{-3} M).
   Fig. 1 demonstrates the effects of all these tested
substances on pyrimidine biosynthesis in cultures from
Neurospora crassa. Allopurinol and oxipurinol have
marked inhibitory effects. Allopurinol-1-ribonucleoside
inhibits pyrimidine biosynthesis in the order of 55 to
6o percent in concentrations from 10^{-5} to 10^{-3} during
our incubation period of 45 minutes. Allopurinol-1-
ribonucleotide shows influences comparable to those of
allopurinol. The inhibitory effect of oxipirinol-7-
ribonucleoside is in straight parallelity to the effect
of oxipurinol, whereas oxipurinol-7-ribonucleotide be-
gins with its marked effect (8o - 9o percent inhibition)
at concentrations between 10^{-4} to 10^{-3} M.

II. Influence of allopurinol and oxipurinol, their
respective ribonucleosides and ribonucleotides
on purine and pyrimidine biosynthesis in normal
human skin fibroblasts.

Allopurinol and oxipurinol inhibit purine biosynthesis
in human skin fibroblasts nearly in the same range.
There is a gradual decrease in ^{14}C-FGAR formation in

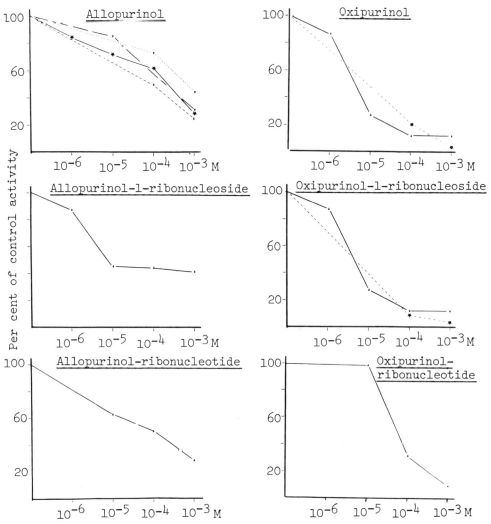

Fig.1. The effects of allopurinol, allopurinol-
1-ribonucleoside, allopurinol-1-ribo-
nucleotide, oxipurinol, oxipurinol-7-
ribonucleoside and oxipurinol-7-ribo-
nucleotide on pyrimidine biosynthesis
in cultures from Neurospora crassa.

the order of 4o percent at a concentration of 10^{-4} M
to 6o percent at 10^{-3} M. Allopurinol-1-ribonucleoside
and allopurinol-1-ribonucleotide show essentially no
significant differences concerning the inhibitory effect, furthermore, the influences demonstrate clear-

cut parallelity to the values obtained by allopurinol.
Oxipurinol-7-ribonucleoside exerts a more pronounced
effect than oxipurinol, oxipurinol-7-ribonucleotide
shows marked inhibitory effects, however only at the
highest concentrations (6o percent inhibition at 10^{-3}M).

The influences of allopurinol and allopurinol-1-
ribonucleoside on pyrimidine biosynthesis in human
fibroblasts show parallelity, there is a gradual decrease of pyrimidine biosynthesis in the order of 35
to 4o percent at concentrations from 10^{-4} to 10^{-3} M.
The effect of allopurinol-1-ribonucleotide is minor and
only at high concentrations (15 to 2o percent inhibition
at 1o $^{-3}$ M). Oxipurinol inhibits pyrimidine biosynthesis gradually, to an extent of 3o percent at 10^{-3} M,
in comparison to allopurinol the effect is insignificant smaller. With oxipurinol-7-ribonucleoside we get
inhibitory effects (4o percent at 10^{-4} M and 10^{-3} M)
which are to be comparedt to those of allopurinol-1-
ribonucleoside. Oxipurinol-7-ribonucleotide shows no
influences on pyrimidine biosynthesis up to concentrations of 10^{-4} M, whereas at 10^{-3} M there is a marked
inhibitory effect of 45 percent.

DISCUSSION

The inhibitory effect of allopurinol on the first
three steps of purine biosynthesis in Neurospora crassa
cultures could be due 1. to the formation of allopurin-
ol-1-ribonucleotide (by HGPRTase; 1,19 nMol/mg protein/
hr in our Neurospora crassa cultures) 2. due to an
oxipurinol-7-ribonucleotide effect, namely the formation of oxipurinol (by xanthine oxidase) and then
oxipurinol-7-ribonucleotide (by OPRTase) and 3. by
formation of allopurinol-1-ribotide through allopurin-
ol-1-riboside (by a purine nucleoside phosphorylase)
to allopurinol-1-ribonucleotide (by a nucleoside kinase).
This latter pathway seems rather improbable because
we could not demonstrate an effect on purine biosynthesis by direct addition of allopurinol-1-ribonucleoside
to the incubation medium. Allopurinol-1-ribonucleotide
inhibits significantly at higher concentrations. This
is to explain if there is free access of the ribonucleotide form through the cell membrane (at least in these
high concentrations) or if there is conversion to the
ribonucleoside (which then enters the cell) by a nucleotidase located on the cell membrane. This latter mechanism however seems more improbable bcause there was no
effect by allopurinol-1-ribonucleoside (which excludes
the possibility for subsequent formation of allopurin-
ol-1-ribonucleotide by a nucleoside kinase, at least

during the short incubation time of 30 minutes).
The inhibitory effect of oxipurinol on the first three
steps of purine biosynthesis could be due to the formation of oxipurinol-7-ribonucleotide by OPRTase. The possibility for conversion to oxipurinol-7-ribonucleoside
(by uridine phosphorylase) with subsequent formation of
oxipurinol-7-ribonucleotide (by an uridine specific
kinase) cannot be excluded. This seems even probable
on account of the significant inhibitory effects of
oxipurinol-7-ribonucleoside which otherwise (if there
exists no uridine specific kinase to form directly
oxipurinol-7-ribonucleotide) had to be explained by
conversion to oxipurinol with subsequent formation of
oxipurinol-7-ribonucleotide by OPRTase. In all cases
we cannot however exclude other direct or indirect
effects of allopurinol- or oxipurinolribonucleosides in
these high concentrations on purine or pyrimidine biosynthesis. The slight inhibitory effect of oxipurinol-
7-ribonucleotide in high concentrations (10^{-3} M) would
be best explained ba transport of this form through the
cell membranes. Whether this would be of physiologic
importance during therapy in normal persons cannot be
decided at the moment.

So, our results would be consistent with the view
that the ribonucleotides of allopurinol and oxipurinol
are to be accused for the inhibitory effects on purine
and pyrimidine biosynthesis in Neurospora crassa.

The inhibitory effects of allopurinol and oxipurinol on purine and pyrimidine biosynthesis in human fibroblasts could be explained by formation of their respective ribonucleotides either by HGPRTase (allopurinol)
or OPRTase (oxipurinol). The effects of allopurinol-1-
ribonucleoside are hardly to be explained. Theoretically the allopurinol-1-ribonucleoside can be converted
either to the free base (catalyzed by a purine nucleoside phosphorylase) or to allopurinol-1-ribonucleotide
(catalyzed by a nucleoside kinase). According to Utter
et al. (4) there seems however to be a lack of kinases
in mammals which effectively phosphorylate inosine.
Furthermore, indirect experiments of Elion et al. (5),
where no detectable nucleotide formation or incorporation into nucleic acids was observed in vivo with ^{14}C-
allopurinol would support this. Nevertheless, our results would claim for a direct conversion of allopurin-
ol-1-ribonucleoside to allopurinol-1-ribonucleotide in
human fibroblast, at least at doses from 10^{-5} to 10^{-3} M.
Otherwise we would have to explain the inhibitory effects
of allopurinol-1-ribonucleoside by direct influences on
purine synthesis, for the other possibility theoretical-

ly existing (namely that allopurinol-1-ribonucleoside
would be catalyzed to allopurinol and this directly to
allopurinol-1-ribonucleotide by HGPRTase) is more improbable because the results of Kelley and Wyngaarden
(1) with fibroblasts showed that, at higher doses, the
effect of allopurinol could not be related on the presence of HGPRTase.

After intravenous injection of allopurinol-1-ribo-
nucleoside (100 mg daily/5 days) into normal persons
we found however no significant changes in serum uric
acid levels. This we would expect if direct conversion
of allopurinol-1-ribonucleoside to allopurinol-1-ribo-
nucleotide would exist in significant amounts during
physiological conditions.

The effects of allopurinol-1-ribonucleotide during
our in vitro experiments would again be best explained
with passage of this form through the cell membranes.
Oxipurinol-7-ribonucleoside has to be converted to its
nucleotide by an uridine kinase to exert the demonstrated effects (on the basis of the view that the ribonucleotide forms are to be accused for the inhibitory
influences) and for oxipurinol-7-ribonucleotide we
have again to suggest passage of this form through the
cell membranes or dephosphorylation, entry of ribonucleoside with subsequent intracellular phosphorylation.

The discussion of the inhibitory influences of
these tested substances on pyrimidine biosynthesis in
human fibroblasts brings about the same problems as did
purine biosynthesis. The differences lie only in the
percent range of inhibition by the different substances
and not in the interpretation of the molecular mechanisms.

Our results could not conclusively explain the
mechanisms for allopurinol- and oxipurinol-mediated
inhibition of purine and pyrimidine biosynthesis. Most
of our results are in accordance with the view that
allopurinol-1-ribonucleotide and oxipurinol-7-ribonucleo-
tide are to be accounted for the inhibitory effects on
both purine and pyrimidine biosynthesis. Relatively
high concentrations of allopurinol-1-ribonucleoside
and oxipurinol-7-ribonucleoside have inhibitory effects
which are difficult to explain with the view that there
exist no inosinic or uridine specific kinases in fibroblasts, catalyzing to the respective ribonucleotides.

We hope that time kinetic studies, the studies
with cell homogenates and combination of inhibitory
substances to see potentiating effects will give additional clue to the molecular action of these substances
on purine and pyrimidine biosynthesis.

REFERENCES

1. Kelley,W.N., and J.B.Wyngaarden
   J.Clin.Invest. 49,602 (1970)
2. Kelley,W.N.,I.H.Fox,T.D.Beardmore and J.C.Meade
   Annals New York Acad.Sciences 179,588 (1971)
3. Rosenbloom,F.M.,J.F.Henderson,I.C.Caldwell,
   W.N.Kelley and J.E. Seegmiller
   J.Biol.Chemistry 243,1166 (1968)
4. Utter,M.F.,in P.D. Boyer,H.Lardy and K.Myrbäck
   (Editors),The Enzymes,Vol II.,
   Academic Press,New York, 78 (1960)
5. Elion,G.B.,Kovensky,A., and Hitchings,G.H.
   Biochem.Pharmacol. 15,863 (1966)

EFFECTS OF ALLOPURINOL AND OXONIC ACID ON

PYRIMIDINE METABOLISM IN THE PIG

P.J. Hatfield, H.A. Simmonds, J.S. Cameron,
A.S. Jones, and A. Cadenhead

Guy's Hospital and Rowett Research Institute
London, S.E.1 9RT, U.K. and Bucksburn, Aberdeenshire, U.K.

Pigs given allopurinol showed an increase in urinary orotic acid and orotidine excretion from a mean of 5 mg to a mean of 50 mg/24 hours. Although the dose of allopurinol was constant the levels of orotic acid and orotidine fell gradually from their initial peak. When the drug was stopped levels returned rapidly to normal. These results are similar to those obtained in rat and man (1), where this effect has been attributed to inhibition of the enzyme orotidylic decarboxylase (2).

When guanine was given together with the allopurinol, the increased urinary excretion of orotic acid and orotidine remained unchanged at 50 mg/24 hours. This is in contrast to the findings in man, where exogenous RNA fed together with allopurinol eliminated the increase in excretion of orotic acid and orotidine (3).

The combination of guanine and allopurinol produced an acute crystal nephropathy in the pig and permanent renal damage (4). As renal function deteriorated there was a proportional increase in urinary levels of orotic acid and orotidine. This increased excretion would appear to be related to the increase in plasma oxipurinol levels which have been reported in renal failure (5).

Oxonic acid is a uricase inhibitor (6) and has been given to rodents to produce an animal model for gout (7). When given to pigs urinary allantoin excretion decreased, as uric acid excretion increased. These findings indicate inhibition of the enzyme uricase. However, oxonic acid also produced a great increase in orotic acid and orotidine excretion. Urinary levels increased from 5 mg to 900 mg/24 hours, and this would imply a complete or

637

nearly complete block in pyridimine nucleotide de novo synthesis.

These findings cast considerable doubt on the suitability of oxonic acid as an agent for the production of an animal model in the study of gout.

REFERENCES

1. Brown, G.K., Fox, R.M., O'Sullivan, W.J. (1972). Biochem. Pharmacol., 21, 2469.

2. Foster, D.M., See, C.S., O'Sullivan, W.J. (1972). Biochem. Med., 7, 61.

3. Zöllner, N., Gröbner, W. (1971). Ges. Exp. Med., 156, 317.

4. Farebrother, D.A., Hatfield, P.J., Simmonds, H.A., Cameron, J.S., Jones, A.S., Cadenhead, A. In preparation.

5. Wood, M.H., Sebel, E., O'Sullivan, W.J. (1972). Lancet, 1, 761.

6. Johnson, W.J., Stavric, B., Chartbrand, A. (1969). Proc. Soc. Exp. Biol. and Med., 131, 8.

7. Klinenberg, J.R., Bluestone, R., Schlosstein, L., Waisman, J., Whitehouse, M.W. (1973). Ann. Int. Med., 78, 99.

RIBONUCLEOTIDES OF ALLOPURINOL AND OXIPURINOL IN RAT TISSUES AND

THEIR SIGNIFICANCE IN PURINE METABOLISM

Gertrude B. Elion and Donald J. Nelson

Wellcome Research Laboratories

Research Triangle Park, N.C. 27709

INTRODUCTION

In the early studies on the metabolism of allopurinol in animals and in man [1,2], it was not possible to find any allopurinol or oxipurinol ribonucleotides in acid-soluble extracts of tissues with methods sensitive enough to detect 10^{-6}M concentrations. Similarly, no such ribonucleotides were detectable in the acid-soluble metabolites of allopurinol in Ehrlich Ascites cells [3] or in human fibroblasts [4]. Although allopurinol was a substrate for hypoxanthine-guanine phosphoribosyltransferase (HGPRT) in vitro, the kinetic parameters of allopurinol with human red cell HGPRT [5] indicated that the conditions were not favorable for the formation of this nucleotide in vivo (Table 1). Thus, the K_m for allopurinol was 1 mM, whereas a level of 0.01 mM is the highest level attained after therapeutic doses. The binding of oxipurinol to this enzyme was even poorer, K_i = > 10 mM, and the velocity so low that it was not measurable at the 0.125 mM concentration used in the assay [5], nor at 0.5 mM, used subsequently. Moreover, the relatively low K_m values of hypoxanthine, and xanthine, and the increased level of these bases resulting from xanthine oxidase inhibition, increased the probability that hypoxanthine and xanthine would successfully compete with their respective analogues for HGPRT under in vivo conditions. Nevertheless, the work of Fox [6] and of Kelley [7,8] indicated that the nucleotides of allopurinol and oxipurinol might indeed be formed and be responsible for the orotic aciduria and orotidinuria seen in allopurinol-treated patients. In order to attempt the quantification of such nucleotides, the present studies in rats were undertaken with large doses of ^{14}C-allopurinol of high specific activity. By the extraction of large amounts of tissue, it was

possible to isolate, identify, and quantify the radioactive
metabolites. A more detailed account of these investigations will
be published elsewhere [9].

Table 1

	Human Red Cell HGPRT	
	K'_m (K_i), mM	V_m
Hypoxanthine	0.0024	55
Guanine	0.0018	66
Xanthine	0.25	1
Allopurinol	1.0	13.5
Oxipurinol	> 10.	

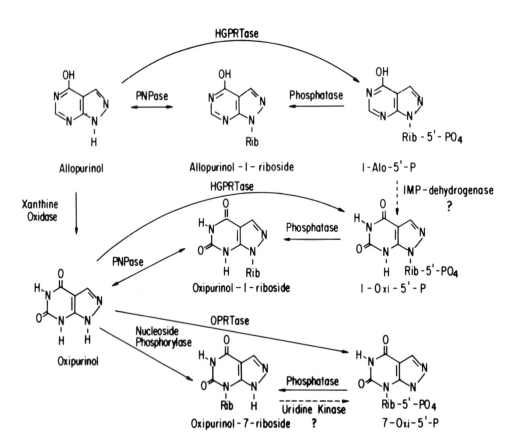

Figure 1.   Biochemical transformations. [9]

Table 2

DOSE ^{14}C-ALLOPURINOL	TISSUE	METABOLITE CONC. IN TISSUES		
(mg/kg)		1-ALO-5'-P	7-OXI-5'-P μM	1-OXI-5'-P
50, i.v.	LIVER	4.18	0.076	0.73
(3 hr)	KIDNEY	4.06	2.51	0.56
	RBC	0.27	< 0.01	< 0.01
50, p.o.	LIVER	0.28	0.09	0.22
(4 hr)	KIDNEY	0.21	0.12	< 0.03
	RBC	0.12	< 0.01	< 0.01

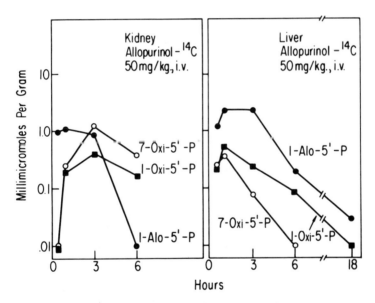

Figure 3.   Time course of the concentrations of allopurinol and oxipurinol ribonucleotides after a single dose of [6-^{14}C]allopurinol, 50 mg/kg., i.v., in the rat. [9]

IDENTIFICATION AND QUANTIFICATION OF NUCLEOTIDES

    The pathways for the formation of allopurinol-1-ribonucleotide
(1-Alo-5'-P) and the two ribonucleotides of oxipurinol (1-Oxi-5'-P
and 7-Oxi-5'-P) are shown in Figure 1. The nucleotides required as
standards for comparison with the metabolites of allopurinol were
synthesized enzymatically <u>in vitro</u> from their respective bases,
utilizing phosphoribosyltransferases from <u>E</u>. <u>coli</u> and beef erythrocytes [10]. The conversion of 1-Alo-5'-P to 1-Oxi-5'-P did not occur
with the inosinate dehydrogenase from Sarcoma 180 cells [9]. It is
not known whether oxipurinol-7-riboside is a substrate for uridine
kinase.

    Rats were given intravenous or oral doses of ^{14}C-allopurinol,
and were sacrificed at various time intervals thereafter. Individual
tissues were extracted with perchloric acid and the extracts were
chromatographed on large DEAE Sephadex A-25 columns. A typical
profile of the radioactive peaks in a liver extract is shown in
Figure 2. The free bases and nucleosides were present in Peak 1.

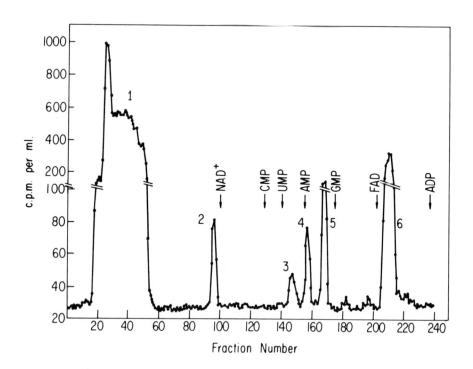

Figure 2.   Chromatography on DEAE Sephadex A-25 of the perchloric
acid extract of 19.6 g. of rat liver, 6 hours after the administration of [6-^{14}C]allopurinol (2.72 mCi/mmole), 5 mg/kg. i.p. [9]

Peak 2 has not yet been identified. Peak 3 represented a trace radioactive impurity present in the original ^{14}C-allopurinol. The major peaks were 4,5, and 6, the latter sometimes occurring as two peaks, 6A and 6B, which were clearly separable on a Dowex-1 formate column. Peaks 4,5 and 6B were unequivocally identified as 1-Alo-5'-P, 7-Oxi-5'-P and 1-Oxi-5'-P respectively, by comparison with the synthesized standards, with respect to retention times on the LCS-1000 high-pressure liquid chromatograph, elution volumes from Sephadex A-25 and Biorad AG-1, conversion to their respective ribonucleosides by phosphatase treatment and hydrolysis to their respective bases by acid [9]. The tissue concentration of 1-Alo-5'-P in peak 4 (Fig. 2) was ca. 10^{-8}M. No radioactivity was detected in fractions 218 -450, the region where the di- and triphosphates of adenine and guanine appeared (ADP eluted at fraction 250, GDP at 295, ATP at 400, and GTP at 435), indicating that ribonucleoside di- or triphosphates of allopurinol or oxipurinol were not present at the 10^{-9}M level. This finding is in agreement with the absence of enzymes in mammalian tissues of kinases for inosinic and xanthylic acids. No radioactivity was found in fraction 175, where 4-amino-6-hydroxypyrazolo(3,4-d)pyrimidine-1-ribonucleotide (prepared via the E. coli enzyme) was known to elute. Thus, no conversion of 1-Oxi-5'-P to the guanine analogue had occurred at a level > 10^{-9}M.

Table 2 shows the amounts of the three nucleotides found in rat liver kidney and red blood cells 3 hours after a high i.v. dose (50 mg/kg) of allopurinol. Liver and kidney showed 4-5 µM levels of 1-Alo-5'-P, whereas RBC showed only 0.27 µM. The principal oxipurinol ribonucleotide in liver was the 1-Oxi-5'-P; in kidney it was 7-Oxi-5'-P. Neither oxipurinol ribonucleotide was detectable in red cells at the 10^{-8}M level. The absence of 1-Oxi-5'-P in the red cells is of interest and correlates with the lack of activity of the human red cell HGPRT with oxipurinol (Table 1). Whether the liver HGPRT is qualitatively or quantitatively different from the red cell enzyme is not known. When a 50 mg/kg dose was given p.o., and an extra hour was allowed to permit absorption, the concentrations of 1-Alo-5'-P in liver and kidney were about 5% of the levels seen after i.v. administration, and the level in RBC was about half. The levels of the oxipurinol nucleotides in kidney were also much lower after p.o. than after i.v. administration.

The concentrations of the nucleotides were studied as a function of time after a 50 mg/kg i.v. dose of ^{14}C-allopurinol (Figure 3). The concentration of 1-Alo-5'-P remained at the 1-4 µM level for 3 hours and then fell rapidly. The oxipurinol nucleotides fell more rapidly in liver than in kidney and were below 10^{-8}M after 18 hours.

In general, the concentrations of the nucleotides in liver and kidney correlated with the concentrations of the respective free bases in the tissues (Figure 4). When the concentration of allopurinol in the tissue was 10 µM, the level of 1-Alo-5'-P was

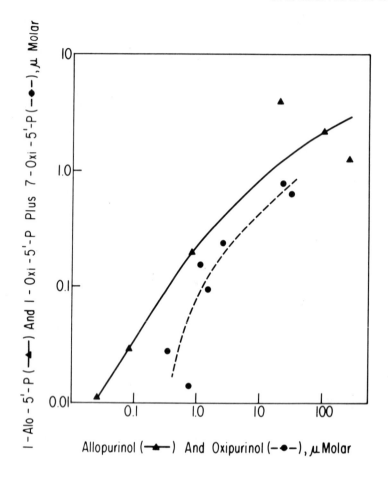

Figure 4.   Relationship between concentrations of free allopurinol
and oxipurinol and the amounts of the respective ribonucleotides
in rat liver and kidney.

about 1μM, when the oxipurinol concentration was 30 μM, the 1-and
7- ribonucleotides totaled about 0.8 μM.

        This sort of correlation should enable one to make some
predictions about man, assuming the rat and human enzymes are
similar in their specificities, turnover numbers and amounts.
The plasma levels of allopurinol and oxipurinol in the rats at
various times were determined (Table 3).

Table 3

DOSE	Time,	Plasma		Liver	
Rat	hrs.	Allo	Oxi	Allo	Oxi
		µM		µM	
50 mg/kg, i.v.	1	180	78	104	31
	3	55	44	20	22
	6	2	5		
50 mg/kg, p.o.	4	2	16		
Man					
300 mg. p.o.	3	< 10	65		
	6	< 1	65		

After a single 50 mg/kg i.v. dose, the allopurinol concentration
was 180 µM at 1 hour but fell to 2 µM at 6 hours. In man, after a
single oral dose of 300 mg, the plasma allopurinol remains below
10 µM at 1 - 3 hours and is undetectable at 6 hours [11]; after a
100 mg. dose, it is 1-2 µM at 3 hours [1,2]. One might, therefore,
expect the maximum 1-Alo-5'-P levels in man to be approximately 10^{-7}M
and to fall rapidly thereafter. The plasma oxipurinol levels, on
the other hand, are generally maintained at a constant level of about
65 µM in man on a daily 300 mg dose, because of the reabsorption
of oxipurinol in the kidney tubule [12] and the consequent slow rate
of clearance. Thus, oxipurinol ribonucleotide might be comparable in
man to that seen 1-3 hours after the high i.v. dose in the rat, i.e.
10^{-6}M in liver and kidney, but undetectable in red blood cells. This
assumes, however, that the levels of competing natural substrates
are similar in the two species and that the enzyme levels are
comparable.

EFFECTS ON PURINE METABOLISM

What, then, is the significance of the allopurinol and oxipurinol ribonucleotides? Their relationship to pyrimidine metabolism
will be discussed by Dr. Hitchings [13]. Feedback inhibition of
de novo purine synthesis has been found to occur in man [13] and
animals [3,15] treated with allopurinol and oxipurinol. From the
quantitative data presented here, it seems unlikely that allopurinol
ribonucleotide can be responsible for this effect. In Table 4 are
shown the pool sizes of some of the naturally occurring nucleotides,
as well as the expected levels of allopurinol ribonucleotide.

Table 4

	Pool Sizes	PRPP Amidotransferase	
	Rat Liver mM	Pigeon K_i mM	Human $[I]_{0.5}$
AMP	0.3	0.1 - 2	1.8
ADP	1.2	0.04 - 0.6	
ATP	3.1	0.04 - 1	
IMP	0.08	0.2 - 3	
GMP	0.08	0.1 - 1	0.5
GDP	0.2	0.4 - 5	
1-Alo-5'-P	< 0.004	0.6	
(Human	< 0.0001)		

Since allopurinol has a half-life of only about 1-1/2 hours in man, due to its rapid oxidation, the level of allopurinol ribonucleotide in human liver is probably < 0.0001 mM. If one uses, as a first approximation, the K_i values reported for the pigeon liver PRPP-amidotransferase, K_i = 0.6 mM, [16] the levels of allopurinol ribonucleotide in liver are far below the amount required to in-hibit this enzyme appreciably. On the other hand, the levels of the natural purine ribonucleotides [9] lie much closer to the K_i values for the pigeon liver enzyme [16,17] or to the $[I]_{0.5}$ values in human lymphoblasts [18]. Hypoxanthine and xanthine levels are increased by the inhibition of xanthine oxidase produced by allopurinol and oxipurinol. The reutilization of these oxypurines is very efficient under these conditions [19,20]. This could result in feedback inhibition of de novo synthesis through temporary increases in IMP and XMP, and subsequently AMP and GMP levels. This effect would be enhanced by the cooperative effect of AMP and GMP in the inhibition of PRPP - amidotransferase [17].

It has been suggested that allopurinol may exert an inhibition of de novo purine synthesis by lowering PRPP levels [21]. Such a decrease has been reported in human red cells during a 5-hour period following a dose of allopurinol (2-4 mg/kg, p.o.), but not following oxipurinol administration [21]. These data may not be relevant since red cells do not have a de novo pathway for purine biosynthesis. In human fibroblasts, in which the de novo pathway is presumably functional, there was no significant drop in PRPP levels when the allopurinol concentration was 0.1 mM [7], which is 10 - 100 times higher than the plasma levels in man. In rat liver, which has approximately 10 μM PRPP concentrations [23], it is conceivable that at the high dose of allopurinol (50 mg/kg, i.v.) enough

1-Alo-5'-P and 1-Oxi-5'-P would be formed to cause a temporary decrease of PRPP levels. However, more will need to be known about the functional capacity of PRPP synthetase, and the amounts and relative turnover rates of all the enzymes utilizing PRPP before the significance of PRPP levels can be interpreted.

Attempts were made to determine whether temporary changes in purine nucleotide pool sizes could indeed be detected in tissues following allopurinol administration. The nucleotide pool sizes in rat liver were measured using the LC1000 liquid chromatograph, 1,3 and 24 hours after a single i.v. dose (20 mg/kg) of allopurinol. A transient decrease in AMP, GMP, GDP and ATP levels was seen at 1 hour (Figure 5), but by 3 and 24 hours these had been restored to normal, or slightly higher than normal levels. It is not clear whether such temporary decreases were due to feedback inhibition or possibly to decreased PRPP levels. In kidney, on the other hand, the levels of AMP and ATP were slightly elevated after 1 hour (Figure 6) but were normal at 24 hours, the next time period at which they were examined. However, GDP levels in the kidney remained elevated at 24 hours and, indeed, after 7 days of allopurinol administration (20 mg/kg i.p.). Levels of GMP could not be measured accurately because orotic acid and GMP eluted simultaneously, and orotic acid was elevated. It is apparent that pool sizes of purine nucleotide can fluctuate rapidly and that they may fluctuate differently in different tissues. In general, whatever small changes occurred in the purine nucleotide pools after a large dose of allopurinol, these pools appeared to return to normal quickly, presumably because of the efficient regulatory mechanisms of purine biosynthesis.

The steady state purine nucleotide levels in rat liver were examined one month after chronic feeding with either 0.05% or 0.1% allopurinol in the diet. At these levels of allopurinol, the excretion of allantoin, the principal purine end-product in rats, was decreased 40% and 50% respectively. Urinary allantoin was only partially replaced by urinary xanthine and hypoxanthine. Thus, at the 0.05% level of allopurinol in the diet, allantoin excretion dropped by 13.3 μmoles/mg creatinine while oxypurines rose by 1.6 μmoles/mg creatinine, indicating that an inhibition of de novo purine synthesis had occurred. The purine nucleotide levels of such treated animals showed no net change in the mono, di-or triphosphates of adenine or guanine (Figure 7). There was, however, a slight elevation of IMP levels, as one might have expected, although the statistical significance is doubtful. The fact that purine nucleotide levels remained constant while de novo purine synthesis was inhibited is readily explained by the reutilization of hypoxanthine and xanthine for nucleotide synthesis.

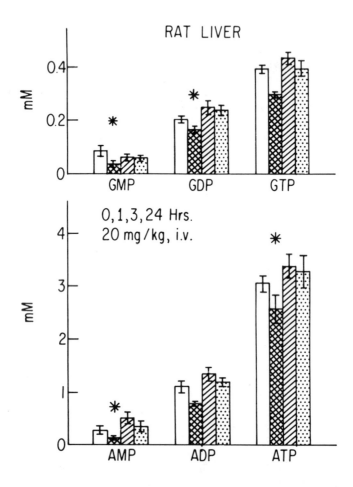

Figure 5.   Acid-soluble ribonucleotide levels in rat liver 0, 1, 3 and 24 hours after a single 20 mg/kg i.v. dose of allopurinol.

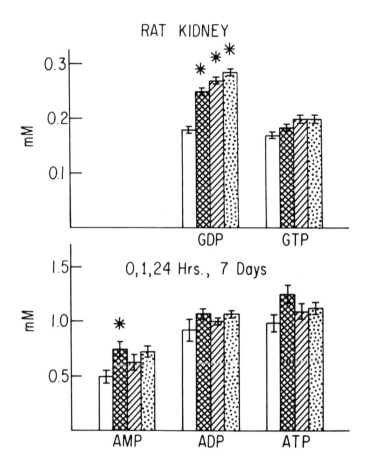

Figure 6.   Acid-soluble ribonucleotide levels in rat kidney 0, 1, and 24 hours after a single 20 mg/kg i.p. dose of allopurinol, and 24 hours following 7 days of administration of the same dose.

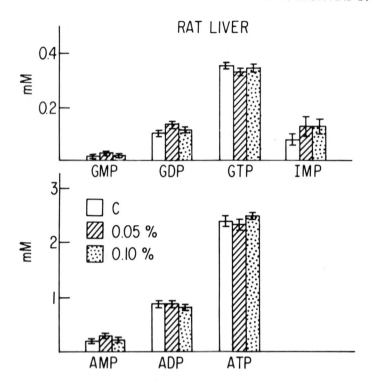

Figure 7.   Acid-soluble ribonucleotide levels in rat liver after
4 weeks of feeding a diet containing 0, 0.05% or 0.10% allopurinol.

SUMMARY

    The ribonucleotide of allopurinol and both the 1- and 7-
ribonucleotides of oxipurinol have been identified and quantified
in rat tissues under a variety of conditions. The amounts of these
nucleotides formed were shown to be related to the levels of individual bases, but must also depend on the levels of PRPP and of the
appropriate enzymes in the different tissues.

    On the basis of the quantitative data, the feedback inhibition
of de novo purine biosynthesis appears to be a function of the
naturally occurring ribonucleotides, rather than of 1-Alo-5'-P.
As a result of the reutilization of hypoxanthine and xanthine,
normal ribonucleotide pools are maintained.

REFERENCES

1.   G.B. Elion, A. Kovensky, G.H. Hitchings, E. Metz and R. W. Rundles (1966) *Biochem. Pharmacol.*, 15: 863.

2.   G.B. Elion (1966) *Ann. Rheum. Dis.*, 25: 608.

3.   G. H. Hitchings (1969) In: Biochemical Aspects of Antimetabolites and of Drug Hydroxylation. *FEBS Symposium*, 16: 11.

4.   W.N. Kelley and J.B. Wyngaarden (1970) *J. Clin. Invest.*, 49: 602.

5.   T.A. Krenitsky, R. Papaioannou and G.B. Elion (1969) *J. Biol. Chem.*, 244: 1263.

6.   R.M. Fox, M.H. Wood and W.J. O'Sullivan (1971) *J. Clin. Invest.*, 50: 1050.

7.   W.N. Kelley, T.D. Beardmore, I.H. Fox and J.C. Meade (1971) *Biochem. Pharmacol.*, 20: 1471.

8.   T.D. Beardmore and W.N. Kelley (1971) *J. Lab. Clin. Med.*, 78: 696.

9.   D.J. Nelson, C.J.L. Bugge, H.C. Krasny and G.B. Elion (1973) *Biochem. Pharmacol.*, in press.

10. J.A. Fyfe, R.L. Miller and T.A. Krenitsky (1973) *J. Biol. Chem.*, in press.

11. D.J. Nelson and G.B. Elion, unpublished.

12. G.B. Elion, T.F. Yü, A.B. Gutman, and G.H. Hitchings (1968) *Amer. J. Med.*, 45: 69.

13. G.H. Hitchings, J.A. Fyfe and D.J. Nelson (1973) accompanying manuscript.

14. R.W. Rundles, J.B. Wyngaarden, G.B. Elion, G.H. Hitchings and H.R. Silberman (1963) *Trans. Assoc. Amer. Physicians*, 76: 126.

15. G.H. Hitchings (1966) *Ann. Rheum. Dis.*, 25: 601.

16. R.J. McCollister, W.R. Gilbert, Jr., D.M. Ashton and J.B. Wyngaarden (1964) *J. Biol. Chem.*, 239: 1560.

17. C.T. Caskey, D.M. Ashton and J.B. Wyngaarden (1964) *J. Biol. Chem.* 239: 2570.

18. A.W. Wood and J.E. Seegmiller (1973) *J. Biol. Chem.*, 248: 138.

19. R. Pomales, S. Bieber, R. Friedman and G.H. Hitchings (1963) *Biochim. Biophys. Acta*, 72: 119.

20. R. Pomales, G.B. Elion and G.H. Hitchings (1964) *Biochim. Biophys. Acta*, 95: 505.

21. I.H. Fox, J.B. Wyngaarden and W.N. Kelley (1970) *New Eng. J. Med.*, 283: 1177.

22. A. Clifford, J.A. Riumallo, B.S. Baliga, H.N. Munro and P.R. Brown (1972) *Biochim. Biophys. Acta*, 277: 443.

THE EFFECT OF ALLOPURINOL ON ORAL PURINE ABSORPTION

AND EXCRETION IN THE PIG

H.A. Simmonds, A. Cadenhead, A.S. Jones,
P.J. Hatfield, and J.S. Cameron

Guy's Hospital and Rowett Research Institute
London, S.E.1 9RT, U.K. and Bucksburn, Aberdeenshire, U.K.

Allopurinol therapy in hyperuricaemic man has been shown to be advantageous from two points of view. Firstly, it reduces urinary uric acid excretion and increases the excretion of the precursor purines xanthine and, to a lesser extent, hypoxanthine. In addition, total urinary purine excretion (the sum of these three) may be reduced by as much as 50% during allopurinol therapy (1). This latter effect has been attributed to the formation of nucleotides of either hypoxanthine (1) or allopurinol itself (2), which in turn exert a "feed back" inhibitary effect on the first enzyme of de novo purine synthesis.

In this paper we present results of experiments in which we have investigated the effect of allopurinol in the pig, firstly on a low purine diet and then given an oral purine supplement in the form of guanine (3).

In the pig on a purine free diet allopurinol had little effect on total urinary purine excretion, even at dosages of 300 mg/kg/day.

Pigs given an oral purine supplement of guanine (150 mg/kg/day) absorbed and excreted up to 50% of the dose in the urine. When the guanine diet was supplemented with allopurinol (300 mg/kg/day) the increase in total urinary purine excretion resulting from the guanine alone was greatly reduced. Other workers have likewise found that in man allopurinol reduces substantially the increased purine excretion resulting from exogenous yeast purine (RNA) (4,5). This finding is difficult to explain on the basis of feedback inhibition of de novo purine synthesis alone.

653

These observations have been investigated by following the
relative rates of incorporation of $(8-^{14}C)$ guanine into body
fluids and tissues during the oral administration of guanine, and
guanine supplemented with allopurinol at the above dosages (6).
Incorporation of radioactivity into tissues was minimal when ^{14}C
guanine was given orally during the feeding of either guanine
alone or guanine plus allopurinol.

The results with oral ^{14}C guanine showed that during
guanine feeding alone approximately 43% of the administered
radioactivity was excreted in the urine and 5% in the faeces. It
is presumed that the 50% of radioactivity unaccounted for was
excreted via the lungs as CO2 following bacterial degradation in
the gut, since Sorensen (7) has reported the recovery of 55.5% of
orally administered ^{14}C uric acid as $^{14}CO2$ in man.

When dietary guanine was supplemented with allopurinol the
excretion patterns changed considerably, only 14% of the radioactivity being excreted in the urine, the remainder (85%) being
found in the faeces. The total recovery of radioactivity in
these latter experiments suggests that little degradation of
purines by gut bacteria has taken place during allopurinol
therapy.

These experiments show that in the pig during concomitant
allopurinol therapy a large proportion of an exogenous purine
load is eliminated via the gut.

Similar experiments using intravenous ^{14}C guanine have been
carried out to determine whether allopurinol produces this effect
by blocking absorption or enhancing purine excretion into the gut.
In these experiments 13% of the intravenously administered label
was excreted into the gut during allopurinol therapy, and less
than 1% was found in the faeces when guanine was given alone.
However, respiratory CO2 was measured for only the first twenty-
four hours in these experiments, and these results therefore
require to be substantiated.

CONCLUSION

1.   Allopurinol therapy in pigs produces a net reduction in the
      absorption of exogenous purine from the gut.

2.   This is reflected in a reduction in total urinary purine
      excretion. This reduction therefore is not the result of a
      reduction in de novo purine synthesis, or increased tissue
      incorporation of exogenous purine.

3. Excretion of purine into the gut in the presence of allo-
    purinol has been demonstrated, but block of purine absorp-
    tion by allopurinol has not been excluded.

4. In patients not on a low purine diet, at least part of what
    has hitherto been considered feedback inhibition of de novo
    purine synthesis is probably due to decreased absorption of
    dietary purine.

5. Rigorous dietary purine restriction does therefore not
    appear necessary during allopurinol therapy.

6. Allopurinol may exert an additional effect in reducing
    purine degradation by gut bacteria. This effect on nitrogen
    recycling has yet to be evaluated.

REFERENCES

1. Rundles, R.W., Wyngaarden, J.B., Hitchings, G.H., Elion, G.B.
    (1969). Ann. Rev. Pharmacol., 9, 345.

2. Fox, I.H., Wyngaarden, J.B., Kelley, W.N. (1970). N. Eng.
    J. Med., 283, 1177.

3. Simmonds, H.A., Hatfield, P.H., Cameron, J.S., Jones, A.S.,
    Cadenhead, A. (1973). Biochem. Pharmacol. In press.

4. Bowering, J., Margen, S., Calloway, D.H., Rhyne, A. (1969).
    Am. J. Clin. Nutr., 22, 1426.

5. Zöllner, N., Griebsch, A., Gröbner, W. (1972). Ernahr. -
    Umschau, 19, 79.

6. Simmonds, H.A., Rising, T.J., Cadenhead, A., Hatfield, P.J.,
    Jones, A.S., Cameron, J.S. (1973). Biochem. Pharmacol. In
    press.

7. Sorensen, L.B. (1960). Scand. J. Clin. Invest., 12, 1,
    (Supplement 54).

ALLOPURINOL AND THIOPURINOL : EFFECT IN VIVO ON URINARY OXYPURINE EXCRETION AND RATE OF SYNTHESIS OF THEIR RIBONUCLEOTIDES IN DIFFERENT ENZYMATIC DEFICIENCIES.

AUSCHER C., MERCIER N., PASQUIER C.,DELBARRE F.

Centre de Recherches sur les Maladies Ostéo-articulaires
Unité N°5 INSERM. ERA 337 CNRS. Clinique de Rhumatologie
Médicale et Sociale Hôpital Cochin 27, rue du Fg St
Jacques 75014 PARIS   FRANCE

At the present time, we just report some experimental results
of a study on the mechanism of action of allopurinol (4-hydroxy-
pyrazolo (3,4-d) pyrimidine) and thiopurinol (4 thiopyrazolo
(3,4-d) pyrimidine) on de novo biosynthesis of uric acid. In this
present work, we have compared effect of allo and thiopurinol on
oxypurine (xanthine and hypoxanthine) urinary excretion with their
rate of synthesis of ribonucleotides in vitro by erythrocyte
hemolysate in some particular enzymatic deficiencies (hypoxanthine-
guanine phosphoribosyltransferase : HGPRT, adenine phosphoribosyl-
transférase : APRT and xanthinuria).

I-EFFECT OF ALLOPURINOL AND THIOPURINOL ON URINARY OXYPURINE EXCRETION (table 1)

In gout with normal renal function and normal HGPRT activity,
we have shown that thiopurinol like allopurinol reduces uric acid
synthesis. However, while allopurinol reduces urinary uric acid
excretion with partly balanced increase of oxypurine, thiopurinol
has no significant effect after prolonged period administration on
urinary excretion and plasma oxypurines (DELBARRE et al 1967,1972).

In patients with nearly complete deficiency of erythrocyte
HGPRT activity (either Gout or LESH NYHAN syndrome) thiopurinol has
no effect on plasma and urinary excretion of uric acid (DELBARRE
et al. 1970), while in the same patients treated with allopurinol
there is a rapid andimportant decrease of uric acid balanced by
nearly stochiometric increase of oxypurines. Gouty patients with
HGPRT deficiency have higher urinary oxypurine excretion with a
more important contribution of hypoxanthine (H/X = 2,36) than

Table I

EFFECT OF ALLO AND THIOPURINOL ON URINARY
OXYPURINE(1) EXCRETION - GOUTY PATIENTS -

Treatment	HGPRT normal (2)						HGPRT < 2% of normal (3)					
	none	allo	var^n	none	thio	var^n	none	allo	var^n	none	thio	var^n
Uric acid	560	363	-197	603	413	-190	801	371	-430	801	788	-13
Uric acid + Oxypurines	591	513	- 78	634	445	-189	872	1007	+135	872	846	-26
Oxypurines	31	150	+119	31	32	+ 1	71	638	+567	71	58	-13
H/X (4)	1,08	0,48					2,36	1,02				
n	7	4					3	3				
± s	0,17	0,07					0,43	0,22				

(1) values are expressed in mg/day uric acid
(2) Gout with normal renal function (from DELBARRE et Al. 1972)
(3) 3 gouty patients presented in this symposium by AMOR
(4) personnal results.
var^n = difference between treatment and none

normal patient(H/X = 1,08). On allopurinol therapy the averaged
contribution of hypoxanthine in increased oxypurines excretion is
about 50% (H/X = 1,02) in HGPRT deficiency and 32 % (H/X = 0,48)
in normal activity.

II - RATE OF SYNTHESIS OF AMP, IMP, GMP , ALLO AND THIOPURINOL
RIBONUCLEOTIDE IN VITRO BY HEMOLYSATE OF ERYTHROCYTES.

A - Control Values (table II)

Activity of APRT , HGPRT and the rate of synthesis of allo and
thiopurinol nucleotide are determined by isotopic method (CARTIER
et al 1968). For routine study PRT activity is expressed in nM/mg
hemoglobin /hour (unwashed erythrocytes) but for more precise work
content of protein in hemolysate of washed erythrocytes (modification
of LOWRY method with albumine as standard) is determined. Synthesis
of nucleotide was confirmed both by effect of 5' nucleotidase

Table II

RATE OF SYNTHESIS OF RIBONUCLEOTIDES OF ALLO AND
THIOPURINOL BY LYSATES OF ERYTHROCYTES – CONTROL SUBJECTS

Substrat	mM in reaction	nM/mg hemoglobin/h		nM/mg protein/h	
		n	mean + s	n	mean + s
Adénine	2,4	24	22,6 ± 5,1	9	15,1 ± 3,4
Hypoxanthine	2,4	24	109,1 ± 12,7	9	80,2 ± 6,3
Guanine	0,6	24	139,1 ± 21,5	9	106,5 ± 9,6
Allopurinol	0,33	5	1,92 ± 0,23	5	1,54 ± 0,27
Thiopurinol	0,29	5	1,43 ± 0,23	5	1,14 ± 0,21

(from Sigma) on the product of reaction and by depletion of PRPP
(phosphoribosylpyrophosphate) when erythrocytes with normal HGPRT
are incubated either with allo or with thiopurinol in presence
of glucose (SPERLING et al 1972).

B – Enzymatic Deficiencies (table III)

     Clinical observations of patients and their family are report-
ed in this symposium by DELBARRE and AMOR.

     In patients NAD., BRE., (son) MON., (m) whatever the importance
of the deficiency there is a correlation between rate of synthesis
of IMP, GMP, and allo and thiopurinol nucleotide. The gouty patient
LUG., has nearly normal rate of synthesis of IMP (73 % of normal)
but an important deficiency when guanine is substrat (26 %). KELLEY
et al (1969) mentioned such anomalies but found more important
loss of activity with both substrates (10 % of control for hypoxanthine and 0,5 % with guanine). In this particular enzymatic
mutation rate of synthesis of allo and thiopurinol nucleotides
in vitro are about 50 % of normal. The mother of LESH-NYHAN
syndrome BRE.,(m), GIR., the gouty patient with partial APRT deficiency (38 %) and MOR., the xanthinuric man have normal rate of
synthesis of IMP, GMP, allopurinol and thiopurinol nucleotides.

Table III

RATE OF SYNTHESIS OF PURINE, ALLO, AND THIOPURINOL
NUCLEOTIDES BY LYSATES OF ERYTHROCYTES (% of control values) IN ENZYMATIC DEFICIENCIES

Patient	Diagnostic	Sex	Substrats: Adénine	Hypoxanthine	Guanine	Allopurinol	Thiopurinol
NAD.	Gout	M	129	~2	~2	~2	~2
LUG.	Gout	M	129	73	26	53	56
BRE (son	Lesh-Nyhan	M	254	~1	~1	~1	~1
(mother		F	108	103	92	113	101
MON (son	Gout	M	113	2	2		
(mother	Gout [1]	F	128	60	54	58	65
GIR.	Gout	M	38	109	107	110	110
MOR.	Xanthinuria	M	89	88	95	88	91

Underlined values are outside 95 % range limits (mean ± 2s)
(1) Onset 48 years.

III - EFFECT OF ALLO AND THIOPURINOL ON URINARY OXYPURINE
EXCRETION AND THEIR RATE OF SYNTHESIS IN VITRO OF RIBONU-
CLEOTIDE IN ENZYMATIC DEFICIENCIES (table IV)

The important effect on the oxypurine excretion (with high
ratio of hypoxanthine/xanthine)during allopurinol therapy and the
inefficacy of thiopurinol, in both patients LUG., and MON., (m)
are quite the same as in patient with complete deficiency. They
have 50 % of control values of rate of synthesis of allo and
thiopurinol nucleotide in vitro. (PRPP content of erythrocytes
is normal.) The response of allopurinol therapy on uric acid
and oxypurine excretion in the mother of LESCH-NYHAN BRE., (m)
is the same as other gouty patient with normal HGPRT and normal
rate of synthesis of allo and thiopurinol nucleotide. The gouty
patient GIR., with partial deficiency of APRT (38% of normal)
has an important overproduction of uric acid . This was confirmed
by incorporation of labeled glycine into uric acid (DELBARRE et al
1969). Allopurinol and thiopurinol has an explosive effect on uric
acid biosynthesis de novo (-40%). With allopurinol therapy there

Table IV

EFFECT OF ALLO AND THIOPURINOL ON URINARY
PURINE EXCRETION IN ENZYMATIC DEFICIENCY

Patient	Treatment	PRPP[1]	Uric acid mg/day	Oxypurines + uric ac. mg/day	Oxypurines[2] mg/day	Hypoxanthine / Xanthine
LUG.	None	3,70	680	724	44	2,45
	Allo		180	694	514	1,00
	Thio		668	708	43	2,23
MON.(3)	None		441	463	22	
	Allo	8,52	255	442	187	1,30
	Thio		420	449	29	
BRE(4)	None	6,69	686	732	46	
	Allo		382	515	133	0,53
GIR.	None	2,78	1024	1052	28	0,78
	Allo		301	577	276	0,85
	Thio		586	610	24	0,82
MOR.	None	6,70	17	347	330	0,41
	Allo		20	385	365	0,35
	Thio		17	403	386	0,40

(1) PRPP : phosphoribosylpyrophosphate Control values (8) mean ± s : 6,64 ± 1,90 mM/ml
    erythrocytes (Sperling method)
(2) oxypurines = xanthine + hypoxanthine expressed in mg uric acid.
(3) mother of gout (tab. III)
(4) mother of LESH-NYHAN - PRPP of the son 52 nM/ml erythrocytes (see tab.III)

is a supplementary decrease of uric acid balanced by increase of
oxypurines. There is no significant modification of the ratio
H/X. Effect of thiopurinol on the decrease of uric acid synthesis
de novo has been confirmed by the diminushion of incorporation
of labeled 1 -14 C glycine into uric acid. More detailed study of
this gouty patient will be reported in this symposium by DELBARRE.
Neither allopurinol nor thiopurinol have any effect on the amount
of oxypurine excretion and on the ration H/X in xanthinuric patient
MOR. Nevertheless, the rate of synthesis of nucleotide is normal
with all substrats (purines and pyrazolopyrimidines). On the
other hand, ratio H/X is in the 95 % range limits of that of
gouty patient with normal HGPRT treated by allopurinol (table IV).
More detailled study of this case will be reported in an other
communication of this symposium.

SUMMARY

1. In the present study we have shown that in vitro with hemo-lysate of erythrocytes there is a conversion of allo and thiopuri-nol to allo and thiopurinol ribonucleotide. The rate of synthesis of both nucleotides is slow (about 1.5 to 2 % that of IMP).

The per cent of control values in partial HGPRT deficiency is the same as well as with hypoxanthine and guanine or with allo'and thiopurinol as substrat. In the particular enzymatic deficiency LUG, who has nearly normal PRT activity with hypoxanthine as substrat, rate of synthesis of allo and thiopurinol are about 50 % of normal value.

Rate of synthesis of allo and thiopurinol ribonucleotides is normal with hemolysate of erythrocyte of the partial APRT deficien-cy patient, the mother of LESH-NYHAN syndrome and the xanthinuric man.

2. Urinary oxypurine excretion is higher in patient with com-plete or partial HGPRT deficiency. The contribution of hypoxanthine is also higher in HGPRT deficiency as in untreated patient as in the increased oxypurines produced by allopurinol. Whatever is the importance of the deficiency there is a great effect of allopurinol on decreased of uric acid balanced by oxypurines, it is the same as in complete HGPRT deficiency as well as inefficacy of thiopurinol. The response of allopurinol on decrease of uric acid synthesis in the mother of LESH-NYHAN syndrome is the same as in normal HGPRT activity gouty patient.

3. APRT activity of hemolysate erythrocytes does not contribute in the in vitro synthesis of allo and thiopurinol nucleotide. More-over there is a very important reduction of biosynthesis de novo of purine (-40 %) in this gouty patient treated either with allo or thiopurinol (100 to 200 mg/day).

4. As well as in vitro APRT, HGPRT activity and rate of syn-thesis of allo and thiopurinol nucleotide are normal in the xanthi-nuric man. But there is no effect of allo (200 mg/day) and thiopu-rinol (500 mg/day), on oxypurine excretion, and on the ratio hypo-xanthine to xanthine.

The contribution of hypoxanthine in oxypurine excretion is nearly the same in xanthinuric patient that in gouty patient with normal renal function and HGPRT activity treated by allopurinol.

URINARY EXCRETION OF 6 HYDROXYLATED METABOLITE AND OXYPURINES IN

A XANTHINURIC MAN GIVEN ALLOPURINOL OR THIOPURINOL

AUSCHER C., PASQUIER C., MERCIER N., DELBARRE F.

Centre de Recherches sur les Maladies Ostéo-articulaires
Unité n°5 INSERM. ERA 337 CNRS. Clinique de Rhumatologie
Médicale et Sociale Hôpital Cochin, 27, rue du Fg St
Jacques 75014 PARIS   FRANCE

Xanthinuria is characterized by a large urinary excretion of
oxypurine (xanthine + hypoxanthine) which replaces uric acid at
the end product of purine metabolism. Patients with xanthinuria
are very deficient in xanthine oxidase activity. This rare metabolic disorder may be of interest for both information :
1) the contribution of xanthine oxidase in the 6 hydroxylation of
4 -hydroxy-pyrazolo (3,4-d) pyrimidine [(Pyrazolo (3,4-d) pyrimidine = PP)] : allopurinol and 4-thio-pyrazolo (3,4-d) pyrimidine :
thiopurinol. 2) the contribution of their nucleotides (allo and
thiopurinol ribonucleotide) to the reduction of biosynthesis de
novo of purine.

We have given allopurinol and thiopurinol to a xanthinuric
man and in this communication we report on the urinary excretion
amount of 4-6 dihydroxy PP : oxipurinol, or 4-thio-6-hydroxy PP :
oxithiopurinol and the concomitant values of xanthine and hypoxanthine.

PATIENT AND EXPERIMENTS
The clinical case of this xanthinuric man has been reported
by DELBARRE in a previous paper (to be published). This man is a
pastry cook; since he could not be in the hospital more than one or
two days, most determinations were done without control diet. This
explains the great difference of daily urinary oxypurine values
that were found. This patient has normal renal function (creatinine
clearance = 99 ml/min). His averaged uric acid and oxypurines
(xanthine and hypoxanthine) plasma concentration were 1.2 and
0.58 mg per 100 ml. His 95 % range (mean $\pm$ 2s) urinary excretion
of uric acid is 7 to 27 mg per day and of oxypurine 310 to 618 mg

per day or 230 to 430 mg per gramme of creatinine. The averaged
contribution of xanthine and hypoxanthine are 71 % and 29 % that is
to say that the averaged ratio of hypoxanthine to xanthine is
0,41± 0,05. His urinary excretion was normal for adenine (1,73 mg/
day) and for guanine (0,93 mg/day). We dit not find any measurable
amount of 8 hydroxy-7-methyl guanine in daily urine but a concomitant
increase of 7-methyl guanine (9,10 mg/day: control :6,9 mg/day).
SKUPP and AYVAZIAN (1969) have demonstrated that 8-hydroxy-7- methyl
guanine is the product of oxidation of7-methyl guanine but only by
human liver xanthine oxidase. This xanthinuric man has a family, but
up to now we have only seen one of his sisters who also has
xanthinuria.

The amount of urinary excretion of each PP and its 6-hydroxy-
lated metabolite as well as the concomitant excretion of oxypurine
were determined :1) in daily urine after prolonged period of oral
administration. 2) in the hours following a single dose of oral
administration. The same experiment was carried out in a non-
xanthinuric man with normal renal function.

RESULTS

I-Urinary Excretion Amount of Unchanged PP and of their
   6-Hydroxylated Metabolite

Prolonged period administration. a) After 3 and 8 days
of 200 mg per day of allopurinol we have found 26 mg of unchanged
allopurinol (13 %) in the daily urine, and 146 (73 %) and 113 mg
(56,7 %) of oxipurinol of the daily dose of allopurinol. b) After
two periods of 8 days of oral administration of thiopurinol
(400 mg per day) we found only traces of unchanged thiopurinol
(less than 0,05 %) ; 224 mg (73 %) and 163 mg (56 %) of oxithiopurinol excreted in the daily urine.

Single dose administration (7 mg/kg). Urines were collected 2,4,6,10 and 24 hours after oral administration of 400 mg to
the xanthinuric subject and 600 mg to the control subject either of
allopurinol or of thiopurinol. As for the control subject 58 mg
of unchanged allopurinol (15 %) was excreted within the six hours
following administration. The cumulative urinary excretion amount
of oxipurinol was equal to 1,6 % of the dose at the end of two
hours, 5, 2 % at 4 h, 8,7 % at 6 h, 12,5 % at 10 h and 32,5 % at
24 hours. These values are quite similar to those obtained with
the control subject. (fig 1) and agreed with those found by ELION
et al. (1966) in non xanthinuric subject by using an isotope dilution method. Thiopurinol was also given in the same fashion. As for
the control subject we found only traces or unchanged thiopurinol

(less than 0,05 %) in the first hours following administration and a rapid excretion of oxithiopurinol. The cumulative excretion amount of this 6-hydroxylated metabolite is 13 % at 2 hours, 24,2% at 4 h, 43,3 % at 6 h, 54,2 % at 10 h and 63,7 % at 24 h. We obtained quite the same values in the control subjet. (fig.2).

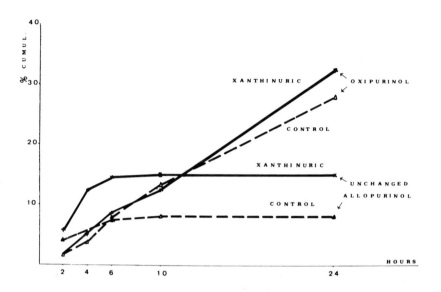

Fig. 1.   Allopurinol - single dose.

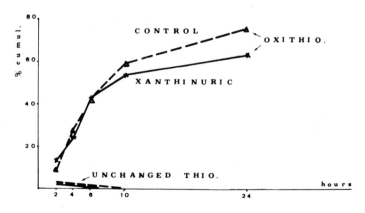

Fig. 2.   Thiopurinol - single dose.

II –Urinary Excretion of Oxypurines : Hypoxanthine and
Xanthine.

We have found no modification of urinary excretion amount
of oxypurines or of the ratio of hypoxanthine to xanthine after
3,7 and 57 days allopurinol (200 mg/days) and 3,5,14 and 25 days
of thiopurinol(500 mg/days) administration, nor during the 24 hours
following administration of a single dose of every PP. All the
values found were in the 95 % range limits of the mean values
obtained without treatment.

DISCUSSION

1) After a prolonged period administration of allopurinol or
of thiopurinol as well as in the hours following administration of
a single dose of every PP, we have found that the amount of the
6-hydroxylated metabolite :(oxipurinol or oxithiopurinol) is quite
similar in the xanthinuric man as in the control subject. There
is only a small amount of unchanged PP in both cases. It is
thought that the enzyme responsible for 6 hydroxylation of PP
would be xanthine oxidase. However, in vivo, the rate of 6 hydro-
lation of each PP is as rapid in the xanthinuric subject as in the
non-xanthinuric one, while, in vitro, 6-hydroxylation is extremely
slow. Owing to the rarity of xanthinuria, very few studies could
have been carried out up to now. ELION et al. (1966) have found
only 2 % of oxipurinol excreted in the urine of a xanthinuric
patient given allopurinol for one day. In contrast CHALMERS et
al (1969) who have given xanthinuric man 10 days allopurinol
have found 63 % of the daily dose excreted in the 24 hours urine
as oxipurinol.

As for our results we do not think that xanthine oxidase
could be the enzyme responsible for the in vivo hydroxylation of
PP. Which is the enzyme responsible for this rapid and important
hydroxylation? We cannot say at the moment. CHALMERS et al (1969)
have proposed a scheme involving oxidation of allopurinol ribonucleotide into oxipurinol ribonucleotide and the formation of oxipurinol via oxipurinol ribonucleoside. That may be an explanation,
and as we know (AUSCHER et al Tel Aviv 1973) the rate of synthesis
in vitro of allo and thiopurinol nucleotide is slow.

2) We have shown that allopurinol and thiopurinol (about 200 mg
per day) have no significant effet on the decrease of total urinary
oxypurine excretion and on the ratio : hypoxanthine to xanthine,
either after prolonged period administration or after a single
dose. It is well known that thiopurinol like allopurinol reduces
the synthesis of uric acid in non xanthinuric subjects ; this effect
needs normal activity of HGPRT.

In our xanthinuric man activity of adenine and hypoxanthine -
guanine phosphoribosyl transferase in erythrocytes is normal as
well as rate or synthesis of ribonucleotide of allo and thiopurinol by hemolysate of erythrocytes but as we have shown, it is slow
(about 1,5 to 2 % of that of IMP). PRPP content of erytrocytes is
also normal : 6,70 nM/ml erythrocytes (mean± s : 6,64 ±1,90 nM/ml
erythrocytes).

BRADFORD (1968) has shown by isotopic method that in xanthinuric female hypoxanthine is a very active metabolite and only
5,7 % of its daily turnover is excreted in urine. In contrast,
xanthine would be an end product(61 % of its daily turnover are
excreted in urine). It is known that there is a strong regulation
of synthesis of AMP ans GMP from IMP. Each of them regulates their
own synthesis, moreover HGPRT is inhibited and regulated by IMP
and GMP. The great turnover of the active metabolite hypoxanthine
in xanthinuria and the rapid synthesis of IMP afford probably an
optimal balance in the regulation of endogenous nucleotide AMP
and GMP and therefore a very favorable ration between them, that
is to say that in our xanthinuric man, glutamine-phosphoribosyl-
amido-transferase would be in optimal condition of regulation.
Whether or not the nucleotide of thio or of allopurinol is formed
in vivo in our xanthinuric man, we cannot answer at the present
time. The results presented above show that even if it is formed,
it does not contribute to the regulation of the synthesis de novo
of purine.

From our study and from the results of BRADFORD (1968) on the
important daily turnover of hypoxanthine in xanthine oxidase deficiency, we would like to say that deficiency (or inhibition by
allo or oxipurinol) of xanthine oxidase activity should be an
important factor in the regulation of de novo biosynthesis of uric
acid. (over producter with normal HGPRT activity).

Adenine Therapy in
the Lesch-Nyhan Syndrome

ACUTE RENAL FAILURE DURING ADENINE THERAPY

IN LESCH–NYHAN SYNDROME

Mario Ceccarelli, Maria Laura Ciompi, Giampiero Pasero

Deptm. of Pediatrics and Deptm. of Internal Medicine

University of Pisa, Italy

Additional adenine and folic acid are required for optimal growth of hypoxanthine–guaine phosphoribosyltransferase (HGPRT) deficient fibroblasts in tissue culture [1]. On this ground adenine therapy was advised in Lesch–Nyhan (LN) syndrome and resulted in a prompt correction of megaloblastic anemia in two cases [2]. In mammalian tissue adenine is easily converted into adenosine monophosphat by adenine phosphoribosyltransferase, utilizing 5–phosphoribosyl-pyrophosphate (PRPP). "Inappropriate" purine biosynthesis "de novo" in LN syndrome can be just accounted for an increased availability of PRPP due to missing competition between HGPRT and glutamine-PRPP amidotransferase for this substrate [3-5]. Consequently PRPP consumption from adenine is likely to result in a decreased purine biosynthesis, as proved by lesser incorporation of 14C-glycine in urinary uric acid (UA) [2]. Nevertheless adenine therapy showed conflicting evidence of inhibiting purine biosynthesis, since a decrease of urinary UA was not evidenced in two cases in spite of a reduction of erythrocyte PRPP [6]. Therefore we think that any further contribution on this topic is of some interest.

We have treated with adenine a 4-years old boy (G.B.), the first case of LN syndrome observed in Italy [7]. The weight at delivery was 3.550 gm, but growth became soon slower and inadequate, so that at age 4 the patient weighted 13 kg. Early objective evidence of illness was the presence of orange flecks in the diapers, pointed out by the mother to the pediatrician but disregarded by him (we mean emphasize this warning sign, noticed in a further case of LN syndrome recently observed by us [8]). Later, about at the end of the first year, the patient developed the classical pattern of the disease: spastic cerebral palsy, choreo-a-thetosis, growth retardation, pronounced dysarthria and self-mutilating behavior. Growth standstill should be accounted for nutritional problems, such as dysphagia, food refusal and vomiting. Self-destructive behavior, chiefly represented by biting his fingers and lower lip, seemed to be compulsive in its development, since the patient wanted to be protected against himself with gloves and bandages. The diagnosis of LN syndrome was advised at 3 years and half, when the patient was referred to us by the Pediatric Neuropsychiatric Service of our Hospital for a metabolic study. Blood UA was 10.5 mg%, urinary UA excretion at purine-free diet averaged 600 mg/day, that is 46 mg/kg/day. HGPRT activity was undetectable in hemolysates, while erythrocyte PRPP resulted 11 nmol/ml and 14C-glycine cumulative incorporation in urinary UA exceeded 0.4% in 7 days.

Adenine therapy was started in march 1972 at a dose of 1 g/day, that is 80 mg/kg/day by mouth. An attempt to rise protein intake almost up to 30 gm/day of animal proteins was unsuccessful owing to the lack of cooperation by the patient. Frequent control of UA excretion was hard to perform for difficulties in urine collection, moreover since the child was treated as an out-patient. An immediate objective finding of reduced UA excretion was however the disappearance of orange flecks on the diapers and underclothes; the reliability of this finding was testified by its prompt reappearance when treatment was discontinued temporarily for lack of adenine. Whenever adequate urine collection was obtained, UA excretion resulted unequivocally reduced; the lessening of purine biosynthesis was associated with a decrease in erythrocyte PRPP and blood UA; a representative pattern of metabolic changes after two months of treatment showed an erythrocyte PRPP of 7 nmol/ml, with blood UA of 7.8 mg% and urinary UA excretion of 380 mg/day. Spasticity, athetoid movements and solf-mutilation exhibited no definite improvement, while mental retardation seemed to display a tendency towards the correction; in any case behavioral changes are detailed elsewhere.

In july 1972 the child showed progressive lethargy, while urine output fell till almost complete anuria. At the admission in our service (july 15 th) blood urea was 328 mg% (one month earlier it was 36 mg%), blood creatinine 7.8 mg%, blood UA 48 mg% with an uricase method, blood actual pH 7.10, while plasma electrolytes were in normal range. Alkaline and high-caloric fluid intake with Ringer solution, sodium bicarbonate and L-gulitol and administration of allopurinol released the diuresis and allowed a progressive decline in blood levels of nitrogen compounds (urea, creatinine and UA). Overcorrection of acid-base balance with alkalosis and generalized seizures and urinary tract infection with fever worsened the course of illness, but after three weeks (august 7th) a satisfactory metabolic balance was restored: blood urea was 44 mg%, blood actual pH 7.33 (Fig. 1). The recovery of renal function after acute renal failure was good: on september 21, glomerular filtration rate, as measured by clearance of radiohypaque and corrected for body surface, resulted 8 ml/min.

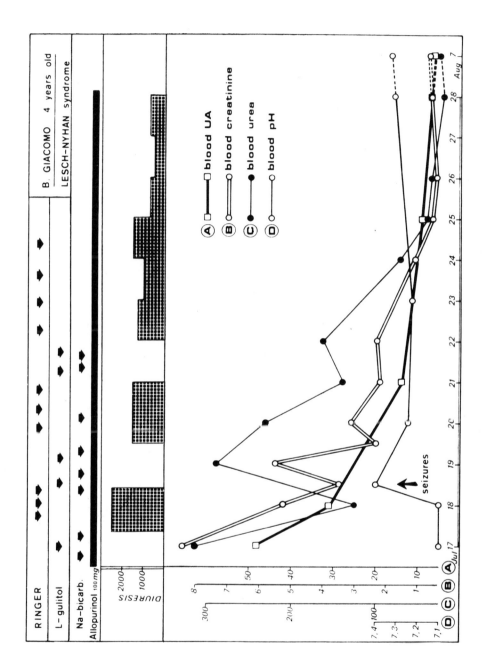

Fig. 1

Clinical picture was consistent with tubular obstruction by UA and/or oxidized derivatives of adenine. Adenine is oxidized by xanthine oxidase to 2,8-dioxyadenine, a compound with a very low solubility[9] and a definite toxic effect on kidney[6,10]. Therefore it is likely that dioxyadenine was the initiating factor in tubular obstruction; however very high blood UA level and recovery of urine output by alkalinization pointed out the role of urate crystallization. We were not able to measure the ratio of absorbances of solution of urinary sediment at 305 m/280 m, as an indication of the content of dioxyadenine in urine[6] and we didn't know the solubility pattern of dioxyadenine according to pH. Ultimately the problem remaine unresolved.

After normalization of nitrogen and acid-base balance adenine was re-administered at a dose of 500 mg/day (40 mg/kg/day), associated with allopurinol (150 and later 200 mg/day). This time the efficacy of the treatment was very poor: the mother noticed the reappearence and the persistence of the orange flecks on the underclothes; blood UA level rosed to 13.5 mg% in january 1973 and to 14.5 mg% in april 1973, when urinary UA excretion was respectively 900 and 710 mg/day. This was an unexpected finding, since it must almost be observed the UA level reducing effect of allopurinol. No interference of adenine on allopurinol activity has been demonstrated[6]. It is admissible a reduced intake of allopurinol due to the vomiting (adenine was administered as a powder mixed with the foods, while allopurinol was taken as a crumbled tablet), but it remains to explain the different activity of adenine, unless it was believed to be dose-related.

No further renal change was experienced still june 1973, when urine output suddenly stopped. Anuria lasted 48 h (june 3-5) and in the meanwhile blood urea rose to 87 mg%, blood creatinine to 4 mg% and blood UA to 26 mg%, whereas blood actual pH decreased to 7.29. Once again by stopping adenine, fluid and alkali intake allowed to restore urine output and normalize the blood chemistry. In the first urine specimen after anuria (300 ml), pH was 6.5, UA excretion 210 mg and the ratio of the absorbances of urine sediment at 305/280 mμ was 0.39: this means the presence of considerable amount of dioxyadenine[6].

In Summary adenine therapy gives conflicting evidence of inhibiting purine biosynthesis, fails to ameliorate the neurologic pattern of the disease, with a possible exception for the behavior, but undoubtedly induces a toxic effect on kidney. Allopurinol doesn't seem to prevent this risk; consequently adenine therapy doesn't appear a reliable treatment for LN syndrome.

REFERENCES

(1)   DE MARS R., SARTO G., FELIX J.S. BENKE P.: Lesch-Nyhan mutation: prenatal
      detection with amniotic fluid cells. Science **164**, 1303, 1969.

(2)   VAN DER ZEE S.P.M., LOMMEN E.J.P., TRIJBELS J.M.F., SCHRETLEN E.D.A.M.:
      The influece of adenine on the clinical features and purine metabolism in the
      Lesch-Nyhan syndrome. Acta Paediat. Scand. **59**, 259, 1970.

(3)   ROSENBLOOM F.M., HENDERSON J.F., CALDWELL I.C., KELLEY W.N., SEEGMILLER
      J.E.: Biochemical bases of accelerated purine biosynthesis "de novo" in
      human fibroblasts lacking hypoxanthine-guanine phosphoribosyl-transferase.
      J.Biol.Chem. **243**, 1166, 1968.

(4)   GREENE M.L., SEEGMILLER J.E.: Elevated erythrocyte phosphoribosylpyrophosphate
      in X-linked uric aciduria: importance of PRPP concentration in the regulation
      of human purine biosynthesis. J.Clin.Invest. **48**, 32a, 1969.

(5)   PASERO G., CIOMPI M.L.: L'importanza del fosforibosil-pirofosfato (PRPP) nella bio-
      sintesi purinica. Atti Simposio int. su Malattie reumatiche e alterazioni meta-
      boliche. (to be published).

(6)   SCHULMAN J.D., GREENE M.L., FUJIMOTO W.Y., SEEGMILLER J.E.: Adenine
      therapy for Lesch-Nyhan syndrome. Pediat. Res. **5**, 77, 1971.

(7)   NISSIM S., CIOMPI M.L., BARZAN L., PASERO G.: Coreo-atetosi, insufficienza menta-
      le, automutilazione, alterazioni del metabolismo purinico: la sindrome di
      Lesch-Nyhan. Prima osservazione in Italia. Neuropsich. Infant. **135**, 709, 1972.

(8)   CECCARELLI M., CIOMPI M.L., TADDEUCCI G., PASERO G.: Nuova osservazione di
      sindrome di Lesch-Nyhan. Reumatismo (to be published).

(9)   BENDICH A., BROWN G.B., PHILIPS F.S., THIERSCH J.B.: Direct oxidation of adenine
      "in vivo". J.Biol.Chem. **183**, 267, 1950.

(10) PHILIPS F.S., THIERSCH J.B., BENDICH A.: Adenine intoxication in relation to "in
      vivo" formation and deposition of 2,8-dioxyadenine in renal tubules.
      J.Pharmacol. **104**, 20, 1952.

BEHAVIOURAL CHANGES DURING ADENINE THERAPY
IN LESCH–NYHAN SYNDROME

Simonetta Nissim, Maria Laura Ciompi, Laura Barzan, Giampiero Pasero

Deptm. of Child Neuropsychiatry, S. Chiara Hospital and Deptm.

of Internal Medicine, University Pisa, Italy

It seems useful to us to report here the changes observed in a child affected by the Lesch–Nyhan syndrome when given adenine treatment for 15 months.

The aim of this treatment was to reach the regression of the metabolic disturbances which underlie this illness. The changes produced in the metabolism and the complications consequent on adenine therapy are described elsewhere.

When the patient was first diagnosed, in November 1971, his neurological condition clearly fitted the criteria for the Lesch–Nyhan syndrome reported in the literature: hypertonic state, spastic cerebral palsy, accompanied by choreoathetoid movements. These hindered the completion of any pattern of action. When lying, the patient was in opisthotonus. He could only be kept erect by holding his under the armpits, and both thighs were turned inwards, with legs in a scissored position. No spontaneous or automatic walking movements were observed.

The speech was drastically impaired. Dysarthria and slurring together of incompletely pronounced words were marked. These features were partly due to lack of coordination of lips and tongue, dependent on choreoathetoid movements.

By comparison, verbal comprehension was greater, especially when meanings were given in clear, simple words in an emotionally favourable situation.

There was a clear intention to communicate at the pre-verbal level of mime and gesture.

On the psychological plane, it was striking that the boy's attentive, smiling expression when in his mother's arms was replaced by a state of tension as soon as he was placed in new situations. And tension turned to alarm, with serious crises of anguish and panic, when any attempt was made to establish direct contact between the investigator and the patient.

The boy showed a strong, compulsively self-mutilative tendency. He used to bite his fingers hard when his hands were left free from bandages. Subsequently, he started to bite also his lower lip, producing a large injury on both sides.

Pharmacological treatment.

The initial dose of adenine was 1 gr (90 mg/Kg – per day).

After a period of severe renal insufficiency, treatment was resumed at a lower dose of 500 mg, together with 150-200 mg of allopurinol.

The boy was observed at weekly clinical sessions, in a free situation. Changes affecting inter-personal relationships began to emerge with increasing clearness. The child's willingness to take part in social situation increased, while his anxiety lessened, so allowing him to cope with new situations and to do without his mother for short periods.

The child's interest in his surroundings increased, even if intermittently, and there were incipient attempts to move deliberately towards toys and to involve adults in game-situations.

At this stage the greater cooperativeness of the child allowed specific treatments, such as physiotherapy and language therapy which were practised twice a week.

Little or no results were observed from the neurological point of view; there was a slightly less stiffness and a greater ability of grasping. If supported, the boy was able to maintain a standing position. He seemed to have acquired control of this acquired control of his head movements.

Clear progress was observed in all language skills.

The first development noted, was a greater willingness to indulge in pre-verbal communication, especially in terms of mime and gesture. A marked advance in verbal skills followed. After 20 sessions it was possible to summarise the situation as follows:

Phonologically, the patient's laguage was dysarthric. This resulted in a distortion of most phonemes, especially when sequential articulation was attempted. This did not, however, prevent the boy from being understood reasonably well.

Semantically, the boy's vocabulary of words understood and words used has developed very slowly. He is able to use about 100 words. His motivation towards spontaneus verbal communication, has been encouraged by his family; this has been another factor leading to improvement.

Syntactically, the patient is now getting beyond the stage when only phrasewords are used. He has recently begun to use phrases containing two elements.

Self-mutilation was prevented by extracting all the boy's first set of teeth (as suggested in the literature). So far, the autoaggressive behaviour has not been resumed, although he now has his incipient second set of teeth. Nor does he now experience crises of anguish when looking at his own hands.

Conclusions

There is no doubt that during the 15 months of treatment the child has shown clear improvement in general behaviour and specific skills, despite the lack of marked neurological progress.

The clinical picture has revealed a pattern of coexistent improvement and regression in different fields. While progress was being made in verbal skills, for instance, there was a temporary regression in behaviour, or, instead of outward-directed aggression, there were compulsive outbursts of swearing and abusive language triggered of by the tension felt in stressing situations.

It has not so far been possible to measure the child's I.Q., but his expressive vivacity, his adaptability to changing situations, and his emotive involvement, which, though still incomplete, is tinged with cunning and a sense of humour, all go to indicate that his intellectual capabilities are not severely impaired.

According to the literature, the Central Nervous System satisfies its need for nucleotides at the expense of the "salvage–pathway; as in this disorder the neurological symptoms can be attributed to lack of GMP, it seems probable that the general improvement observed, may be due to a higher availability of GMP, as the end-result of the conversion $AMP \rightarrow IMP \rightarrow GMP$.

It is our clinical impression that the role of adenine was to help foreward the appearance of certain behaviour patterns representing a greater degree of maturity, thus making therapeutic action possible in specific areas and releasing a stable situation of non-response.

Treatment with adenine does, however, involve serious risks to health, dependent on the low solubility of its oxidised metabolytes.

These risks do not seem to be eliminated by the concomitant administration of allopurinol, and are so dangerous, that they appear to be out of proportion to the results we have achieved.

Therefore, we decided to stop this treatment, after a second acute fit of renal failure.

Benzbromaron, 3-Hydroxy Purines, 3-Butylazathioprine

TREATMENT OF GOUT AND HYPERURICAEMIA BY BENZBROMARONE

ETHYL 2 (DIBROMO -3,5 HYDROXY - 4 BENZOYL) - 3 BENZOFURAN

A. de GERY, C. AUSCHER, L. SAPORTA, F. DELBARRE

Institut de Rhumatologie - Unité 5 INSERM-ERA 337 CNRS

Hôpital COCHIN - PARIS I4ème

Benziodarone has been largely used for the treatment of gout since the discovery of its action to reduce plasma uric acid level, in I965 ; its uricosuric effect have been proved later (3-4).

Nevertheless some restriction have been placed on its prolonged use, due to its iodine content. Some cases of hypo or hyperthyroïdism have been reported (7), more particularly by Camus France who related two cases of hyperthyroïdism with muscle disease (7) that is the reason why, as early as I967, we have employed and prefered a bromide by product (fig.I) éthyl-2 (dibromo-3,5, hydroxy-4) ben-zoyl benzofuranne (4,5) (benzbromarone).

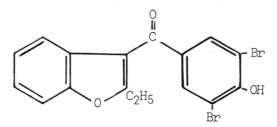

ETHYL 2(DIBROMO-3,5 HYDROXY-4BENZOYL)-3

BENZOFURAN

Fig. 1

Our first results have been confirmed by our experiment of this drug on fifty gouty and five control subjects.

RESULTS

We have used benzbromarone at a daily dose of one hundred mg, sometimes fifty, even twenty five mg have been effective.

Uricaemia always decreased, on an average it fell from 9.20 to 4.80 mg per cent(a decrease of 48.I per cent) (fig.2).

The average of daily urinary uric acid was increased from 520 mg to 746 mg, that is to say i.e an increase of 43,2 per cent (fig.3).

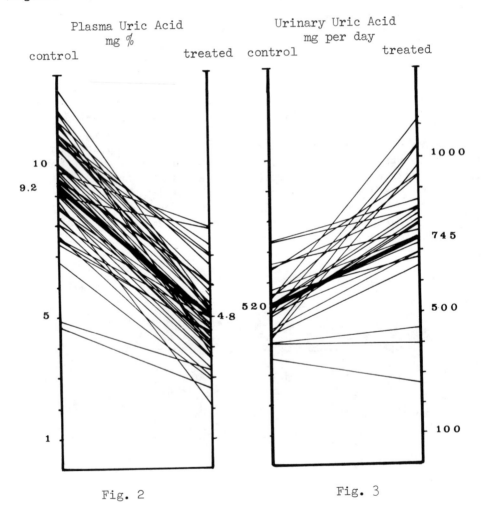

Fig. 2   Fig. 3

The clearance of uric acid measured one two and three hours after a single oral dose of the drug given to a man is always increased, so is the ratio uric acid clearance to creatinine clearance and we have compared, in the same subject, benzbromarone to probenecid and a placebo. We just report here the results in 4 control subjects, it has been proportionnally the same in gouty others - figure 4.

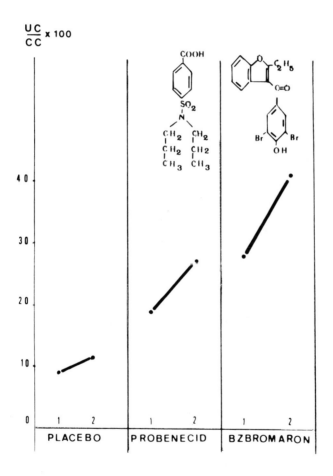

Fig. 4. Compared variations of uric acid clearance to creatinine clearance, ratio UC/CC, one and two hours after an oral dose of a uricosuric drug in four men.

The urinary output of xanthine plus hypoxanthine remains un-changed (tableau I).

TABLEAU I : URINARY EXCRETION (XANTHINE + HYPOXANTHINE) mg/24H.

	BEFORE TREATMENT	UNDER TREATMENT by 100 mg per day of BENZ-BROMARONE.
I	18,I	24,0
2	29,0	34,0
3	3I,9	29,8
4	25,7	20,2
5	34,5	33,0
6	29,9	40,5
7	24,I	22,6
8	I6,7	29,0
9	23,0	I6,0
IO	I5,0	I5,0
II	I9,8	I8,4
AVERAGE	24,2	25,7

This excretion has been measured in eleven gouty subjects treated by hundred mg per day.

DISCUSSION

These results show that benzbromarone is a strong uricosuric drug and has no inhibitory effect on purine synthesis more over on xanthine oxidase activity, as proved by unchanged levels of plasma and urine xanthine plus hypoxanthine.

The action is thus been compared to benziodarone, both have the same contra indications (over excretion of uric acid of renal stones).

In 7 patients benzbromarone and benziodarone have been given at the same daily dose that is to say I00mg per day. Plasma uric acid decreased in the same way, which proves that the uricosuric activity of the bromide byproduct is as powerful as that of the iodine one. These results have been corroborated by others authors (8 - 9 - I0 - II - I2 - I3 - I4 - I5).

Like benziodarone (3) benzbromarone remains affective in case of mild renal failure : in eight of our cases, blood creatinine level was over 2 mg per cent, and this, is of practical interest, in some of our patients with nephropathy who became intolerant either to allopurinol or to thiopurinol and developed cutaneous rashes.

TOLERANCE

In our experience one has to insist on the excellent tolerance of this drug which is a point of importance.

A great number of our patients have been treated for many years; two of them have been treated for over six years.

There has been no cutaneous rash, no hematologic renal or hepatic side effects, and let us un derligned the particular gastric tolerance, only one of our patients had had gastralgic and diarrhea 48 hours after the beginning of the treatment which had thus to be stopped.

The possible action which have been saught in relation to the deposits due to an other benzofuran hyproduct, Amiodarone, has not been detected, either in patients under prolonged treatment or in chicken embryo after the injection of the drug into the eggs Yolk (I6).

CONCLUSION

The activity of benzbromarone is as powerful as benziodarone and so more effective than other classical uricosuric drugs.

The tolerance is particularly appreciate and this byproduct with bromide does not modify thyroid functions.

REFERENCES

I - NIVET M, MARCOVICI J, LARUELLE P, FARAH M.
     Note préliminaire sur l'action d'un Benzofuranne sur l'uricémie.
     Bull. Soc. méd. Hôp. Paris. 1965, 73, II87.

2 - DELBARRE F, AUSCHER C, AMOR B.
     Action uricosurique et antigoutteuse de certains dérivés du Ben-
zofuranne.
     Presse méd. I965, 73, 2725.

3 - RICHET G, COTTET J, AMIEL C, LEROUX-ROBERT C, PODEVIN R.
     Traitement de l'hyperuricémie des goutteux et des insuffisants
rénaux par la Benziodarone.
     Presse méd. I966, 74, I247.

4 - DELBARRE F, AUSCHER C, OLIVIER J.L, ROSE A.
     Traitement des hyperuricémies et de la goutte par les dérivés
du Benzofuranne.
     Sem. Hôp. Paris, I967, 24, II27.

5 -DANCHOT J.
     Les hypo-uricémiants de structure benzofurannique.
     Thèse Médecine, Paris, I968.

6 - HARRISON M.T, CAMERON A.J.V.
     Iodine-induced hypothiroïdism due to benziodarone.(Cardivix)
     Brit. Med. J. I965, I, 840.

7 - CAMUS J.P., PRIER A, KARTUN P, MAUGEIS de BOURGUESDON J.
     Thyréotoxicose et benziodarone.
     Rev. Rhum, I973, 40, 2, I48-I50.

8 - STERNON J, KOCHELEFF P, COUTURIER E, BALASSE E, VAN DEN ABEELE P.
     Effet hypo-uricémiant de la benzbromarone. Etude de 24 cas (Ré-
sultats préliminaires).
     Acta clin. belgica, I967, 5, 285.

9 - VAN BOGAERT P.
     Essai clinique du L 22I4 dans le traitement de la goutte ou des
syndromes rhumatologiques associés à une hyperuricémie.
     J. belge Rhum. Méd. phys. I969, 24, 295.

IO - MERTZ D.P.
     Veränderungen der Serumkonzentration von Harnsaure unter der
Wirkung von Benzbromaron. Munsch. med. Wschr, I969 , 9,49I.

II - ZOLLNER N, DOFEL W, GROBNER W,
    Die Wirkung von Benzbromaron's auf die renale Harnsaureausscheidung
der Gesunder.
    Klin. Wschr, I970, 7, 246.

I2 - KROPP R.
    Zur urikosurischen Wirkung von Benzbromaronum and Modell der
pyrazinamidbedingten Hyperurikämie.
    Méd. Klin. I970, 32/33, I4448.

I3 - ZOLLNER N, GRIEBSCH A, KINK J.K.
    Uber die Wirkung von Benzbromaron um auf den Serumharnsaurespiegel
und die Harnsaureausscheidung des Gichtkranken.
    Dtsch. med. Wschr, I970, 48, 2405.

I4 - FAMEY J.P, VAN DEN ABEELE G.
    Analyse de l'action hypo-uricémiante de la Benzbromarone dans
40 cas d'hyperuricémie goutteuse et non goutteuse.
    J. belg. Rhum. Med. phys. I970, 25, 5.

I5 - MASBERNARD A, GUILBAUD J, DRONIOU J.
    Experience clinique du traitement au long cours de la maladie
goutteuse par les dérivés iodé et bromé du benzofuranne.
    Soc. méd. chir. Hôp. Form. san. Armées, I97I, 5, 393.

I6 - BROEKHUYSEN J, PACCO M, SION R, DEMEULENAERE L, VAN HEE W.
    Metabolism of benzbromarone in man.
    Europ. J. clin. pharmacol. I972, 4, I25-I30.

THE PIG AS AN ANIMAL MODEL FOR PURINE

METABOLIC STUDIES

J.S. Cameron, H.A. Simmonds, P.J. Hatfield,
A.S. Jones, and A. Cadenhead

Guy's Hospital and Rowett Research Institute
London, S.E.1 9RT, U.K. and Bucksburn, Aberdeenshire, U.K.

Small laboratory rodents have many disadvantages for the study of purine metabolism in relation to human disease. They normally pass their purine metabolites in much smaller volumes of urine both in relation to bodyweight and filtration rate (1); in addition, the distribution of enzymes of purine catabolism in these species differs from man (2,3). The pig in contrast has a kidney which structurally and functionally resembles human kidneys closely (4,5), and the distribution of xanthine oxidase and guanase is similar to man (2,3). Because of these advantages, and the finding of "guanine gout" in pigs (6), we studied purine metabolism in this animal (7). The pig has proved a much better model.

Urinary total purine excretion in pigs was found to be comparable to that in man and was excreted in comparable volumes of urine (7). The chief purine end product was allantoin, together with small amounts of uric acid and other purines. Oral loading with guanine showed that pigs possess enzymes capable of metabolising large amounts of guanine.

Allopurinol was given orally up to fifty times maximum human therapeutic doses without side effects. Allopurinol and oxipurinol appeared to be handled similarly by pig (7,8) and human kidney (in contrast to rat, mouse and dog) (9). Xanthine oxidase was maximally inhibited by allopurinol at 2.2 mmol (300 mg)/kg/day with no further increase in oxipurinol and xanthine excretion at higher doses. No limit, however, was found for the formation of allopurinol riboside. 94% of radioactivity from (6-^{14}C) allopurinol was recovered from the urine and no ^{14}C incorporation was detected in tissue nucleotides.

REFERENCES

1.   Hitchings, G.H. (1966). Ann. Rheum. Dis., 25, 601.

2.   Al-Khalidi, U.A.S., Chaglassian, T.H. (1965). Biochem. J.,
     97, 318.

3.   Levine, R., Hall, T.C., Harris, C.A. (1963). Cancer, 16, 269.

4.   Nielsen, T.W., Maaske, C.A., Booth, N.H. (1966). In: Swine
     in Biochemical Research, Bustad, L.K., McLellan, R.O. (Eds).
     Bartel Memorial Trust, p.529.

5.   Farebrother, D.A., Hatfield, P.J., Simmonds, H.A.,
     Cameron, J.S., Jones, A.S., Cadenhead, A. In preparation.

6.   Virchow, R. (1866). Virchow's Arch. Path. Anat. Physiol.,
     35, 358.

7.   Simmonds, H.A., Hatfield, P.J., Cameron, J.S., Jones, A.S.,
     Cadenhead, A. (1973). Biochem. Pharmacol. In press.

8.   Simmonds, H.A., Rising, T.J., Cadenhead, A., Hatfield, P.J.,
     Jones, A.S., Cameron, J.S. (1973). Biochem. Pharmacol. In
     press.

9.   Elion, G.B., Kovensky, A., Hitchings, G.H., Metz, E.,
     Rundles, R.W. (1966). Biochem. Pharmacol., 15, 863.

INFLUENCE OF A URICOSURIC DRUG ON CONNECTIVE TISSUE METABOLISM

H. Greiling and M. Kaneko

Klinisch-chemisches Zentrallaboratorium der Rheinisch-Westfälischen Techn. Hochschule Aachen

Benzbromaron acts as an uricosuric substance, but there is a quantitative discrepancy between uric acid excretion and the depression of uric acid level in blood serum. In figure 1 you see that there is an increased excretion of uric acid but there remains an amount of uric acid, which is not due to the increased uric acid excretion. Benzbromaron is in vitro a xanthine-oxydase inhibitor, but as you see from these experiments there is no decrease in oxypurine excretion after benzbromaron application.

An explanation for this phenomenon would be a decreased biosynthesis of uric acid by benzbromaron, for instance by activating the hypoxanthine-guanine-phosphoribosyl transferase. In vitro studies have shown, that benzbromaron inhibits the hypoxanthine-guanine-phosphoribosyl transferase in erythrocytes. With $5 \cdot 10^{-4}$ M benzbromaron there is an inhibition of ca. 40%. Müller and Bresnik on the other hand have found by in vivo studies an increase of hypoxanthine-guanine-phosphoribosyl transferase activity in erythrocytes following benziodaron, the jodine analogue of benzbromaron (1).

Secondly we postulate that an accumulation of uric acid in other organs or tissues occurs. In arthritis urica the connective tissue is involved and also a predilected deposition of monosodium urate in connective tissues as for instance in joint cartilage is observed. Therefore we have studied the influence of benz-

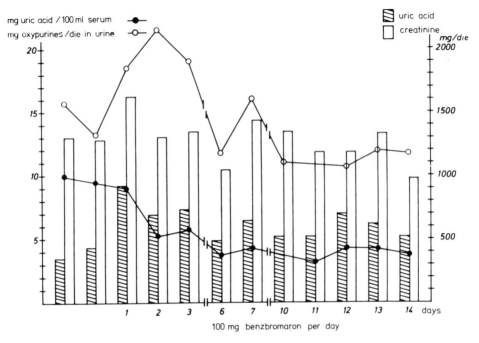

Fig. 1: Uric acid level in blood serum and uric acid,
oxypurine and creatinine excretion before and
after application of benzbromarone by a patient
with gout.

	cpm/μMol GAG	per cent inhibition
control	38 480	—
10^{-4} M benzbromaron	11 440	70
10^{-5} M benzbromaron	30 720	18

Fig. 2: Inhibition of $^{35}SO_4$-incorporation into the glycosaminoglycans (GAG) of cornea by benzbro-
maron

bromaron on connective tissue metabolism. The incorporation
of radioactive sulfate into the bovine cornea is inhibited by benzbromaron. This means, that there is an inhibition of sulfatation
of proteokeratan sulfate and proteochondroitin sulfate in cornea
(Figure 2).

Most antiinflammatory drugs, for instance acetylsalicylic
acid, phenylbutazone, indomethacin, cortison, flufenaminic
acid etc. are inhibitors of sulfate metabolism in connective
tissue (2). These antiinflammatory drugs are inhibitors of
oxydative phosphorylation and therefore in presence of these
substances the content of tissue ATP and its reaction product
with sulfate: 3'-phosphoadenylylsulfate is decreased.

We have found, that benzbromaron inhibits the 3'-phospho-
adenylylsulfat:chondroitin-6-sulfotransferase, an enzyme which
we have isolated from liver (3). This enzyme is competitively
inhibited by benzbromaron ($K_i = 2 \cdot 10^{-5}$ M (Figure 3) and the
inhibition of sulfate incorporation into proteoglycans is probably
due to this enzyme inhibition (Figure 4).

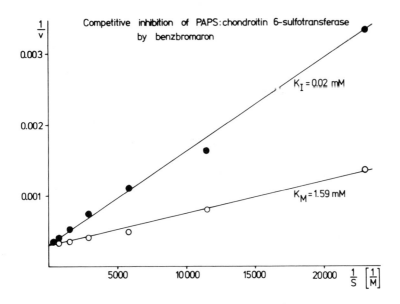

Fig. 3: Influence of benzbromaron on enzymatic
sulfatation of chondroitin

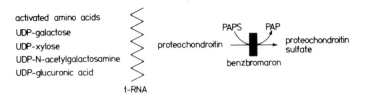

Fig. 4: Postulated mechanism of inhibition by benz-
         bromaron

If we postulate that the proteoglycans are regulators for the
influx of electrolytes and also for the penetration of monosodium urate into connective tissue, than it could be possible,
that a decreased synthesis of proteoglycans causes an increased entrance of monosodium urate into the connective tissue.
In the synovial fluids of patients with gout is an increased lysosomal activity. The origin of the lysosomal enzymes
are the polymorphnuclear leucocytes, which are increased in
an acute gouty attack in the synovial fluid. This increased enzyme activity is not specific for gout, but is also found in other
inflammatory joint diseases as for instance in rheumatoid arthritis. Conform with the increase of lysosomal enzymes is the increase of glycolytic enzymes in the synovial fluid. For example
a high lactate-dehydrogenase activity in synovial fluid is accompanied by a high lactate concentration. This biochemical parameters are indicators for the grade of inflammation. From
earlier studies we have experimentally found, that chondroitin
sulfate can transform monosodium urate into free uric acid by
a cation exchange function. Free uric acid is markedly more
insoluble than monosodium urate. An increased lactate concentration and a decrease of pH in connective tissue would promote
these cation exchange reaction and therefore the precipitation
of uric acid (4). This view is also supported by Katz and Schubert (5), who showed, that the depolymerized proteoglycans
decrease the solubility of monosodium urate. Lysosomal proteoglycan degrading enzymes depolymerize proteochondroitin sulfate. The activity of these lysosomal enzymes is increased,
when the pH is lowered in the inflamed joint cartilage. The

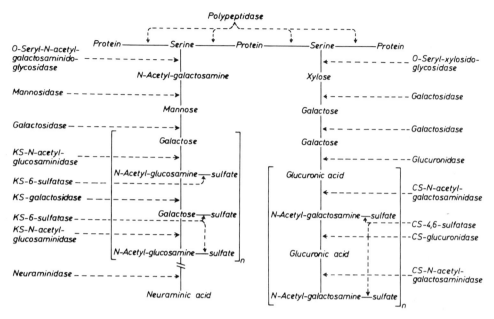

Fig. 5: Sequential lysosomal degradation of the proteo
chondroitin sulfate - keratan sulfate hybrid of
human knee joint cartilage

polymorphnuclear leucocytes contain lysosomal enzymes, which
degrade also proteoglycans. In figure 5 schematically the enzymatic degradation of the main proteoglycan of the knee joint cartilage is presented. In cartilage only those polypeptidases,
which probably extrude the peptido glycosaminoglycans from
the extracellular matrix can be experimentally proved today.
The other lysosomal enzymes listed here, we have demonstrated in liver and kidney lysosomes (6). In conclusion: The lactate production and therefore the pH depression in the inflamed
synovial system of the gouty joint and also the increased activity of lysosomal proteoglycan degrading enzymes enhance the
process of deposition of monosodium urate in joint cartilage.

References:

1.   Müller, M.M., and W. Bresnik, Klin. Wschr. 50, 1015
       (1972)

2.   Whitehouse, M.W., in Progress in Drug Research, Vol.
       8, 320, Birkhäuser Verlag Basel und Stuttgart (1965)

3.   Momburg, M., H.W. Stuhlsatz, R. Kisters and H. Grei-
       ling, Hoppe-Seyler`s Z. physiol. Chem. 353, 1351 (1972)

4.   Greiling, H., Th. Herbertz, B. Schuler and H.W. Stuhl-
       satz, Z. Rheumaforsch. 21, 50 (1962)

5.   Katz, W.A. and M. Schubert, J. Clin. Invest. 49, 1783
       (1970)

6.   Greiling, H. in Connective Tissue and Ageing, S. 168,
       Excerpta Medica Amsterdam (1973)

THE SYNTHESIS AND IMMUNO-ENHANCING ACTIVITY OF 3-BUTYLAZATHIOPRINE

A.H. Chalmers, A.W. Murray and P. Verakalasa

School of Biological Sciences, Flinders University of

South Australia, Bedford Park, South Australia, 5042

INTRODUCTION

As part of a programme of the design and testing of potential immunosuppressive and anticancer drugs, 9-butylazathioprine was synthesised (Chalmers, Knight & Atkinson, 1969) and shown to be an immunosuppressive agent (Chalmers, Gotjamanos, Rao, Knight & Atkinson, 1971; Chalmers, Burdorf & Murray, 1972). During the chemical synthesis of this compound from azathioprine and 1-iodobutane (Chalmers et al., 1969) small amounts of the 3-isomer (3-butylaza-thioprine) were formed. This paper describes the isolation and characterisation of this compound and reports the ability of the analogue to enhance the antibody response of mice to sheep red blood cells.

MATERIALS AND METHODS

Antibody Response of Mice to Sheep Red Blood Cells

The numbers of spleen cells producing antibody against sheep red blood cells were estimated by the Jerne-plaque technique (Jerne & Nordin, 1963). Both Balb/c and Swiss albino mice were used as described in the individual experiments. Details of antigen and drug administration are given in the text.

Synthesis of 3-Butyl-6-(1-Methyl-4-Nitroimidazol-5-yl)-Thiopurine
(3-Butylazathioprine)

In a large scale preparation of 50 g 9-butylazathioprine as
described by Chalmers et al. (1969), 19 g of a viscous, yellow oil
was obtained from the chloroform extract which after recrystallization
from isobutyl methyl ketone (150-200 ml) gave small, granular,
yellow crystals (5% overall yield). These crystals after
recrystallization from isobutyl methyl ketone and drying over P_2O_5
had m.p. 108-110° (decomp.) and gave the following analysis.
Found : C, 47.0; H, 4.6; N, 29.4; O, 9.7; S, 9.6; $C_{13}H_{15}N_7O_2S$
requires C, 46.9; H, 4.6; N, 29.5; O, 9.7; S, 9.7.

The n.m.r. spectrum in deuterated chloroform showed two
1-proton singlets at 1.7 τ and 1.8 τ, a two proton triplet at 5.5 τ,
a singlet at 2.3 τ, a three proton singlet at 6.25 τ and a seven
proton multiplet from 7.7 - 9.3 τ.

Characterization of 3-Butylazathioprine

Conversion into 3-butyl-6-diethylaminopurine. 3-Butylazathioprine
(0.2 g) was reacted with diethylamine in a sealed glass tube at
100° for 7 hours. After cooling overnight, 3 ml of water was added
and the reaction mixture acidified to pH 1 with HCl. 5-Mercapto-1-
methyl-4-nitroimidazole (70 mg) was filtered off and the mother
liquor adjusted to pH 9-10 with 5 N KOH and extracted with chloroform
(4 x 10 ml). The chloroform was dried over anhydrous sodium sulphate
and evaporated to give 0.2 g of a light, yellow-coloured, viscous
oil. Thin layer chromatography on ECTEOLA cellulose with 0.9% (w/v)
NaCl as the mobile phase, gave a single purple fluorescing spot at
R_f 0.28 when viewed under ultra-violet light (c.f. 3-butylazathioprine,
R_f 0.56)

The spectra of this compound under acid and alkali conditions
(Table 1) were characteristic of 3-alkyl-N,N-dimethyladenine
derivatives (Townsend et al., 1964).

Conversion into 3-butyl-6-mercaptopurine. To 3-butylazathioprine
(1 μmole) in 3 ml of 0.05 M Tris (pH 7.4) was added 2-mercaptoethanol
(30 μmoles). The thiolysis was allowed to proceed for 2 hours.
The reaction products were separated by chromatography on Whatman
3MM paper using n-butanol saturated with water as the mobile phase.
The atmosphere in the tank was made ammoniacal by placing 200 ml
n-butanol-2.5% aq. ammonium hydroxide (1:1) in the bottom of the
tank. 5-Mercapto-1-methyl-4-nitroimidazole was located at R_f 0.09
and a purple fluorescence at R_f 0.50. After elution into water the
spectral properties summarised in Table 1 were obtained. These
absorbances are characteristic of 3-alkyl-6-mercaptopurine (Elion,
1957). The presence of mercuric ions with the compound in 0.1 N HCl

TABLE 1

The absorbance characteristics of N-butylated purine analogues.

Values given in parentheses are the millimolar extinction coefficients at the wavelength indicated. The buffer used was 0.05 M potassium phosphate, pH 7.4.

Analogue	Solvent	$\lambda_{max.}$ (nm)	$\lambda_{min.}$ (nm)
9-Butylazathioprine	0.1 N HCl Buffer	279-280 (18.7)	244 (7.7)
	0.1 N KOH	279-280 (18.7)	245-246 (9.1)
3-Butylazathioprine	0.1 N HCl Buffer	300-301 (21.9) 297 (21.3)	258 (6.7) 250 (7.1)
3-Butyl-6-diethyl adenine	0.1 N HCl Buffer	290-291	242-243
	0.1 N KOH	295	250
3-Butyl-6-mercapto- purine	0.1 N HCl	334-335 243-244	
	0.1 N KOH	334 243	
3-Butyl-6-benzyl- adenine	0.1 N HCl	286-287	241
	0.1 N KOH	289	248
	Buffer	287	249

shifted the maximum absorbance to 322 nm, thus confirming the presence of a free SH group in the purine ring.

Conversion into 3-butyl-6-benzylaminopurine. 3-Butylazathioprine (50 mg) was heated at 100° for 15 min. with 0.5 ml of benzylamine. Ethyl alcohol (5 ml) was added to the cooled residue, followed by 30 ml of water. This solution was refrigerated at 2° overnight and a crystalline residue filtered off and recrystallised from alcohol-water. The spectral properties of this compound are summarised in Table 1 and are characteristic for 3-alkyl-6-benzyladenine derivatives (Skoog et al., 1967).

RESULTS

All of the evidence for the location of the butyl group in the 3-position of the new compound is indirect and is based on the reported absorption characteristics of 3-methyl purine derivatives. Unequivocal confirmation of its structure will require an unambiguous synthesis from simpler precursors.

TABLE 2

Effect of 3-butylazathioprine on the antibody response of mice to sheep red blood cells.

Female mice (6 to each group) were injected i.p. with 3-butylazathioprine (3-BA) or 9-butylazathioprine (9-BA) on day 0 (immediately following injection of 1.5×10^7 sheep red blood cells) and on days 1 and 2; animals were sacrificed and plaque forming cells titred on day 5.

Experimental Animal	Drug	Dose (mg/kg body weight)	Plaque forming cells/10^6 spleen cells	Standard error of the mean	P
Balb/c	-	-	70	21	
	3-BA	45	255	32	<0.02
	3-BA	90	175	30	<0.05
Swiss albino	-	-	28	25	
	3-BA	30	177	17	<0.01
	9-BA	30	30	19	N.S.

3-Butylazathioprine increased the number of spleen cells producing antibodies against sheep red cells in both Balb/c and Swiss albino mice (see Table 2). At a comparable dose 9-butylaza-thioprine had no effect on the antibody response. It has already been shown that at higher doseage levels the 9-alkyl analogues will suppress this response (Chalmers et al., 1972). Although detailed dose-response experiments have not been done, only a marginal enhancement was observed when three injections of 8 mg/kg body weight were given to Swiss albino mice; there was no effect with 0.8 mg/kg body weight.

With a low antigenic challenge (1.5×10^7 sheep red blood cells per mouse) the maximum stimulatory response to three daily doses of 3-butylazathioprine was obtained five days after challenge (see Table 3). A single dose of 3-butylazathioprine given immediately following the administration of sheep red cells also resulted in an enhancement of the number of plaque forming spleen cells after five days. The enhancement by 3-butylazathioprine at five days was less pronounced (on a percentage basis) when mice were given a greater antigenic challenge (1.5×10^8 sheep red blood cells per mouse; Table 3). In some experiments, no stimulation, or even slight inhibition, was observed at the higher level of antigen. However, in these experiments the development of the immune response with time was not followed.

TABLE 3

Immunoenhancement by 3-butylazathioprine at different times after antigenic challenge.

Female Swiss albino mice (5 to each group) were injected i.p. with 3-butylazathioprine (3-BA; 30 mg per kg body weight) on day 0 (immediately following injection of 1.5×10^7 sheep red blood cells) and on days 1 and 2; animals were sacrificed and plaque forming cells titred on the days indicated.

Time after challenge (days)	Treatment	Plaque forming cells/10^6 spleen cells	Standard error of the mean	P
4	Control	30	11	
	3-BA	222	103	<0.1
5	Control	27	10	
	3-BA	243	46	<0.002
	3-BA*	200*	70	<0.05
7	Control	11	3	
	3-BA	20	6	N.S.[†]
5[‡]	Control[‡]	230	58	
	3-BA[‡]	391	32	<0.05

*Animals received one injection of 3-BA (30 mg/kg body weight) immediately following the injection of sheep red blood cells.

[‡]In these groups (6 animals each) mice received a larger antigenic challenge (1.5×10^8 sheep red blood cells).

[†]Not significant.

DISCUSSION

These results have clearly shown that, with the conditions of administration used, 3-butylazathioprine will enhance the antibody response of mice to sheep red blood cells. Similar non-specific enhancement has been obtained with chemically defined polynucleotides (Braun, Ishizuka, Yajimi, Webb & Winchurch, 1971) and with the thiazoles tetramisole and levamisole (Renoux & Renoux, 1972). Such compounds, together with a variety of chemically undefined preparations (Hiu, 1972), may augment natural surveillance against malignant cells.

The nature of the enhancement of antibody production by 3-butylazathioprine has not been established. Enhancement has been obtained in animals pretreated with 6-mercaptopurine and other commonly used immunosuppressive drugs (Schwartz, 1965). Schwartz proposed that the cytotoxic drugs resulted in cell damage with subsequent release of 'stimulatory' polynucleotides. The observed effects of 3-butylazathioprine could be such an indirect result of cytotoxicity or alternatively may reflect a direct interaction of the drug with cells involved in the immune response.

Acknowledgements. We thank Mr. M. Froscio for excellent technical assistance. This work was supported by grants from the University of Adelaide Anti-Cancer Foundation and the Australian Research Grants Committee.

REFERENCES

Braun, W., Ishizuka, M., Yajima, Y., Webb, D. & Winchurch, R. (1971). In 'Biological effects of polynucleotides' p. 139. R.F. Beers & W. Braun (editors), Springer-Verlag, Berlin, Heidelberg, New York.
Chalmers, A.H., Burdorf, T. & Murray, A.W. (1972). Biochem. Pharmacol. 21, 2662.
Chalmers, A.H., Gotjamanos, M.M., Rao, M.M., Knight, P.R. & Atkinson, M.R. (1971). J. surg. Res. 11, 284.
Chalmers, A.H., Knight, P.R. & Atkinson, M.R. (1969). Aust. J. exp. Biol. med. Sci. 47, 263.
Elion, G.B. (1957). In 'The chemistry and biology of purines' p. 46. G.E.W. Wolstenholme and C.M. O'Connor (editors), Ciba Foundation.
Hiu, I.J. (1972). Nature, New Biology 238, 241.
Jerne, N.K. & Nordin, A.A. (1963). Science 140, 405.
Renoux, G. & Renoux, M. (1972). Nature, New Biology 240, 217.
Schwartz, R.S. (1968). In 'Human transplantation' p. 440. F. Rappaport and J. Dausset (editors), Gurne & Stratton, Inc., New York.
Skoog, F., Hamzi, H.Q., Szweykowska, A.M., Leonard, N.J., Carraway, K.L., Fujii, T., Helgeson, J.P. & Leoppky, R.N. (1967). Phytochem. 6, 1169.
Townsend, L.B., Robins, R.K., Leoppky, R.N. & Leonard, J. (1964). J. Am. Chem. Soc. 86, 5320.

RENAL HANDLING OF URATE

Hypouricemia and
the Effect of Pyrazinamide
on Renal Urate Excretion

THE CLINICAL SIGNIFICANCE OF HYPOURICEMIA

C. M. Ramsdell and W. N. Kelley

Department of Medicine, Duke University Medical Center

Durham, North Carolina   27710

The development of automated techniques in the clinical laboratory which routinely include the measurement of uric acid has
raised important questions concerning the clinical significance
of abnormal serum urate values. Although numerous studies have
focused on the relevance of hyperuricemia, the possible significance
of hypouricemia has not been investigated. Hypouricemia has been
reported as a manifestation of a number of relatively rare diseases
including xanthinuria, Fanconi's syndrome due to any cause,
acute intermittent porphyria and an isolated renal defect in transport of uric acid.   In addition, many drugs can lower the serum
urate concentration and could potentially lead to the development
of hypouricemia. The reported and potential causes of hypouricemia
are summarized in Table 1. The purpose of the present prospective
investigation was to determine the etiology, pathogenesis and
clinical significance of hypouricemia in hospitalized patients.

A total of 6,629 consecutive serum urate determinations were
monitored at two large hospitals in Durham, North Carolina. Hypouricemia was defined as a serum urate concentration of 2 mg/100 ml
or less. Periodic checks showed that a value in this range was
always more than two standard deviations below the mean of measurements made at either hospital on a given day. When significant
hypouricemia was noted, the patient and his record was examined
to determine the cause. If an obvious cause could not be ascertained by this method, another serum sample was obtained for
measurement of uric acid, creatinine, phosphate, and salicylate,
and a 24 hour urine specimen was collected in order to estimate
uric acid, oxypurine, creatinine, phosphate, glucose, amino acid,
and porphobilinogen excretion.

TABLE 1

POSSIBLE CAUSES OF HYPOURICEMIA

I.   Decreased Production
     A.   Xanthine oxidase deficiency
     B.   Allopurinol administration

II. Increased Excretion
     A.   Isolated defect in renal transport of uric acid.
          1.   Idiopathic
          2.   Hodgkins Disease
     B.   Generalized defect in renal tubular transport (Fanconi's
          syndrome).
          1.   Idiopathic
          2.   Wilson's Disease
          3.   Cystinosis
          4.   Multiple myeloma
          5.   Heavy metals
          6.   Type I Glycogen Storage Disease
          7.   Galactosemia
          8.   Hereditary Fructose Intolerance
          9.   Outdated tetracyclines
         10.   Bronchogenic carcinoma
     C.   Drugs
          1.   Salicylates (large doses)
          2.   Sulfinpyrazone
          3.   Probenecid
          4.   Phenylbutazone
          5.   X-ray contrast agents
          6.   Chlorprothixine
          7.   6-azauridine
          8.   Halofenate
          9.   Benzbromarone

III. Mechanism Unknown
     A.   Pernicious anemia
     B.   Liver Disease
     C.   Acute Intermittent Porphyria

From the total of 6,629 uric acid measurements, 0.97% including 66 men and two women, had a serum urate concentration of 2 mg/100 ml or less. In 4 of these 68 patients, the serum urate was found to be falsely low, since a value of greater than 2 mg/100 ml was obtained on the same serum sample by the uricase assay. The other 64 patients were separated into two groups, those with hypouricemia of known cause (53 patients) and those with hypouricemia of unknown cause (11 patients).

In the group of 53 patients with hypouricemia of known cause, drugs seemed to be responsible in 44 subjects (Table 2). The drugs implicated were aspirin, allopurinol, and X-ray contrast agents. In addition, six hypouricemic subjects were receiving glycerol guaiacolate, and although this drug was not known to alter uric acid

TABLE 2

HYPOURICEMIA OF KNOWN CAUSE

	Number of Patients	Mean Serum Urate Level
		mg/100 ml
Drugs	44	
Aspirin		
Taken as directed	18	1.8
Taken surreptiously	3	1.5
Allopurinol	2	1.9
X-ray contrast agents		
Iopanoic acid (Telepaque)	8	1.6
Iodipamide meglumine (Cholografin)	1	1.5
Diatrizoate sodium (Hypaque)	6	1.9
Glyceryl guaiacholate	6	1.6
Neoplastic Disease	9	
Glioblastoma	1	2.0
Carcinoma of the cervix	1	2.0
Carcinoma of the tongue	2	2.0
Hodgkins disease (IV-B)	2	1.7
Carcinoma of unknown origin	1	2.0
Sarcoma of unknown origin	1	2.0
Carcinoma of the pancreas	1	2.0

(From Ramsdell and Kelley, 1973).

metabolism, this finding stimulated us to examine its effect on the serum urate concentration. In four subjects studied prospectively, we found that glyceryl guaiacolate had a modest hypouricemic effect. With therapeutic doses of this drug for 3 days, a decrease in the serum urate concentration of up to 3 mg/100 ml was observed. Although each of the drugs implicated above can cause a reduction in the serum urate concentration, we cannot exclude the possibility that other factors, such as low dietary purine intake, hemodilution, or severe illness may have contributed to the development of hypouricemia in some of the patients.

Nine of the 53 patients had neoplastic disease with evidence of metastases or extensive dissemination. The types or locations of the neoplastic lesions are summarized in Table 2. Hypouricemia appeared to be persistent in seven of the nine patients and transient in two. Four of the nine patients also had evidence of severe obstructive hepatic disease. Seven of these subjects had a poor dietary purine and protein intake, and four were noted to be hypoosmolar, with osmolarity values ranging from 267 to 273 milliosmols/liter.

Four of the subjects with neoplastic disease were studied in more detail (Table 3 and Table 4). In each subject, the ratio of C_{urate} to $C_{creatinine}$ was found to be increased (Table 3). There was no evidence of renal glucosuria, hyperphosphaturia, or bicarbonate wasting, which would suggest a generalized defect of tubular transport in these patients (Table 4). One patient, however, patient W.G., had a generalized amino aciduria and another, patient A.H., had an increase in alanine excretion. Two of the patients, J.L. and W.G., with neoplastic disease and an increased $C_{urate}/C_{creatinine}$ ratio had evidence of obstructive jaundice, with plasma bilirubin levels of 15 and 20 mg/100 ml, respectively. In these same two patients with extensive hepatic involvement there was also an increase in the urinary excretion of oxypurines (Table 4).

TABLE 3

HYPOURICEMIA ASSOCIATED WITH NEOPLASTIC DISEASES

Patient	Diagnosis	Serum Urate	Urine Urate	$\dfrac{C_{urate}}{C_{creatinine}}$
		mg/100 ml	mg/24 hr	
J. L.	Adenocarcinoma with hepatic metastases	1.6	508	0.15
W. G.	Carcinoma of pancreas	2.0	628	0.28
A. H.	Gliobastoma	2.0	235	0.18
J. A.	Carcinoma of tongue	2.0	765	0.40

(From Ramsdell and Kelley, 1973).

TABLE 4

HYPOURICEMIA ASSOCIATED WITH NEOPLASTIC DISEASE

Patient	Diagnosis	Tubular Reabsorption of Phosphate (%)	Urinary Glucose	Urinary Amino Acids	Urinary Oxypurines (μmoles/24 hr)
J. L.	Adenocarcinoma with hepatic metastases	88	Negative	Normal	254
W. G.	Carcinoma of pancreas	90	Negative	Generalized increase	364
A. H.	Glioblastoma	94	Negative	Alanine	<10
J. A.	Carcinoma of tongue	98	Negative	Normal	<10

(From Ramsdell and Kelley, 1973)

TABLE 5

HYPOURICEMIA OF UNKNOWN CAUSE

Patient	Associated Condition	Duration of Hypouricemia	Serum Urate
			mg/100 ml
C. S.	Alcoholism	Not known	1.9
D. L.	Toxic epidermal necrolysis	Persistent	1.9
L. K.	Severe Cirrhosis	Transient	1.0
H. T.	Prolonged diabetic hyperglycemia	Transient	1.2
M. G.	Myocardial ischemia	Not known	1.8
P. I.	Carotid artery obstruction	Transient	1.9
S. G.	Paraplegia with urinary tract infection	Transient	1.9
W. J.	Gastric ulcer with bleeding	Transient	1.9
C. B.	Rocky Mountain spotted fever	Transient	2.0
C. L.	Fever of unknown origin	Transient	1.6
J. S.	Abdominal abscess	Transient	1.7

(From Ramsdell and Kelley, 1973).

In 11 of the 64 hypouricemic patients, no illness or drug known to be associated with hypouricemia could be identified. The diseases found among members of this group are summarized in Table 5. In marked contrast to the hypouricemia described in patients with neoplastic diseases, the hypouricemia in this group was transient in eight of the nine subjects in whom more than one value was obtained. In the only case where the serum urate did not return to normal, the patient, D.L., died soon after the documentation of hypouricemia. One of the patients in this group, C.B., had an elevated $C_{urate}/C_{creatinine}$ ratio (0.24), with no elevation of oxypurine excretion and no evidence of a generalized defect in tubular function. Another patient J.S., had an elevated $C_{urate}/C_{creatinine}$ ratio (0.24), with amino aciduria and an elevated oxypurine excretion of 566 μmols/24 hr. Eight patients in this group also had a reduced dietary intake of purines and protein.

In the 20 patients with hypouricemia not caused by drugs, including 9 with neoplastic disease, a variety of factors may have contributed to the development of this chemical abnormality. A probable defect in tubular transport of uric acid was noted in seven patients; two of these had Hodgkin's disease, and four had carcinoma. In previous studies, hypouricemia in Hodgkin's disease (Bennett, et al., 1972) and in patients with carcinoma (Weinstein, et al., 1965) has been attributed to an abnormality in the renal handling of uric acid. Severe obstructive hepatic disease was observed in four of the hypouricemic patients with malignancy in this study. The association of severe liver disease and hypouricemia has been noted previously (Pasero and Masini, 1958), and it seems to be associated with an increase in the excretion of uric acid (Pasero and Masini, 1958; Ullmann, 1923). Two jaundiced patients who had an elevated $C_{urate}/C_{creatinine}$ ratio also had elevated urinary oxypurine excretion. This suggests that a deficiency or inhibition of hepatic xanthine oxidase may contribute to the development of hypouricemia in some patients. Finally, a low serum osmolarity was noted in eight hypouricemic patients, and many of the hypouricemic patients were consuming a low protein diet.

In summary, it is clear that hypouricemia is a relatively common finding, as would be expected on statistical grounds. Although hypouricemia previously has been reported to be a manifestation of some relatively rare diseases, these illnesses are unusual causes of the hypouricemia found in a general hospital population.

REFERENCES

Bennett, J. S., Bond, J., Singer, I. and Gottlieb, A. J. 1972. Hypouricemia in Hodgkin's disease. Ann. Intern. Med. 76: 751-756.

Pasero, G., and Masini, G.   1958.   L'ipouricemia negli itteri
    colurici.   Minerva Med.   49: 3155-3158.

Ramsdell, C. M. and Kelley, W. N.   1973.   The clinical significance
    of hypouricemia.   Ann. Intern. Med.   78: 239-242.

Ullmann, H.   1923.   Zur frage der harnsaureausscheidung im urin bei
    ikteruskranken.   Klin Wochenschr.   2: 2174-2175.

Weinstein, B., Irreverre, F. and Watkins, D. M.   1965. Lung carcinoma,
    hypouricemia and aminoaciduria.   Amer. J. Med.   39: 520-526.

HYPOURICEMIA, HYPERCALCIURIA AND DECREASED BONE DENSITY.

A NEW HEREDITARY SYNDROME

O. Sperling, A. Weinberger, I. Oliver,
U. A. Liberman and A. de Vries

Rogoff-Wellcome Medical Research Institute and
the Metabolic Unit of Department of Medicine D,
Tel-Aviv University Medical School, Beilinson
Hospital, Petah Tikva, Israel

Genetically determined hypouricemia in man due to increased renal urate clearance is usually associated with additional renal tubular defects (1-3). Renal hypouricemia due to an isolated renal tubular defect has also been described but appears to be a rare condition (4,5). We report a new genetically determined syndrome in man, in which the renal hypouricemia is associated with idiopathic hypercalciuria and decreased bone density.

A 53-year old man, suffering from severe bone pains since seven years, was admitted for metabolic evaluation. The only abnormal findings were hypouricemia with a serum uric acid 0.6-1.1 mg percent, hypercalciuria with a 24 hours urinary calcium excretion of up to 462 mg and on roentgen examination decreased density of the spine, long bones and hands. There was no aminoaciduria, glucosuria or phosphaturia. Serum and urinary copper and serum ceruloplasmin were in the normal range.

More detailed studies revealed a 24 hours urinary uric acid excretion of 405-691 mg, with a renal urate clearance of 46.2 ml/1.73 m^2.

A pyrazinamide suppression test (Table 1) indicated defective renal tubular reabsorption of urate. The clearance of urate and the ratio of the urate clearance

TABLE 1 – PYRAZINAMIDE SUPPRESSION TEST

	Period (min)	C_{urate} (ml/min)	Excreted urate (mg/min)	$\dfrac{C_{urate}}{C_{inulin}}$
Control mean		55.3	0.504	0.657
Pyrazinamide	0– 25	68.9	0.462	0.957
	25– 53	59.1	0.396	0.799
	53– 78	49.0	0.417	0.551
	78–105	50.8	0.318	0.726
	105–128	40.4	0.404	0.469
	128–149	74.6	0.429	0.828

to inulin clearance, which were both markedly elevated
in the control period – 55.3 ml/minute and 65.7 percent
respectively, were only slightly altered by the administration of this drug, decreasing to 40.4 ml/min and
46.9 percent, respectively. The various parameters of
the renal handling of urate in this patient were calculated according to Steele and Rieselbach (6). The
tubular secretion of urate was 1.46 ug/ml. The fractional
excretion of urate during the pyrazinamide test was
markedly elevated, being 46.9% (normal value 2.14±0.54%),
and secreted uric acid amounted to only 36.9% of that
excreted (normal value 79.1%).

The results obtained by studying the effect of
probenecid on the clearance of urate were consistent
with the pyrazinamide test, indicating defective renal
tubular reabsorption of urate. Administration of
probenecid , which normally increases the urate clearance
by two to five- fold (7), affected the uric acid clearance in the patient only slightly, increasing it from
55.3 ml/minute to 74.8 ml/minute. The ratio of the
urate clearance to the inulin clearance increased from
65% to 84%.

Calcium balance studies revealed a positive balance
of 80 mg per day, increased absorption from the gut and
hypercalciuria. Serum calcium and phosphorus and
tubular reabsorption of phosphate were normal. A calcium
infusion test gave a normal response.

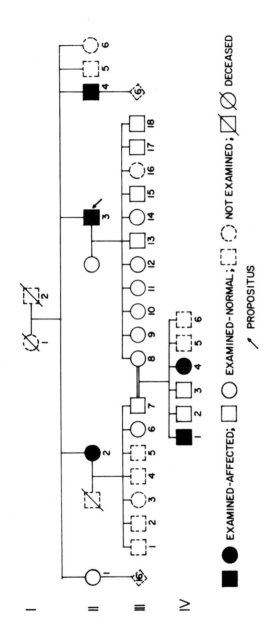

Fig. 1. Pedigree of affected family.

Study of the family (Fig. 1) revealed among the
proband's 5 siblings, of which only 3 were available
for study, 2 additional similarly affected subjects, a
55-year old brother and a 65-year old sister. They had
renal hypouricemia, hypercalciuria and decreased bone
density. Ten out of the eleven children of the propositus and two out of the seven children of his similarly
affected sister were studied and found to be normouricemic and normocalciuric and they had no bone abnormality.

The only consanguineous marriage in the family was
between the daughter of the propositus and the son of
his affected sister. Four out of the six children born
to them could be studied. Two children, a 9-year old
boy and a 4-year old girl were found to have renal hypouricemia and hypercalciuria. Roentgenograms, taken only
of the boy's hands revealed decreased bone density. The
inheritance of the syndrome in this family is autosomal
recessive.

Regarding the hypercalciuria, for lack of a defined
etiology, we classify it for the time being with the
socalled idiopathic hypercalciuria. If correct, this,
to our knowledge, would be the first reported instance
of hereditary familial idiopathic hypercalciuria.

Regarding the decreased bone density, although in
the propositus associated with bone pains and kyphosis,
it is essentially a roentgenological feature, of which
we do not know as yet the pathological substrate. There
is no evidence for any of the known disorders causing
osteomalacia or any hereditary disorder causing decreased
bone mineralization. Furthermore, idiopathic hypercalciuria generally does not cause bone demineralization.
We therefore define the syndrome as hereditary renal
hypouricemia, idiopathic hypercalciuria and decreased
bone density.

The association of idiopathic hypercalciuria, decreased
bone density and renal hypouricemia in all the affected
subjects distinguishes them from the two patients reported
in the literature (4,5), having hypouricemia due to an
isolated renal tubular defect. The family studied by us
presents, to our knowledge, a hitherto undescribed
syndrome. The etiological connection between the three
components of this syndrome - renal hypouricemia, idiopathic hypercalciuria and decreased bone density - is
not understood; their relationships to one mutation are
unclear.

References

1. Bishop C., Zimdahl W.T. and Talbott J.H. Proc. Soc. Exp. Biol. Med. 86:440, 1954.

2. Baron D.N., Dent C.E., Harris H., Hert E.W. and Jepsen J.B. Lancet 271:421, 1956.

3. Watts L.A. and Engle R.L.Jr. Amer. J. Med. 39:520, 1965.

4. Praetorius E. and Kirk J.E. J. Lab. Clin. Med. 35:865, 1950.

5. Greene M.L., Marcus R., Aurbach G.D., Kazam E.S. and Seegmiller J.E. Amer. J. Med. 53:361, 1972.

6. Steele T.H. and Rieselbach R.E. Amer. J. Med., 43:868, 1967.

7. Sirota J.H., Yü T.F. and Gutman A.B. J. Clin. Invest. 31:692, 1952.

SUPPRESSION OF URIC ACID SECRETION IN A PATIENT WITH RENAL HYPOURICEMIA

Peter A. Simkin, Maurice D. Skeith, and L.A. Healey

Division of Rheumatology, Department of Medicine

University of Washington, Seattle, Washington

Despite much recent progress, we remain uncertain as to the precise mechanisms of uric acid excretion and the actions of interfering drugs. We have explored these questions with the assistance of an unusual individual with renal hypouricemia. Our patient is an essentially normal 55 year old woman who first consulted Dr. Skeith for low grade hypertension and mild obesity. Physical examination and laboratory evaluation are otherwise normal with no evidence of aminoaciduria or renal glycosuria. She consistently, however, has been found to have a serum urate level of 0.3 mg/% while her total uric acid excretion is within normal limits. Expressed in the conventional terms of renal physiology, this woman has a clearance of urate which consistently exceeds her glomerular filtration rate. In this respect, she resembles only two other reported humans, although the same property has long been recognized in Dalmatian dogs.

In the first slide the urate to creatinine ratios of random urines collected over portions of 4 weekends have been plotted against the time of day (labelled "hours"). Those of you who have used this parameter clinically will recognize that our patient's ratios are within the normal range. The urate to creatinine ratio of serum however is normally greater than 5.0. In our patient the serum ratio is 0.31 as depicted by the dashed horizontal line. Now, if filtered urate were neither reabsorbed nor secreted (an assumption we make for creatinine) then the urate to creatinine ratio of urine would also be 0.31. Higher ratios can be attained only by tubular secretion of urate. As you can see from the slide, all but three of these specimens showed evidence of secretion and in some the urine ratios are double or more those found in serum indicating secretion of urate at least equal to the amount filtered.

We would also suggest that there is a dirunal pattern of urate

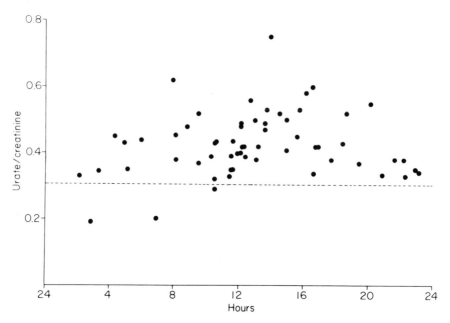

Figure 1.   Urate/creatinine ratios of random urines.

excretion which peaks in the afternoon with a valley in the small
hours of the morning. This same pattern has been observed in normal
individuals. Its presence here suggests, however, that the pattern
is secondary to differences in secretion rather than to varying
filtration or reabsorption of urate.

Lastly, it is also obvious that there is a great deal of scatter
among these experimental points. The patient kept a careful dietary
history during these collections, but the degree of variation in her
urate to creatinine ratio does not correlate with differences in her
diet including moderate use of alcoholic beverages.

We have tried to put these observations in perspective in the
next slide. On the left hand is a schematic drawing of the glomerulus
and proximal tubule. In the conventional model, filtration of urate
is assumed to be free and unrestricted and reabsorption of the filtered urate is assumed to be 98% complete. Urate is then resecreted
in the same segment of the nephron and the ultimate excretion is
approximately 10% of that filtered and is composed primarily of secreted urate.

If, on the other hand, filtered urate is not reabsorbed, as in
our patient L.M., all filtered urate can be expected to appear in the
urine with the addition of any urate secreted. If the secretion of
urate were only 8% of that filtered, the total amount excreted would
be 108% of the filtered load. In the terms of the previous slide,
this would correspond to a urate to creatinine ratio of 0.33.

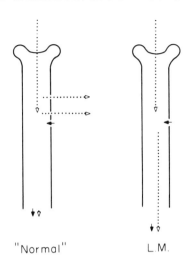

"Normal"   L.M.

Figure 2.   Schema of urate excretion.

Clearly the consistantly higher ratios we observed indicate that
urate secretion in our patient is almost always greater than 8% of
glomerular filtration and may reach 100% or more of that filtered.
Similar data have been reported from the two other known cases of
renal hypouricemia.
     We were able to conduct three different studies of agents which
interfere with renal transport of urate during continuous inulin
infusions. The first such study (a pyrazinamide suppression test),
is shown on the next slide. The data from these experiments will be
shown as the ratio of urate clearance to inulin clearance. A ratio
greater than 1.0 reflects tubular secretion while a lower ratio
would suggest reabsorption. The time of day is on the horizontal
axis labelled "hours".
     In the first study, appropriate control periods were collected
and the patient then took 3 gm. of pyrazinamide. After allowing one
hour for absorption, we obtained three experimental urine specimens.
The baseline clearance ratio of 1.30 is relatively low for our patient
at this time of day, but clearly appears to fall in response to the
pyrazinamide reaching values as low as 1.06. These findings indicate
that pyrazinamide substantially interferes with urate secretion.
Bear in mind that on the scale of this figure the control ratio in
normal individuals would be less than 0.1 and that this would drop
to 0.02 after pyrazinamide. The data in our patient supports the use
of pyrazinamide as an agent to suppress the secretion of urate but
we are unable to say whether complete suppression is achieved with
a 3 gram dose or to evaluate any possible independent effects on
uric acid reabsorption.
     In the next study, we infused sodium salicylate at 16 mg/min
to evaluate this drug which has such well known bidirectional effects

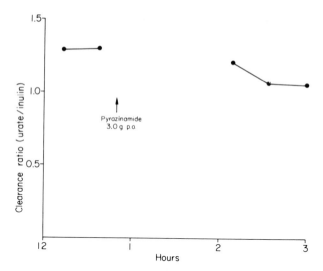

Figure 3.   Pyrazinamide suppression.

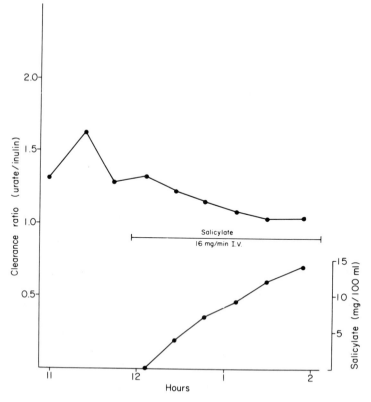

Figure 4.   Salicylate Infusion.

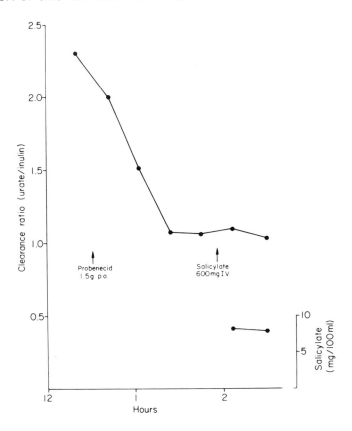

Figure 5.   Oral Probenecid.

on urate excretion. The baseline clearance was again relatively low
on this day, however, the value fell progressively during the infusion
approaching a clearance ratio of 1.0. In a normal individual, the
response to such an infusion is a fall in clearance until a salicylate
concentration (shown here in the lower line) of approximately 5 mg%
is reached. Thereafter, increasing salicylate levels cause progressive
uricosuria.

These findings support the accepted concept that the uricosuric
effects of salicylate result from an independent effect on reabsorption which masks continuing suppression of urate secretion. In fact,
in our patient maximal suppression was not achieved until levels
substantially greater than those causing uricosuria in normal individuals.

During the third study, we observed a much higher baseline
clearance of urate. Probenecid was given orally and a rapid fall in

secretion was observed with a clearance ratio once again approaching 1.0. In normal subjects 1.5 grams of Probenecid would lead to striking uricosuria. Sodium salicylate was infused acutely toward the end of this study and no additional inhibition was recognized. This observation suggests that both agents may inhibit the same site of urate secretion.

In conclusion, we have employed a rare patient with renal hypo-uricemia to study the tubular secretion of urate. We are reporting three principle findings here today. First, we were able to suppress the excretion of urate to, but not below, the glomerular filtration rate. This finding is consistent with the interpretation that renal hypouracemia results from defective reabsorption of filtered urate. Second, secretion is effectively inhibited by the use of pyrazinamide, salicylate, and probenecid with the latter agents being effective inhibitors at levels which are consistently uricosuric in normal people. Last, and perhaps most important, urate secretion is substantially higher in our patient than it has been thought to be in normals. We suspect that tubular reabsorption of secreted urate has masked both the extent and the variability of urate secretion in normal individuals.

EVIDENCE FOR A URATE REABSORPTIVE DEFECT IN PATIENTS WITH WILSON'S DISEASE

D. M. Wilson and N. P. Goldstein

Mayo Clinic and Mayo Foundation

Rochester, Minnesota USA

Most patients with Wilson's disease have hypouricemia. The urate pool is diminished and the clearance of urate is increased two to three times normal. Leu (1) suggested that serum urate returns to normal after treatment with D-penicillamine, but some of our patients had continued hypouricemia after many years of treatment. This prompted us to review urate excretion in a group of patients both before and after drug therapy.

The nature of the renal lesion in Wilson's disease has not been established since net urate excretion reflects filtration, reabsorption and secretion from the tubule in a manner which is not entirely clear. Recently pyrazinamide (PZA) which depresses urate excretion by 95% has been used to examine urate transport (2). At the time these studies were begun it seemed clear that pyrazinamide selectively inhibited urate secretion virtually completely, while not affecting reabsorption. Other studies had suggested that this secretory component was distal to any reabsorptive sites (3). Further analysis of this drug has led to some reservations concerning both of these points so that analysis of the actual mode of action of PZA will be critical to any analysis of the site of altered urate transport in these patients with Wilson's disease.

We have studied filtration and excretion of urate with and without pyrazinamide in 10 patients with Wilson's disease in order to determine if the urate transport defect persists after treatment with D-penicillamine and to make some judgement concerning the site of any defect in urate handling by the renal tubule.

Methods

    Ten patients and 10 control patients were studied. All 10
patients had Wilson's disease based on decreased serum ceruloplasmin levels or copper turnover studies. Eight patients had been on
D-penicillamine from 2.5 to 13 years. All were off drug for at
least two weeks prior to study.

    Standard inulin clearances were obtained. After four adequate
20 minute clearance periods 3.0 g of pyrazinamide were given orally. Fifteen to 20 minute clearance periods were collected for the
next two hours. Uric acid was measured by the uricase method (4)
and inulin by a modification of the Fjelbo autoanalyser technique
(5).

    Secreted urate (Ts urate) was calculated as the amount of
urate excreted per ml GFR before pyrazinamide minus the amount
excreted per ml GFR after the drug was given (during the period
where maximal suppression of urate was apparent). Tubular reabsorption (Tr) was expressed as a percent of filtered load. This
was calculated as the amount of urate filtered (assuming no protein
binding) minus the excreted urate at the point where maximal
suppression was evident. Fractional excretion of urate was expressed as the percent of filtered load excreted during pyrazinamide
administration. Urine samples for amino acid excretion were
collected within three months of the urate studies while patients
were off D-penicillamine. The amino acids were measured on a
Phoenix amino acid analyser.

Results

    A review of Wilson's disease patients at the time first seen
showed that serum uric acid was 2.3 mg% ranging from 1.0 to 5.0 mg%.
After treatment of from 0 to 13 years the mean serum urate was
4.65 mg% ranging from 3.0 to 5.6 mg% (Table I) In the control
group P_{urate} was 5.33 mg% ranging from 3.8 to 6.4 mg%. C_{inulin}
was 91.7 mg/min in the Wilson's disease patients which was lower
than the 109 ml/min for the control group. The clearance of urate
was 15.6 $\pm$ 3.2 ml/min in the Wilson's disease group and 10.7 $\pm$
3.7 ml/min in the controls. C_{urate}/C_{inulin} was .188 $\pm$ .086 and
.098 $\pm$ .044 respectively (Table II).

    Mean Ts urate for the Wilson's disease patients was 1.59 ug/ml
GFR/mg% P_{urate} and only 0.765 $\pm$ 0.01 SE for the controls. In only
one control was the Ts urate higher than in the Wilson's disease
patients. There is clearly a significant difference between the two
groups (p $<$ 0.001) (Fig 1).

    Pyrazinamide virtually abolished the difference in urate
excretion between the two groups. The fractional excretion of

WILSON'S DISEASE

Case	Plasma urate, mg/dl		C_{in}, ml/min/1.73 m^2	Years of treatment
	Initial	Present		
1	1.3	3.0	65	12
2	1.3	4.2	100	12
3	1.7	4.0	51	13
4	2.4	5.3	91	2.5
5	3.4	5.2	107	2.5
6	5.0	3.8	63	0
7	1.0	5.4	95	4
8	1.9	5.1	99	7
9	4.8	4.8	133	0
10	2.5	5.6	113	3
Mean	2.33	4.65	91.7	5.6

Table I.   Clinical data, 10 patients with Wilson's disease

WILSON'S DISEASE
Urate Excretion

Case	C_{ur}/C_{in} (%)		Secreted urate/excreted urate (%)	
	Patient	Control	Patient	Control
1	20.1	11.8	73.7	72.9
2	17.5	9.0	90.6	89.6
3	39.0	8.6	66.5	74.4
4	20.9	8.0	92.8	65.6
5	16.4	6.6	94.3	66.1
6	26.5	7.1	88.5	75.8
7	9.7	16.7	97.5	80.0
8	13.2	14.5	78.4	83.7
9	11.5	8.8	75.3	81.4
10	13.7	7.0	82.2	83.9
Mean ± SD	18.8 ± 8.6	9.8 ± 3.4	84.0 ± 10.3	77.3 ± 7.8

Table II.   Urate excretion in Wilson's disease

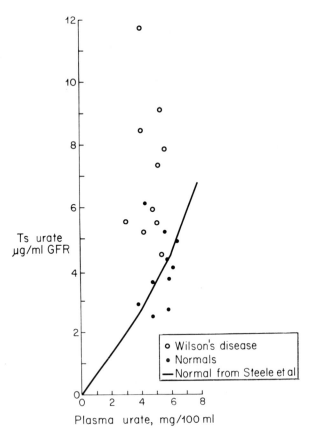

Figure 1.   Ts urate in Wilson's disease

urate after PZA was 2.71 $\pm$ .6 SE in the patients and 2.17 $\pm$ .5 SE in the control subjects. (Fig 2).

Amino acid excretion was elevated generally in these patients. Excretion of neutral, acidic, dibasic and monoamino acids were all elevated.   There were four amino acids of 13 tested in which there appeared to be some correlation between the elevation in Ts urate and the amount of amino acids excreted.   In addition to those seen in Figure 3 valine excretion was correlated in a similar pattern. The r value for these amino acids plotted against Ts urate were: serine 0.65, glutamine 0.61, arginine 0.65 and valine 0.61.   These were all significant correlations (p$<$0.05).   However analysis of Figure 3 shows that two patients with the highest Ts urates had low excretion of all these amino acids.

Discussion

Our data are in agreement with that of Leu in that patients

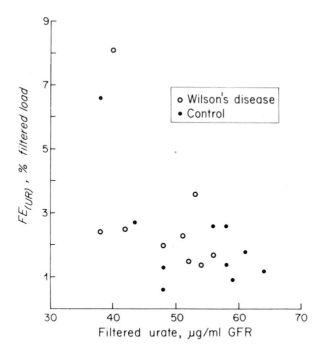

Figure 2.   Fractional excretion of urate after PZA

with Wilson's disease have increased clearances of uric acid. His
untreated patients had a mean clearance of 24 ml/min while normal
persons have a clearance of 9 ml/min (6). Our patients after a
mean treatment period of six years with D-penicillamine still had
a clearance of 15 ml/min. suggesting continued impairment of urate
handling.

D-penicillamine could affect urate transport directly but we
stopped the drug two to four weeks before our studies and Sorenson
has shown that this drug has no effect on net urate excretion (7).

Pyrazinamide was used to help delineate the relationship
between secretion and absorption of urate by the renal tubule. The
mechanism of action of pyrazinamide is controversial however so
the final analysis of urate transport in Wilson's disease using
this drug as a tool will depend on a precise analysis and ultimate
understanding of the action of this drug. The pyrazinamide

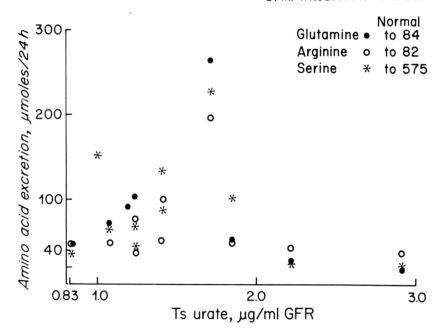

Figure 3.   Amino acid excretion vs Ts urate

suppression test popularized by Steele and Rieselbach (6) was
dependent on two assumptions. The first was that pyrazinamide
selectively and almost quantitatively inhibited urate secretion so
that in the presence of pyrazinamide urate excretion would be a
close estimate of urate filtered and not reabsorbed. The second
assumption for this test would be that any secretion would be
distal to and not coexistant with the reabsorptive site.

Assuming that both of these conditions are true the increased
Ts urate in the Wilson's disease patients could be consistent with
an increased secretion of urate in Wilson's disease. (Fig 4)
Copper is considered to be a toxin in all systems studied in
Wilson's disease and one would hardly expect this result. With
this model a decrease in reabsorption enough to account for the
doubling of urate clearance should have been accompanied by an
increase in the F.E. of urate which we did not observe. (Fig 2)

Many studies have led to the conclusion that the above model
is inappropriate (8). At very high doses pyrazinoic acid is
uricosuric and is therefore similar to several drugs that inhibit
secretion at low doses and reabsorption at high doses. Weiner (9)
and Fanelli (10) have both shown that the dose required to block

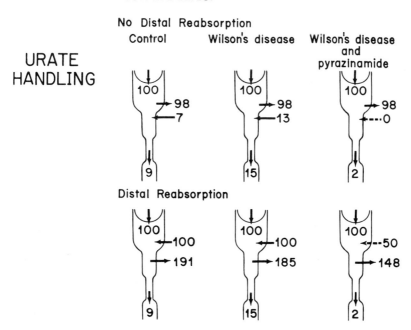

Figure 4.   Models of urate excretion in Wilson's disease:
assuming no reabsorption distal to the secretory site (top panel);
assuming reabsorption coincident with or distal to secretion.

reabsorption is extremely high however and it seems likely that
at the low doses used in our study there was very little inhibition of reabsorption and that we can probably assume selective
blockade of urate secretion. The assumption that the secretory
site is distal to any reabsorptive site now appears to be incorrect
however. Although tubular reabsorption in the distal tubule has
not been demonstrated, microperfusion studies in the rat and
indirect human studies have suggested reabsorption distal to any
secretory site (11)(12). In man mersalyl is a powerful inhibitor
of urate reabsorption. Total excretion of urate in the presence
of this drug is greater than the sum of filtered urate plus
excreted urate before mersalyl. This is strong evidence for reabsorption of some secreted urate implying reabsorption coexistant
with or distal to urate secretion. Fanelli has used this data
(plus the fact that the direct action of pyrazinamide on tubules
is to inhibit transport into or out of cells) to suggest a new
model of urate transport (10). This model is summarized in
Figure 4 (bottom panel). The principal components of this model

include a reabsorptive component with and/or distal to the secretory component. In this model the change in urate excretion after pyrazinamide will reflect how the distalward reabsorptive mechanism responds to a decrease in urate presented after pyrazinamide and therefore not be a quantitative reflection of the secretory rate. In our patients the increased Ts urate could therefore reflect a decrease in the reabsorptive component of transport which is distal to or coexistant with any secretory site.

A complete analysis of other tubular abnormalities was not undertaken in these patients. Glucose reabsorption proceeds to distal portions of the proximal tubule as plasma glucose is raised and becomes uricosuric (12). Only one of our patients had glucosuria at the time of study. Bicarbonate excretion has been studied in eight patients and we have shown an increased splay in Wilson's disease patients (13) with bicarbonatura at low plasmia HCO_3^- levels but there are apparently normal Tm HCO_3^- values. Active phosphate reabsorption is thought to proceed early in the proximal tubule and would therefore be at a different site from urate transport. Phosphate reabsorption may be decreased in untreated Wilson's disease patients but in our patients serum PO_4 was normal and TRP was normal but this data was obtained while on ad libitum diets. Several patients were hypercalciuric.

Amino acid reabsorption proceeds along the whole proximal tubule. In our patients there was a correlation between the excretion of serine, arginine, glutamine and valine and the increase in Ts urate. These amino acids are transported by different transport mechanisms and it seems unlikely that a single amino acid transport system is implicated in altered urate transport. Glycine loading is known to be uricosuric (14) but we saw no particular correlation of the increased urate clearance to glycine excretion in these patients. Glycine loading however causes an increased serine excretion and we did see a correlation here. It is possible that the amino aciduria in some way results in the uricosuria but it seems more likely that the lesion in Wilson's disease affects both transport systems simultaneously.

There are other ways that urate excretion could be altered. Plasma urate binding could be different in these patients or altered in a different way by pyrazinamide, but we have no studies relavant to this point.

References

1.   Leu, M.L., Strickland, G.T., Gutman, R.A.: Renal function in
      Wilson's disease: response to penicillamine therapy. Am J
      Med Sci 260:381-98, 1970.

2. Yü, T.F., Berger, L., Gutman, A.B.: Suppression of tubular secretion of urate by pyrazinamide in the dog. Proc Soc Exp Biol Med 107:905-8, 1961.

3. Steele, T.H., Rieselbach, R.E.: The renal mechanism for urate homeostasis in normal man. Am J Med 43:868-75, 1967.

4. Liddle, L., Seegmiller, J.E., Laster, L.: The enzymatic spectrophotometric method for determination of uric acid. J Lab Clin Med 54:903-13, 1959.

5. Fjeldbo, W., Stamey, T.A.: Adapted method for determination of inulin in serum and urine with an auto-analyser. J Lab Clin Med 72:353-58, 1968.

6. Steele, T.H., Rieselbach, R.E.: The renal mechanism for urate homeostasis in normal man. Am J Med 43:868-75, 1967.

7. Sorensen, L.B., Kappas, A.: The effects of penicillamine therapy on uric acid metabolism in Wilson's disease. Trans Assoc Am Phys 79:157-62, 1966.

8. Holmes, E.W.: Uric acid excretion in man. Kidney Int. 2:115-18, 1972.

9. Weiner, I.M., Tinker, J.P.: Pharmacology of pyrazinamide: metabolic and renal function studies related to the mechanism of drug-induced urate retention. J Pharmacol Exp Ther 180:411-34, 1972.

10. Fanelli, G.M., Weiner, I.M.: Pyrazinoic excretion in the chimpanze. Relation to urate disposition and the actions of uricosuric drugs. J Clin Invest. In press.

11. Lang, R.G.F., Deetjen, P.: Handling of uric acid by the rat kidney. I. Microanalysis of uric acid in proximal tubular fluid. Pflugers Arch 324:279-87, 1971.

12. Steele, T.H., Boner, G.: Secretion and reabsorption of uric acid in man. In Abstrs of the 5th International Congress of Nephrology, 1972, p 131

13. Wilson, D.M., Goldstein, N.P.: Acid base excretion in patients with Wilson's disease. Abstr for consideration III International Symposium on Wilson's Disease, Paris, France, Sept 20-21, 1973.

14. Friedman, M.: The effect of glycine on the production and excretion of uric acid. J Clin Invest 26:815-19, 1947.

REEVALUATION OF THE PYRAZINAMIDE SUPPRESSION TEST

E. W. Holmes and W. N. Kelley

Department of Medicine, Duke University Medical Center

Durham, North Carolina 27710

In 1961 Gutman and Yu proposed a three component system for the regulation of the renal excretion of uric acid in man. The first component of this system is filtration of plasma urate at the glomerulus. While this process is certain to be operative in the human kidney, its quantitative role in the renal excretion of uric acid in man depends upon the extent of urate binding to plasma proteins in vivo. This is a subject that is being discussed in another section of this symposium and will not be considered further in this paper. The second and third component of this system relate to uric acid reabsorption and secretion by the human nephron. Ample data is available to document that both of those processes are operable in the human kidney (Gutman and Yu, 1957; Gutman, et al., 1959), but the relative contribution of each to the final excretion of uric acid has been difficult to determine with conventional clearance techniques. However, a potential solution to this problem of bidirectional uric acid transport appeared in 1967 when Steele and Rieselbach introduced the "pyrazinamide suppression test".

The test is performed by determining uric acid excretion before and after the administration of pyrazinamide or its metabolite pyrazinoic acid (PZA) in a dosage sufficient to give maximal suppression of uric acid excretion. The decrement in uric acid excretion after PZA administration is taken as a quantitative measure of uric acid secretion. In addition the difference between filtered load of uric acid and uric acid excretion at the time of maximal PZA effect is taken as a quantitative measure of uric acid reabsorption. These estimates of uric acid secretion and reabsorption assume the following conditions (Table 1):

TABLE 1

ASSUMPTIONS UNDERLYING THE PYRAZINAMIDE SUPPRESSION TEST

---

1. Secretion of uric acid must occur distal to uric acid reabsorption in the nephron.

2. PZA must completely block uric acid secretion without altering uric acid reabsorption.

3. To be of use in determining the mechanism of action of a uricosuric drug, PZA does not interfere with the metabolism or transport of that drug.

---

1) All uric acid secretion occurs distal to uric acid reabsorption in the human nephron. 2) PZA completely blocks uric acid secretion and it has no effect on uric acid reabsorption and 3) If PZA is to be used to determine the mechanism responsible for a change in uric acid excretion following the administration of a drug, PZA does not interfere with the metabolism or transport of that drug.

Since there was no data available in 1967 which conflicted with these assumptions, the "pyrazinamide suppression test" appeared to be a useful clinical tool for quantitating uric acid secretion and reabsorption in man. However, recent studies of pyrazinamide and uric acid handling by the kidney provide data that have led us to question all three of the assumptions underlying the "pyrazinamide suppression test" (Holmes, Wyngaarden, and Kelley, 1972). Therefore, we would like to consider each assumption and the data relevant to it.

Assumption number one states that uric acid secretion occurs distal to the site of uric acid reabsorption. This is an essential prerequisite if uric acid excretion, the parameter measured in the "pyrazinamide suppression test", is to reflect a change in uric acid secretion alone. However, if uric acid secretion occurred proximal to or coextensive with reabsorption, then any change in uric acid secretion could be subsequently modified by a change in uric acid reabsorption. In addition the change in secretion might secondarily modify uric acid reabsorption through the alteration in distal delivery of uric acid. In this circumstance any difference in uric acid excretion following the administration of PZA could be the result of not one but three factors: 1) The effect of PZA on secretion, 2) The existing rate of uric acid reabsorption distal to the secretory site and 3) The secondary response of the distal reabsorptive mechanism to the change in uric acid load presented to it. Greger, Lang and Deetjen (1971) have

demonstrated by micropuncture that uric acid secretion occurs
proximal to the site for uric acid reabsorption in the rat nephron.
Abramson and co-workers (1973) have reported that net uric acid
reabsorption occurs in the early proximal tubule of the rat and
secretion becomes evident in the late proximal tubule. In the
only direct study available in man, Podevin and co-workers (1968)
have shown that the proximal, but not the distal, tubular epithelium
can transport uric acid from the plasma to the urine. Interpretation of stop-flow and clearance studies in the dog (Zins and
Weiner, 1968; Nolan and Foulkes, 1971), rabbit (Moller, 1965),
guinea pig (Mudge, et al., 1968), and man (Diamond, et al., 1972;
Steele and Boner, 1972) have led to the proposal that uric acid
secretion occurs in the proximal tubule in each of these species.
In the next paper on this program Dr. Diamond and colleagues will
present data supporting a model for a post-secretory reabsorptive
site for uric acid in man. In view of this conflicting data and
the absence of any direct data that indicates uric acid secretion
occurs at a site totally distal to reabsorption in the human
nephron, we believe that assumption number one of the "pyrazinamide
suppression test" as it now stands, cannot be validated.

Assumption number two states that pyrazinamide completely
blocks uric acid secretion. If this requirement were not met,
then the change in uric acid excretion produced by PZA could not
be compared in different patients or the same patient under
different circumstances because the inhibition of uric acid
secretion produced by PZA might be different in each circumstance.
It has been assumed that the inhibition of secretion is virtually
complete in man because uric acid excretion reaches a very low
level with PZA administration. However, this assumption is based
on the prior assumption that uric acid secretion occurs distal to
reabsorption and as already discussed this may not be valid. For
example, the marked reduction in uric acid excretion following PZA
administration may be due to more complete reabsorption distal to
the secretory site. In the absence of any direct data to the
contrary, it seems invalid to conclude that PZA inhibits secretion
completely. More importantly it must also be assumed that PZA
does not alter uric acid reabsorption. If PZA were to alter uric
acid reabsorption in addition to its effect on uric acid secretion,
then the change in uric acid excretion following the administration
of PZA would reflect the balance between these two opposing
processes. Obviously it would be impossible to quantitate uric
acid secretion and reabsorption under these circumstances. Recently,
Weiner and Tinker (1972) have shown that very large doses of PZA
cause a uricosuric response in the dog and Chimpanzee. These
results suggest that in large doses PZA inhibits uric acid reabsorption as well as secretion. Kramp, Lassiter and Gottschalk (1971)
have demonstrated directly by micropuncture that PZA inhibits
uric acid reabsorption in the proximal nephron of the rat. On the

other hand Fanelli, Bohn and Stafford (1970) have shown that the
tubular reabsorptive maximum for uric acid in the Cebus monkey is
increased following the administration of PZA. Although the
directional effect of PZA on uric acid reabsorption is not clear
from these studies, it is difficult to assume that uric acid
reabsorption in the human kidney is not altered during the
"pyrazinamide suppression test".

The third assumption relates to the use of PZA in determining
the mechanism by which a uricosuric drug produces a change in uric
acid excretion. For example, probenecid is a potent uricosuric
drug and its effect is abolished by the administration of PZA
(Diamond, et al., 1972; Steele and Boner, 1972; Fanelli, et al.,
1971). The classical interpretation based on the "pyrazinamide
suppression test" would attribute the uricosuric properties of
probenecid to an increase in uric acid secretion. However, probenecid is known to inhibit not stimulate uric acid secretion from other
more direct studies and its uricosuric effect results from a decrease
in uric acid reabsorption (Nolan and Foulkes, 1971; Moller, 1965;
Mudge, et al., 1968; Kramp, et al., 1971; Fanelli, et al., 1970;
Fanelli, et al., 1971). How then does PZA reverse the uricosuric
effect of probenecid? One explanation is that PZA has a second
effect on the renal handling of uric acid in addition to blocking
secretion. This may be of special concern when several drugs are
involved. Other explanations can be synthesized but each requires
the proposal of another model for the renal handling of uric acid
and this seems injudicious in the absence of any independent data
to support such models. Irrespective of what the explanation may
be, it seems clear from this specific analysis that the "pyrazinamide suppression test" may be misleading.

Unfortunately, at the present time there appears to be no
alternative method for evaluating the renal transport of uric
acid in man which is superior to the PZA suppression test. It
would therefore appear reasonable to determine if a process is
or is not blocked by PZA pretreatment. However, we would recommend
that interpretation of the response be limited and that one
not attempt to associate any response with a specific abnormality
of transport. Perhaps, an effect blocked by PZA should be considered a Type A response and an effect not blocked by this agent a
Type B response. Hopefully, in the future we will have the
additional data necessary to determine the molecular or at least
physiological basis for a Type A and Type B response. Indeed,
with the recent demonstrations that the New World Monkey, a species
that does not have uricase activity, handles uric acid in a
similar if not identical manner to man, it may be possible to
design micropuncture, isolated tubule and other in vitro experiments
that will provide direct data on the sites and mechanisms concerned
with uric acid transport in the kidney.

REFERENCES

Abramson, R. G., Katz, J. H., Maesaka, J. K. and Levitt, M. F. 1973. Uric acid transport in rat kidney. J. Clin. Invest. 52: 1a.

Diamond, H. S., Paolino, J. S. and Kaplin, D. 1972. Evidence for a distal post-secretory reabsorptive site for uric acid. Clin. Res. (abstract) 20: 508.

Fanelli, G. M., Bohn, D. and Stafford, S. 1970. Functional characteristics of renal urate transport in the Cebus Monkey. Amer. J. Physiol. 218: 627-636.

Fanelli, G. M., Bohn, D. L. and Reilly, S. S. 1971. Renal urate transport in the chimpanzee. Amer. J. Physiol. 220: 613-620.

Greger, R., Lang, F. and Deetjen, P. 1971. Handling of uric acid by the rat kidney. Pflugers. Arch. 324: 270-287.

Gutman, A. B. and Yu, T. F. 1957. Renal function in gout with a commentary on the renal regulation of urate excretion, and the role of the kidney in the pathogenesis of gout. Amer. J. Med. 23: 600-622.

Gutman, A. B., Yu, T. F. and Berger, L. 1959. Tubular secretion of urate in man. J. Clin. Invest. 38: 1778-1781.

Gutman, A. B. and Yu, T. F. 1961. A three-component system for regulation of renal excretion of uric acid in man. Trans. Assoc. Amer. Physicians 74: 353-365.

Holmes, E. W., Kelley, W. N. and Wyngaarden, J. B. 1972. The kidney and uric acid excretion in man. Kidney Inter. 2: 115-118.

Kramp, R. A., Lassiter, W. E. and Gottschalk, C. W. 1971. Urate-2-^{14}C transport in the rat nephron. J. Clin. Invest. 50: 35-48.

Moller, J. V. 1965. The tubular site of urate transport in the rabbit kidney and the effect of probenecid on urate secretion. Acta Pharmacol. et Toxicol. 23: 329-336.

Mudge, G. H., McAlary, B. and Berndt, W. O. 1968. Renal transport of uric acid in the guinea pig. Amer. J. Physiol. 214: 875-879.

Nolan, R. P. and Foulkes, E. C. 1971. Studies on renal urate secretion in the dog. J. Pharmacol. Exptl. Ther. 179:

429-437.

Podevin, R., Ardaillou, R., Paillar, F., Fontanelle, J. and Richet, G. 1968. Etude chez l'homme de la cinetique d'apparition dans l'urine de l'acide urique-2-^{14}C. Nephron 5: 134-140.

Steele, T. H. and Rieselbach, R. E. 1967. The renal mechanism for urate homeostasis in normal man. Amer. J. Med. 43: 868-875.

Steele, T. H. and Boner, G. 1972. On the action of uricosuric agents. J. Clin. Invest. 51: 93a.

Weiner, I. M. and Tinker, J. P. 1972. Pharmacology of pyrazinamide: Metabolic and renal function studies related to the mechanism of drug-induced urate retention. J. Pharmacol. Expert. Ther. 180: 411-434.

Zins, G. R. and Weiner, I. M. 1968. Bidirectional urate transport limited to the proximal tubule in dogs. Amer. J. Physiol. 215: 411-422.

EVIDENCE FOR A POST-SECRETORY REABSORPTIVE SITE

FOR URIC ACID IN MAN

Herbert S. Diamond and Ezra Sharon

State University of New York

Downstate Medical Center, Brooklyn, New York

Urinary uric acid is thought to be derived from two sources: uric acid filtered at the glomerulus and incompletely reabsorbed, and uric acid secreted by the renal tubules. The magnitude of the decrease in urinary uric acid which follows administration of pyrazinamide has been widely accepted as an estimate of the secretory component of urinary uric acid. This estimate is dependent upon the assumption that the site of uric acid reabsorption in the renal tubule is proximal to the site of its secretion. Recently, we have reported that total urinary uric acid excretion increases when urine flow rate increases, and have suggested that this was the result of decreased distal tubular reabsorption of uric acid.

If renal tubular reabsorption of uric acid in man occurs at a site distal to the secretion site, part of what is secreted can be reabsorbed downstream. Under these circumstances, the amount of uric acid in the urine that is attributed to secretion is really that which is secreted and escapes reabsorption. Blocking reabsorption would lead to an apparent increase in the secreted fraction. However, if the reabsorption site is proximal to the secretion site, none of the urate added to the urine by secretion can be reabsorbed and blockade of the reabsorption site would not lead to an increase in the apparently secreted fraction.

To test the hypothesis that part or all of renal tubular reabsorption of urate occurs distal to urate secretion, pyrazinamide suppression tests were carried out after inhibition of urate reabsorption. Probenecid or sulfinpyrazone was used to block urate reabsorption.

METHODS

All studies were conducted in the morning following an overnight fast. Medication known to effect the level of serum uric acid or urinary uric acid excretion was withdrawn at least 4 days prior to the studies. All subjects were maintained on a purine-free, low protein, isocaloric diet for at least 3 days prior to and throughout the studies.

Table 1
Effect of Pyrazinamide (PZA) on Probenecid (PB)
Uricosuria in 6 Subjects

	PB	PB+PZA	Decrease (PB-PB+PZA)	% Decrease
Mean urate excretion (µgm/min)	463	135	328	71%
Range	212-638	49-200	163-501	57-79

RESULTS

Uricosuria was established in six subjects by the oral administration of probenecid 2g daily. The acute oral administration of pyrazinamide 3g during probenecid uricosuria decreased urate excretion from a mean of 463 ± 72 µgm/min when probenecid alone was given to 135 ± 21 µgm/min when probenecid was given with pyrazinamide ($p < .01$). Pyrazinamide administration reversed probenecid uricosuria, resulting in urate excretion which was less than control urate excretion when no drug was administered.

Table 2
Effect of Pyrazinamide (PZA) on Probenecid (PB) Uricosuria

	PB 2g		PB 2g + PZA 3g	
	Control UurV	Peak Effect UurV	Control UurV	Peak Effect UurV
Subject	(µgm/min)	(µgm/min)	(µgm/min)	(µgm/min)
L.S.	350	2983	411	691
H.D.	541	3340	530	656
K.P.	330	1261	316	374
MEAN	407	2528	419	574

The decrement in urate excretion produced by pyrazinamide was more pronounced when profound uricosuria was induced by oral administration of a single 2 gram dose of probenecid. Urate excretion was 407 μgm/min on no drug and increased to a peak of 2528 μgm/min after 2 grams of probenecid alone. When 3 grams of pyrazinamide was administered together with probenecid, maximal urate excretion was only 574 μgm/minute, a decrease of 1954 μgm/minute.

Table 3
Effect of Pyrazinamide (PZA) on Sulfinpyrazone (Sf) Uricosuria

	Sf 800 mg		Sf 800 mg + PZA 3g	
	Control	Peak Effect	Control	Peak Effect
	UurV	UurV	UurV	UurV
Subject	(μgm/min)	(μgm/min)	(μgm/min)	(μgm/min)
H.M.	124	867	47	79
H.D.	544	2902	603	871
MEAN	334	1885	325	475

Similarly, urate excretion increased from a mean of 334 μgm/minute when no drug was given to a peak of 1885 μgm/minute following administration of a single 800 mg oral dose of sulfinpyrazone. When 3 grams of pyrazinamide was administered together with sulfinpyrazone, maximum urate excretion was only 475 μgm/minute, a decrease of 1410 μgm/minute. The decrease in urate excretion attributed to pyrazinamide at the time of maximal probenecid or sulfinpyrazone effect, was 4 times as large as urate excretion when no drug was taken. If pyrazinamide effect is attributed to inhibition of renal tubular urate secretion, then urate secretion must be greater than total urinary uric acid on no drug. Thus, apparent renal tubular urate secretion (measured as the decrease in urinary uric acid resulting from pyrazinamide administration), appeared to vary, dependent upon the degree of drug-induced inhibition of uric acid reabsorption.

Table 4
Effect of Aspirin (ASA) on Probenecid (PB) Uricosuria

	PB 2g		PB 2g + ASA 300 mg	
	Control	Peak Effect	Control	Peak Effect
	UurV	UurV	UurV	UurV
Subject	(μgm/min)	(μgm/min)	(μgm/min)	(μgm/min)
H.D.	541	3340	573	1431
L.S.	350	2983	416	873
MEAN	446	3162	495	1152

When urate secretion was inhibited with a 300 mg oral dose

of aspirin, the results were similar to those when urate secretion was inhibited with pyrazinamide. Administration of 300 mg aspirin with 2g probenecid resulted in a decrease in maximal urate excretion from 3162 μgm/min when receiving probenecid alone to 1152 μgm/min when probenecid plus aspirin was given. Again the decrement in urate excretion of 2010 μgm/min, which followed inhibition of urate secretion by aspirin exceeded control urate excretion by almost four-fold.

DISCUSSION

Neither probenecid nor pyrazinamide significantly altered uric acid excretion when administered to patients with serum salicylate levels above 14 mg%.

Thus, the most direct interpretation of the results of the present study is that urate reabsorption occurs, at least in part, distal to the urate secretory site. Under these circumstances, the decrease in urate excretion which follows inhibition of urate secretion by pyrazinamide is a measure of urate secretion less the portion of this secretion that is reabsorbed. Unless reabsorption of secreted urate is inhibited, pyrazinamide administration does not provide a measure of total urate secretion.

Alternative interpretations of these results appear less probable. Pyrazinamide, or its active metabolite, pyrazinoic acid, might competitively inhibit the effect of probenecid and sulfinpyrazone on urate reabsorption. Since the ability to suppress probenecid or sulfinpyrazone induced uricosuria is shown by several drugs which inhibit urate secretion, including aspirin, pyrazinamide, and lactate, the proposal of an additional effect for these drugs is uneconomical. Moreover, this competitive inhibition of drug effects does not explain the apparent excessive urate secretion (as measured by pyrazinamide suppression) observed when uricosuria resulted from water diuresis.

Competition between drugs for binding sites on serum proteins, as has been demonstrated to occur between salicylate and sulfinpyrazone, is possible, but is not likely to be of sufficient magnitude to account for the results here reported.

Probenecid and salicylate decrease urate binding to serum proteins. However, since reabsorption of uric acid is thought to be almost complete at physiologic filtered loads, small changes in filtered urate load should have little or no effect on urate excretion.

The present findings suggest that renal tubular reabsorption of uric acid occurs at least in part at a site in the tubule distal to or co-extensive with the site for uric acid secretion, and that a portion of secreted urate is reabsorbed. During maximum probenecid or sulfinpyrazone induced uricosuria, inhibition of urate secretion with either pyrazinamide or low doses of aspirin resulted in a decrease in uric acid excretion which exceeded total urinary uric acid during control periods by 2 to 4-fold. Thus, tubular secretion of urate appears to be greater than has generally been estimated and may greatly exceed uric acid excretion. A large fraction of secreted urate is reabsorbed.

The pyrazinamide suppression test underestimates urate secretion. Uricosuria induced by probenecid and sulfinpyrazone appears to represent, at least in part, inhibition of post-secretory urate reabsorption.

Various Aspects of
the Renal Handling of Urate

Uric Acid Excretion in Infancy

J. Passwell, S. Orda, M. Brish, H. Boichis

Pediatric Renal Unit and Dept. of Neonatology

The Chaim Sheba Medical Center, Tel-Hashomer, Israel

Studies of the various renal functions in infants have enabled pediatricians to understand their limited capacity to maintain homeostasis. The neonatal glomerular filtration rate is twenty percent of the normal value when corrected for surface area and tubular immaturity is at an even lower functional level than the glomerulus as evident by a lowered renal threshold of bicarbonate and amino acids, a decreased Tm of glucose and a decreased tubular reabsorption of phosphate (1). The morphological factors for this relatively increased tubular immaturity are the increased surface area of glomerular tissue as compared to tubular tissue and the presence of anatomic heterogeneity especially of the tubules. With increasing age the renal growth increases so that the ratio of kidney size to surface area becomes constant. This increase in kidney mass is largely due to the increased growth of the proximal tubules (2). This is an important factor explaining the relative glomerular tubular preponderance of the foetus and newborn.

Urate infarcts in the neonatal kidney and increased amounts of uric acid crystals in a newborn's urine are well known normal findings to neonatalogists. We were therefore prompted to undertake this study to investigate the normal excretion by the kidney of the newborn of uric acid and the progressive maturation of this function during the first year of life.

Our study group included 16 normal newborns who were studied on the third day after birth, 24 infants between the ages of one week and one year and fourteen children aged one to seven years. We also studied 8 low birth weight infants.

The mean serum uric acid in the newborns which was taken on the fourth postnatal day was 3.7 mg/100 ml. This timing is important as the uric acid levels in newborns increase within the first 24 hours of life and reach stable levels at about the third day of life (3). The serum uric acid levels are presented in Table 1.

Table 1

Serum Uric Acid in Term Newborns age 3-4 days

Range   1.7 - 5.2 mg/100 ml

Mean   3.68   mg/100 ml

S.E.   0.08   mg/100 ml

Serum Uric Acid in Infants age 1-52 weeks

Range   1.8 - 5.0 mg/100 ml

Serum Uric Acid in Children age 1-7 years

Range   2.6 - 4.4 mg/100 ml

Urine Collections for clearance studies were started on the third day after delivery and were under constant supervision to avoid spilling of urine. The mean creatinine clearance of the normal infant was 3.65 ml/min, range 1.4 to 8.7 ml/min, and this function progressively increased within the first year of life to approximate those of adults when corrected for surface area. Creatinine clearance at 52 weeks: 19.5-25 ml/min (74-93 ml/min/1.73).

Fig. 1 illustrates the progressive maturation of this function. The creatinine clearance is plotted against the weight of the infant. Our data on premature infants is limited; however it seemed that this function was related more to age than weight.

The fractional excretion of uric acid was calculated from the creatinine and uric acid clearances. Our low birth weight infants showed a large range of 20-63%. The normal range in the newborns was 19-59% and the mean 33%. A gradual maturation of this function is manifest by a decreased % excretion with increasing age. (Infants 1-52 weeks range 26% - 12.5%, children 1-7 years range 16% - 11.5%.)

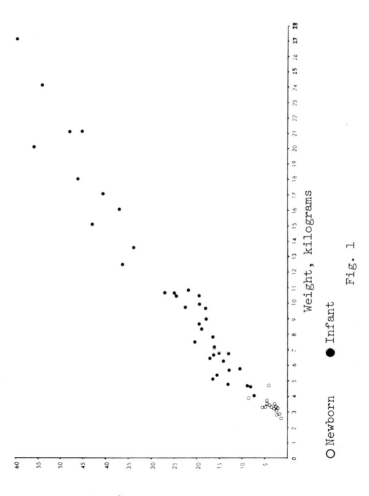

Fig. 1

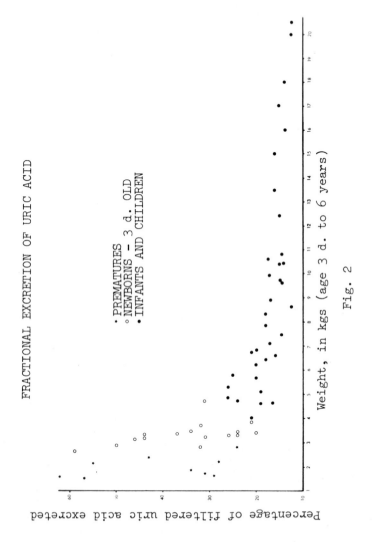

FRACTIONAL EXCRETION OF URIC ACID

· PREMATURES
∘ NEWBORNS – 3 d. OLD
· INFANTS AND CHILDREN

Percentage of filtered uric acid excreted

Weight, in kgs (age 3 d. to 6 years)

Fig. 2

Fig. 2 illustrates this progressive decrease percentage of uric acid excretion. It appears that there is a concomitant increase in the glomerular filtration rate with the progressive decrease in excretion of uric acid.

The urinary uric acid creatinine ratio was of no value as an assessment of maturation of this function.

The mean pH of a fresh urine specimen of our newborns was 5.7 while older infants were capable of normal acidification of their urine. The mean osmolarity of the total urine collection was 310m Osm/kg. These findings are consistent with the known deficiency to acidify and concentrate urine in the first few weeks of life when fed usual infant diets (1).

The increased uric acid load, which the newborn has to cope with is probably due to the physiological fall of white and red blood cell populations seen at this age. It seems that this potential hazard of an excretion of a large uric acid load, as seen in instances of uric acid nephropathy, is facilitated by the concomitant excretion of a relatively alkaline and dilute urine.

The functional interrelationships of reabsorption of the various glomerular solutes are becoming increasingly apparent (4). It is probable that these factors are important aw well as the morphological immaturities in the newborn kidney to explain their increased uric acid excretion.

References :

1. Edelmann C.M., and Spitzer A. : The Maturing Kidney.
    J. of Pediat. 75.509, 1969

2. Fetterman G.H., Shuplock N.A., Philipp F.S., and Gregg H.S. :
    The Growth and Maturation of Human Glomeruli and Proximal
    Convolutions from Term to Adulthood. Studies by Microdissec-
    tion. Pediatrics 35:601, 1965

3. Marks J.F., Kay J., Baum J., Curry L., : Uric Acid Levels in
    Full Term and Low-Birth-Weight Infants. J. of Pediatrics
    73:609-611, 1968

4. Steel T.H. : Control of Uric Acid Excretion. New Eng. J. Med.
    284:1193, 1971

RENAL HANDLING OF URIC ACID IN SICKLE CELL ANEMIA

Herbert Diamond, Ezra Sharon, Dorothy
Holden, and Aurelia Cacatian

State University of New York

Downstate Medical Center, Brooklyn, New York

Hematologic disorders which result in increased cell turnover rates are frequently associated with hyperuricemia and
secondary gout due to increased uric acid synthesis. Although
sickle cell anemia is characterized by a 6-8 fold increase in
red cell turnover, gout is thought to be an uncommon complication
of this disease. Since impaired renal concentrating ability is
a common and early manifestation of sickle cell disease, and
other defects of renal tubular transport such as diminished
tubular maxima for PAH and renal tubular acidosis have been
described, the present study was designed to investigate tubular
transport of urate in sickle cell anemia, and to evaluate the
hypothesis that hyperuricosuria due to defective renal tubular
transport might protect sickle cell anemia patients
against gout.

RESULTS

In 50 adult patients with sickle cell anemia, mean serum
uric acid by the uricase method was 6.2 ± 0.3 mg%. Twenty of
these 50 patients or 40% were hyperuricemic as defined by a serum
uric acid greater than 6.5 mg%. Mean urinary acid/creatinine
ratio was increased to .56 ± .06 in 20 sickle cell anemia patients
compared to .41 ± .04 in 12 normal adults (p < .05).

Renal handling of uric acid was studied in detail in 11 male
and 5 female adults, mean age 25, with sickle cell anemia (Table
1). Mean serum uric acid was 7.0 ± 0.7 mg%. Eight patients were
hyperuricemic and 8 were normouricemic. All but 1 of the 16

patients had a blood urea nitrogen less than 20 mg%. Mean
creatinine clearance was 106 ml/min. Only 3 patients had a
creatinine clearance of less than 75 ml/min. Six of these
patients were considered to be "overproducers" of uric acid on
the basis of 24-hour urinary uric acid excretion of greater than
590 mg after 5 days on an isocaloric purine "free" diet. Three
of the 6 were hyperuricemic and 3 were normouricemic.

Table 1
Clinical Features of 16 Adults
With Sickle Cell Anemia

	Mean	Range
Age	24	15 – 48
Serum Uric Acid (mg%)	7.0 ± .7	3.6 – 13.9
Urine Uric Acid (mg/24 hours)	481 ± 45	268 – 829
Creatinine Clearance (ml/min)	105 ± 12	50 – 245

Thus, there was an equal frequency of uric acid overproduction
among normouricemic and hyperuricemic patients with sickle cell
anemia.

The effect of pyrazinamide on urate excretion was determined
by measurement of urate excretion during 3 consecutive 20-minute
urine collection periods beginning one hour after the oral
administration of 3g of pyrazinamide. The collection period
with the lowest excretion of uric acid was considered to represent
maximum suppression of uric acid excretion. Urate excretion decreased from a mean of 372 ± 38 µgm/min during control periods
to 27 ± 4 µgm/min after pyrazinamide (Table 2). The decrease in
urate excretion which follows pyrazinamide administration has
generally been considered to provide an estimate of urate
secretion. Post-pyrazinamide urate excretion was 7.3 ± 2% of
baseline excretion in the 16 sickle cell anemia patients compared
to 16.8 ± 2.9% in 12 normal subjects (p < .01). Thus, the present
findings are consistent with increased urate secretion in sickle
cell anemia. Since all of the patients studied had renal tubular
disease, at least as indicated by inability to concentrate their
urine normally, impaired post-secretory urate reabsorption is a
more probable interpretation of these results than enhanced
secretion.

Abnormal renal transport of urate may protect some patients
with sickle cell anemia from hyperuricemia. Mean uric acid
excretion was less in hyperuricemic sickle cell anemia patients
than in those with a serum uric acid in the normal range (Table 3).
Mean urate clearance was 8.3 ± 1.1 ml/min in 8 normouricemic
patients compared to 4.0 ± 0.8 ml/min in 8 hyperuricemic patients.
Thus, compared to the hyperuricemic patients, the normouricemic

patients excreted a similar quantity of uric acid at a lesser serum uric acid concentration.

Table 2

Effect of Pyrazinamide (PZA) on Uric Acid Excretion in Sickle Cell Anemia (16 Patients)

	Mean	SEM
Control UurV (mg/min)	372	38
Post-PZA UurV (μg/min)	27	4
Post PZA UurV x 100 (%)		
Control UurV	7.3	2

Table 3

Uric Acid Excretion in Hyperuricemic and Normouricemic Patients with Sickle Cell Anemia

	Normouricemic	Hyperuricemic
# of Patients	8	8
Plasma urate (mg%)	5.0 ± 0.4	9.1 ± 0.8*
UurV (mg/ml)	399 ± 47	351 ± 65
Urate Clearance (ml/min)	8.3 ± 1.1	4.0 ± 0.8*

* $p < .01$

In normal subjects, both pyrazinamide suppressible and total urate excretion increase as serum uric acid concentration increases, particularly when these measures are factored by glomerular filtration rate. In patients with sickle cell anemia, there was a trend toward lesser pyrazinamide suppressible urate excretion in the hyperuricemic patients (Table 4). The decrease in urate excretion after pyrazinamide was 378 ± 46 μgm/minute in the normouricemic patients and 311 ± 63 μgm/minute in the hyperuricemic patients. Glomerular filtration rate, estimated on the basis of creatinine clearance, was similar in the two groups, and the same trend was therefore found when pyrazinamide suppressible urate excretion was factored by creatinine clearance. This ratio, referred to as pyrazinamide suppressible urate excretion per unit of glomerular filtration rate was 4.4 + .9 in the normouricemic patients and 3.0 ± .5 in the hyperuricemic patients. Rieselbach and his co-workers have published 95% tolerance limits for this ratio for normal and hyperuricemis subjects. Pyrazinamide suppressible urate excretion per unit glomerular filtration rate exceeded these tolerance limits in two

of the three normouricemic patients with sickle cell anemia and
uric acid overproduction. This suggests that some sickle cell
anemia patients with uric acid overproduction may be protected
from hyperuricemia by an abnormality in renal transport of urate
which permits enhanced urate excretion at a normal plasma urate
concentration.

Table 4

Effect of Pyrazinamide (PZA)
on Uric Acid Excretion
in Sickle Cell Anemia

	Normouricemic	Hyperuricemic
# of patients	8	8
Sur	378 ± 46	311 ± 63
Ccr	106 ± 22	105 ± 9
Sur/Ccr	4.4 ± 0.9	3.0 ± 0.5

Sur represents the decrease in urate excretion resulting
from pyrazinamide administration. Ccr represents creatinine
clearance.

CONCLUSIONS

Hyperuricemia, uric acid overproduction, and hyperuricosuria
are frequent clinical features of sickle cell anemia. Uric acid
overproduction may result in hyperuricemia in some patients with
sickle cell anemia. However, in other patients, enhanced urate
excretion permits maintenance of a normal serum uric acid
despite uric acid overproduction. Hyperuricosuria in normouricemic patients with sickle cell anemia may result from enhanced
tubular secretion of urate but is more likely to represent
diminished urate reabsorption at a post-secretory site. Hyperuricosuria may protect the young patient with sickle cell anemia
from gouty arthritis and tophi but may also increase the risk of
tubular deposition of urate and nephropathy.

HYPERURICEMIA INDUCED BY ETHAMBUTOL

A. E. Postlethwaite, A. G. Bartel and W. N. Kelley

Department of Medicine, Duke University Medical Center

Durham, North Carolina 27710

Ethambutol is commonly used to treat infections in man caused by Mycobacterium tuberculosis (Fig. 1). We have found that ethambutol also produces a substantial increase in the serum urate concentration in many patients receiving this agent (Postlethwaite, et al., 1972).

In the present study, the effect of prolonged ethambutol therapy on the metabolism of uric acid was assessed in 24 hospitalized tuberculous patients.

Fifteen patients were studied retrospectively. They were treated with doses of ethambutol ranging from 12 to 19 mg/Kg/day. These patients were also receiving isoniazid, pyridoxine, and some were receiving streptomycin in addition to ethambutol. None were on other medications known to alter the metabolism of uric acid. Nine patients with tuberculosis were studied prospectively.

Seven of the 15 patients studied retrospectively, or 47%, developed a mean maximal increase of 4 mg% in their serum urate concentration with a range from 2.4 to 5.6 mg% above pretreatment values during the first 30 days of therapy (Table 1). The remaining eight patients developed a mean decrease of 0.1 mg% in their serum urate concentration. There was no significant change in the serum creatinine concentration in these 15 patients during the period of observation. Isoniazid and pyridoxine administered in the absence of ethambutol was not associated with a change in serum urate concentration.

Eight of 9 patients studied prospectively developed sustained elevations of their serum urate while receiving ethambutol therapy.

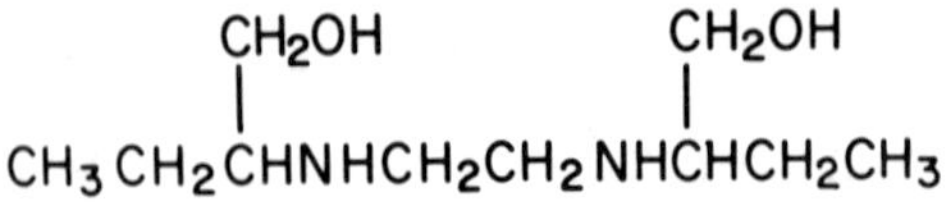

$$\underset{\displaystyle CH_3\,CH_2\,CHNHCH_2CH_2\,NHCHCH_2CH_3}{\overset{\displaystyle CH_2OH \qquad\qquad\quad CH_2OH}{\qquad\; |\qquad\qquad\qquad\quad\; |}}$$

Fig. 1. Structure of ethambutol.

TABLE 1

EFFECT OF ETHAMBUTOL THERAPY FOR 30 DAYS ON SERUM URATE CONCENTRA-
TION IN FIFTEEN PATIENTS STUDIED RETROSPECTIVELY

Subjects	Serum Urate Concentration (mg/100 ml)		Change (mg/100 ml)
	Control	Ethambutol	
Nonresponders (8)	6.4	6.3	-0.1
Responders (7)	5.5	9.6	+4.1
W.R.	8.0	12.0	+4.0
J.T.	4.2	8.0	+3.8
D.B.	5.7	9.7	+4.0
J.A.	4.8	7.2	+2.4
J.D.	5.0	8.3	+2.7
W.S.	6.6	12.2	+5.6
C.H.	4.4	9.7	+5.3

These data are summarized in Fig. 2. Ethambutol was administered
alone for the first 5 days. During these 5 days, all 8 patients
developed a mean maximal elevation of 1 mg% in their serum urate
above control which was accompanied by a mean decrease of 22% in the
$C_{urate}/C_{creatinine}$ X 100. This suggests that ethambutol administered
alone produces an elevation in the serum urate and that this is due,
at least in part, to a decreased renal clearance of uric acid. On
day 6, isoniazid, pyridoxine and in some cases streptomycin were
added to the treatment program. After day 6 there was an additional
increase in the serum urate and a further decrease in the fractional
clearance rate of uric acid.   The possibility that the addition of
isoniazid or pyridoxine on day 6 may have potentiated the hyperuricemic effect of ethambutol cannot be ascertained from our data.

    None of the patients developed an attack of gout while hyperuri-
cemic from ethambutol therapy. Two patients hyperuricemic from ethambutol were treated briefly with allopurinol with a return of the

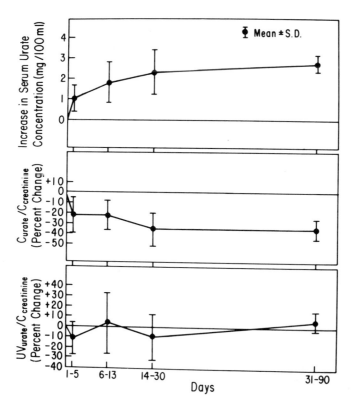

Fig. 2. Effect of ethambutol on serum urate concentration and renal
handling of uric acid.

serum urate concentration to normal.

Relatively few classes of drugs in common clinical use lead to the development of hyperuricemia. Drugs that produce an elevation of serum urate by renal mechanisms are diuretics, salicylates given at a low dose, pyrazinamide and compounds such as alcohol which cause an accumulation of lactic acid. Diuretic agents produce hyperuricemia by causing extracellular fluid volume depletion as well as lactic acid accumulation (Steele and Oppenheimer, 1969; Schirmeister, et al., 1969). Ethanol appears to produce hyperuricemia primarily by causing hyperlacticacidemia although volume depletion may also be a factor (Lieber, et al., 1962). Although the exact basis for the hyperuricemic effect of PZA and salicylates has not been established, each agent has several rather distinctive properties as summarized in Table 2. The hyperuricemic effect of PZA is reversed by salicylates in low dose (Petty and Dalrymple, 1964). Salicylates at any dose characteristically inhibit the effect of uricosuric agents (Gutman and Yu, 1957; Postlethwaite and Kelley, 1972).

In Table 2 we have compared the effect of ethambutol with each of these classes of drugs. Ethambutol did not cause the accumulation of lactic acid in the 3 subjects in whom this was measured.

TABLE 2

CHARACTERISTICS OF DRUGS WHICH PRODUCE HYPERURICEMIA BY ALTERING
THE RENAL HANDLING OF URIC ACID

	Drugs				
	Diuretics	Ethanol	PZA	Salicylates (low dose)	Ethambutol
Properties					
1. Lactic Acid accumulation	+	+	–	–	–
2. ECF Volume Depletion	+	+	–	–	–
3. Refersal by low dose salicylates	–	–	+	–	–
4. Inhibits effect of uricosurics	–	–	±	+	–

\+   Indicates the drug produces the effect described.
–   Indicates the drug does not produce the effect described.

Ethambutol had no effect on the urinary excretion of sodium or potassium and there was no clinical evidence of volume depletion in patients receiving this agent. The hyperuricemia of ethambutol was not reversed by salicylates in 3 patients in whom this was studied. Ethambutol did not inhibit the effect of probenecid, sulfinpyrazone or iopanoic acid in 3 additional patients.

In summary ethambutol produced a substantial increase in the serum urate concentration in 62% of patients receiving this agent for the treatment of tuberculosis. Further studies have shown that this drug produces a decrease in the renal clearance of uric acid. Although the exact basis for this effect is unclear, the effect of ethambutol appears to differ in one or more ways from other agents such as diuretics, ethanol, pyrazinamide and salicylates which have a similar net effect on uric acid clearance.

REFERENCES

Gutman, A. B. and Yu, T. F. 1957. Protracted uricosuric therapy in tophaceous gout. Lancet 2: 1258-1260.

Lieber, C. S., Jones, D. P., Losowsky, M. S. and Davidson, C. S. 1962. Interrelation of uric acid and ethanol metabolism in man. J. Clin. Invest. 41: 1863-1870.

Petty, T. L. and Dalrymple, G. V. 1964. Inhibition of pyrazinamide hyperuricemia by small doses of acetylsalicylic acid. Ann. Intern. Med. 60: 898-900.

Postlethwaite, A. E. and Kelley, W. N. 1972. Radiocontrast agents and aspirin. J.A.M.A. 219: 1479.

Postlethwaite, A. E., Bartel, A. G. and Kelley, W. N. 1972. Hyperuricemia due to ethambutol. New Eng. J. Med. 286: 761-762.

Schirmeister, J., Mau, N. K. and Hallaner, W. 1969. Study on renal and extrarenal factors involved in the hyperuricemia induced by furosemide, p. 59. In G. Peters and F. Roch-Ramel (eds.), Progress in Nephrology. Springer-Verlag, Inc., Berlin.

Steele, T. H. and Oppenheimer, S. 1969. Factors affecting urate excretion following diuretic administration in man. Amer. J. Med. 47: 564-574.

URICOSURIC EFFECT OF AN ANTICHOLINERGIC AGENT IN HYPERURICEMIC

SUBJECTS

A. E. Postlethwaite, C. M. Ramsdell and W. N. Kelley

Department of Medicine, Duke University Medical Center

Durham, North Carolina   27710

In recent years there has been an appreciable expansion of our understanding of secondary hyperuricemia and of certain rare sub-types of gout associated with an excessive production of uric acid. Although the pathogenesis of primary hyperuricemia in most subjects remains obscure, it is clear that many such patients have a dimin-ished renal clearance of uric acid (Wyngaarden and Kelley, 1972). The factor(s) responsible for this abnormal clearance has been difficult to define since our understanding of the manner in which the kidney of normal man regulates uric acid excretion is incom-plete.

A number of pharmacologic agents have been found which alter uric acid excretion in man. Most of the present concepts of uric acid transport in the human kidney are based on an interpretation of the effects of such agents on uric acid excretion (Lieber, et al., 1962; Schirmeister, Mau and Hallaner, 1969; Steele and Oppenheimer, 1969; Gutman, et al., 1969 and Rieselbach, et al., 1970). In the present study, we report that two drugs, Robinul or glycopyrrolate and Pathilon or tridihexethyl chloride, which reduce the transmission of parasympathetic impulses also increase the renal clearance of uric acid in some patients. In addition, we will present evidence which suggests that the characteristics of this uricosuric effect may be different from that observed with other uricosuric drugs.

Robinul 2 mg was administered orally to a total of 34 subjects. Thirteen subjects had gout and had received no hypouricemic therapy for 2 weeks prior to the study. Six subjects had borderline or definite asymptomatic hyperuricemia with serum urate values ranging from 6.7 to 7.1 mg%. The remaining 15 subjects were normouricemic with serum urate values ranging from 3.3 to 6.2 mg%. Care was taken to maintain comparable urine flow during a 2 hour control clearance period before and a 3 hour clearance period after Robinul administration. Robinul Placebo, that is, Robinul not containing glycopyrrolate, was administered to seven selected hyperuricemic subjects. A second anticholinergic, tridihexethyl chloride or Pathilon, was given as a single 50 mg dose orally to 3 gouty patients. The effect of pretreatment with 3 gm of pyrazinamide on the uricosuric response to glycopyrrolate was examined in two patients. The effect of Robinul on the binding of urate to plasma protein in vitro was evaluated by the technique of equilibrium dialysis.

Figure 1 illustrates the change in $C_{urate}/C_{creatinine}$ from control after Robinul administration as a function of the control serum urate concentration in each of the 34 subjects. The open circles represent subjects with gout whereas the closed circles represent subjects without gout. Robinul produced an increase in the $C_{urate}/C_{creatinine}$ in some subjects but not in others. In general, those subjects in whom glycopyrrolate was noted to be uricosuric had a higher serum urate concentration than those who exhibited no uricosuric response to glycopyrrolate. More specifically, an increase in the $C_{urate}/C_{creatinine}$ of greater than 20%

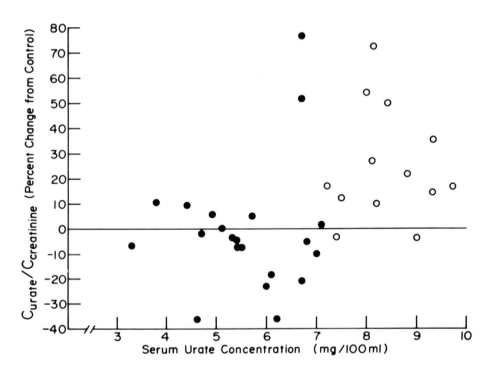

Fig. 1.   Relationship of Robinul response to serum urate concentration.   o Gout,   ● non-gout.

was noted in 8 subjects. Six of these patients had gout whereas the other two had borderline asymptomatic hyperuricemia. The change in $C_{urate}/C_{creatinine}$ was less than 20% in all 15 control subjects as well as in the remaining 11 hyperuricemic or gouty subjects. The failure of some hyperuricemic subjects to exhibit a uricosuric response to glycopyrrolate indicates that such a response cannot be attributed to hyperuricemia per se.

The change in $C_{urate}/C_{creatinine}$ from control after Robinul administration as a function of the control $C_{urate}/C_{creatinine}$ X 100 is illustrated in Figure 2. Normouricemic subjects are indicated by solid dots, hyperuricemic subjects by solid squares and gouty subjects by open squares. A uricosuric response to Robinul was noted predominately, but not exclusively, in those subjects

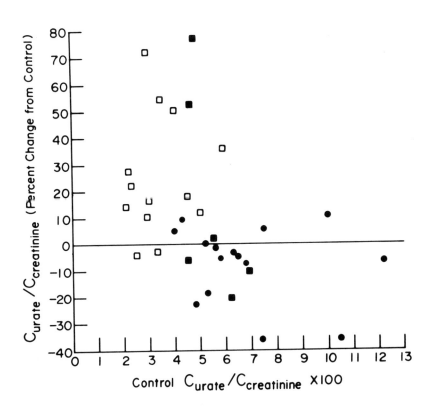

Fig. 2. The percent change in $C_{urate}/C_{creatinine}$ after Robinul administration is plotted as a function of control $C_{urate}/C_{creatinine}$ X 100. ● Normouricemia; ■, Hyperuricemia, □, Gout.

exhibiting a reduced $C_{urate}/C_{creatinine}$.

In the eight subjects exhibiting a uricosuric response to glycopyrrolate the $C_{urate}/C_{creatinine}$, which was initially reduced, returned nearly to the level observed in our normal group of subjects (Table 1). The response of hyperuricemic and normouricemic subjects to glycopyrrolate was significantly different when analyzed by the one or two tailed 2 sample t test with respect to the clearance of uric acid or C_{urate}, the fractional excretion of uric acid or $C_{urate}/C_{creatinine}$ and, the normalized excretion of uric acid or $UV_{urate}/C_{creatinine}$. However, there was no significant difference observed in the creatinine clearance or the serum urate concentration during the period of study in these two groups of subjects.

The change in $C_{urate}/C_{creatinine}$ following the administration of Robinul in 7 selected hyperuricemic patients who are identified with initials is summarized in the center panel of Fig. 3. The percent change in $C_{urate}/C_{creatinine}$ following the administration of Robinul Placebo to these patients is indicated in the right panel of Fig. 3. There was no uricosuric response to the placebo in any of the seven patients. The slight decrease in $C_{urate}/C_{creatinine}$ observed with the placebo was similar to the decrease noted when Robinul containing glycopyrrolate was given to the normouricemic subjects as indicated in the panel to the left.

Additional studies were performed with several subjects who exhibited a uricosuric response to glycopyrrolate. In order to ascertain whether other anticholinergic agents were capable of producing a uricosuric response, a second anticholinergic, tridi-

TABLE 1

RESPONSE OF SUBJECTS TO GLYCOPYRROLATE

Condition	Number Subjects	$C_{urate}/C_{creatinine}$
		Mean
Normouricemic Controls	15	
Control		6.7
Glycopyrrolate		6.2
Hyperuricemic Responders	8	
Control		3.8
Glycopyrrolate		5.7
Hyperuricemic Non-Responders	11	
Control		4.3
Glycopyrrolate		4.3

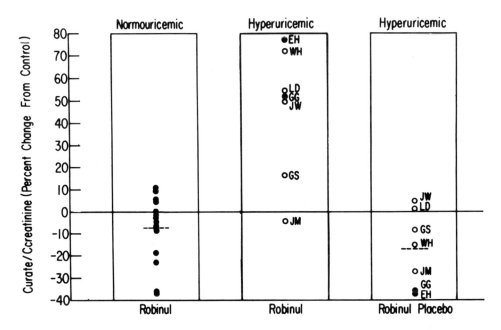

Fig. 3. The percent change in $C_{urate}/C_{creatinine}$. Left panel, response of normouricemic subjects to glycopyrrolate; center panel, response of hyperuricemic subjects to glycopyrrolate; right panel, response of hyperuricemic subjects to Robinul placebo.

hexethyl chloride or Pathilon, was given as a single 50 mg dose orally to 3 gouty patients previously exhibiting a uricosuric response to glycopyrrolate In these 3 subjects, tridihexethyl chloride, increased the $C_{urate}/C_{creatinine}$ by 67 to 83 per cent. The effect of glycopyrrolate and tridihexethyl chloride in patient A.B. is compared in Fig. 4. The effect of PZA pretreatment on the uricosuric response to glycopyrrolate was examined in two patients. The data obtained in patient A.B. are also illustrated in Fig. 4. In patient A.B., as in a second patient studied, the uricosuric effect of glycopyrrolate was completely blocked by the prior administration of pyrazinamide.

Glycopyrrolate did not appear to alter the binding of urate to plasma proteins as determined by the technique of equilibrium dialysis at 4°C (Fig. 5).

In summary, we have made the following observations. 1) Two structurally different anticholinergic agents have a uricosuric effect in some patients. This uricosuric effect was observed only in subjects with a serum urate concentration of 6.7 mg% or greater; however a uricosuric effect was not observed in all hyperuricemic

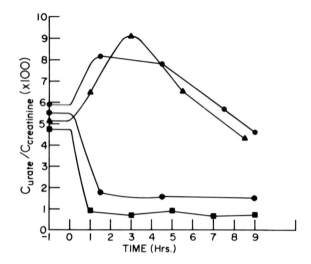

Fig. 4.   Comparison of the effect of Pathilon (▲), Robinul
(●), Pyrazinamide (■) and Pyrazinamide plus Robinul (◆)
on the fractional excretion of uric acid in subject A.B.

subjects.   2)   These anticholinergic agents differ from most
uricosuric drugs in that they are not weak organic carboxylic
acids, they do not reduce urate binding to protein in vitro, and
their effect is completely blocked with pyrazinamide.   3)   Based
on minimal effective dose, glycopyrrolate is more potent than
other uricosuric agents; however, the maximal effect appears to
be limited.

     The mechanism or mechanisms by which these anticholinergic
agents produce their uricosuric effect is not apparent from our
studies.   Although the uricosuric effect of these agents could be
related to their anticholinergic action, several observations do
not appear to support this possibility.   First, a third anticholinergic drug, ℓ-hyoscyamine or Cystospaz, had no uricosuric effect
in two patients who had previously responded to glycopyrrolate.

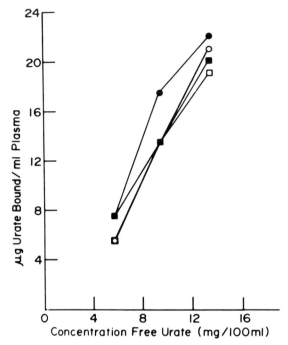

Fig. 5. The binding of urate to plasma proteins at 4°C is plotted
as a function of uric acid concentration in the dialyzing bath.
A.B., control, ● ; glycopyrrolate, ○ . J.W., control, ■ ;
glycopyrrolate, □ .

Second, uric acid retention was not observed after administration
of several parasympathomimetic agents.

REFERENCES

Gutman, A. B., Yu, T. F. and Berger, L. 1969. Renal function in
    gout. III. Estimation of tubular secretion and reabsorption
    of uric acid by use of pyrazinamide. Amer. J. Med. 47: 575-592.

Lieber, C. S., Jones, D. P., Losowsky, M. S. and Davidson, C. S.
    1962. Interrelation of uric acid and ethanol metabolism in
    man. J. Clin. Invest. 41: 1863-1870.

Rieselback, R. E., Sorensen, L. E., Shelp, W. D. and Steele, T. H.
    1970. Diminished renal urate secretion per nephron as a basis
    for primary gout. Ann. Intern. Med. 73: 359-366.

Schirmeister, J., Mau, N. K. and Hallaner, W. 1969. Study on
    renal and extrarenal factors involved in the hyperuricemia
    induced by furosemide, p. 59. In G. Peters and F. Rach-Ramel

(eds), Progress in Nephrology, Springer-Verlag, Inc., Berlin.

Steele, T. H. and Oppenheimer, S.   1969.   Factors affecting urate
    excretion following diuretic administration in man.   Amer.
    J. Med.   <u>47</u>: 564-574.

Wyngaarden, J. B. and Kelley, W. N.   1972.   Gout, pp. 889-968.   In
    J. B. Stanbury, J. W. Wyngaarden, and D. S. Fredrickson (eds.),
    The metabolic basis of inherited disease, McGraw-Hill Book
    Company, New York.

EFFECTS OF BENZOFURAN   DERIVATIVES ON TUBULAR PERMEABILITY TO

URATE-2-^{14}C IN THE RAT NEPHRON

Ronald A. Kramp, M.D. (with the technical assistance
of Régine Lenoir)
Department of Medicine, University of Louvain
School of Medicine, Louvain, Belgium

It is generally admitted that renal excretion of urate may
involve a three component system : glomerular filtration and tubular reabsorption and/or secretion (1). Tubular microinjection experiments in the rat kidney showed that urate was essentially reabsorbed along the proximal tubule presumably by a carrier-mediat-
ed mechanism. Under these experimental conditions a secretory process for urate could not be shown nor excluded (2). In an attempt
to further unravel some of the uncertainties related to these
transport mechanisms we have investigated the effects of benzofuranne derivatives on urate transport along the rat nephron. Rats
were pretreated with benzbromarone or benziodarone which are two
potent uricosuric drugs in man (3,4).

METHODS

The microinjection technique described by Gottschalk, Morel
and Mylle (5) was applied with some minor modifications. Known
volumes (10 to 25 nl) of a colored solution of buffered saline
containing tritiated inulin and urate labeled with ^{14}C were injected with a calibrated pipette into proximal or distal surface convolutions. Inulin was used as the reference substance (6) and was
totally recovered in the urine of the punctured kidney after technically satisfactory microinjections. It was assumed that the injected inulin remained solely within the lumen of the tubule, i.e.
the luminal membrane was impermeable to inulin. It was further
assumed that labeling of urate did not impair its physiological
properties and that excreted ^{14}C represented urate.
Excretion of injected urate was defined in comparison with the presumed tubular handling of inulin. Total recovery of urate was the

777

per cent of injected urate recovered at the renal pelvis. The per
cent of urate excreted in the same time course as simultaneously
injected inulin was termed direct recovery. The latter presumably
represented the fraction of injected urate which remained within
the lumen during tubular transit of urate. Delayed recovery was
defined as urate excreted more slowly than inulin, i.e. a fraction
of injected urate which left the tubular lumen to subsequently re-
gain access to it. The method and rationale for calculation of
these recoveries were previosuly described in detail (5,7).

Anesthetized rats were prepared for micropuncture as previously reported (2). Osmotic diuresis was induced by infusion of a 5 %
mannitol solution in isotonic saline to permit rapid serial collections of urine necessary to determine fractional recoveries of the
isotopes. Prior to abdominal surgery some rats were slowly infused
i.v. with benzbromarone or benziodarone at a dose of 10 mg/kg/BW.
The sites of microinjection were determined by measurement of lis-
samine green tubular transit times (8). Three puncture sites were
taken into consideration. "Early" and "late" proximal respectively
correspond to the first or second and the last proximal convolution to appear on the renal surface and "distal" was a superficial
convolution of a distal tubule. A microinjection was considered to
be technically satisfactory if the rate of injection was maintained
constant, no reflux and no visible tubular dilatation occurred and
if recovery of injected inulin was 100 ± 5 %. The rate of microinjection was adjusted to approximate single nephron filtration rates.

RESULTS

Three groups of rats were studied : control rats with and
without placebo infusion and animals pretreated with benzbromarone
or benziodarone.
In control rats total (72 and 83 %) and direct (52 and 63 %) recoveries, respectively after early and late proximal injections,
significantly increased along the proximal tubule and were different from recoveries (94 and 69 %) following distal injections.
Delayed recovery approximated 20 to 25 % irrespective of the site
of puncture.
In benzbromarone-treated rats, total (83 and 88 %) and direct (67
and 71 %) recoveries after early and late proximal injections were
significantly higher than in control rats (P < 0.001). Delayed recovery however was not different from control results. Recoveries
after distal injections were comparable to the per cent of urate
excreted in rats without drug pretreatment.
Following early and late proximal injections in benziodarone-infus-
ed rats, total (78 and 83 %) and direct (74 and 80 %) recoveries
were also higher than in control animals (P< 0.005) while delayed
recovery was significantly less than in both control and benzbro-

marone-treated rats. After injections in distal convolutions, direct (87 %) and delayed (9 %) recoveries, but not total, varied markedly (P < 0.001) from results in control and benzbromarone-treated rats.

These results indicate that both drugs significantly decreased urate reabsorption in the proximal tubule. Furthermore, delayed excretion was markedly reduced in benziodarone-treated animals but was unchanged after benzbromarone infusion. These drug effects were not related to changes in tubular transit times, urine flow rates, injection rates and so on... which were similar in the three animal groups.

COMMENTS

In this study the microinjection experiments were undertaken during mannitol diuresis. Urate recoveries however compared to those previously reported in hypertonic saline infusion (2). Under these experimental conditions, urate was essentially reabsorbed along the proximal tubule while a larger volume of distribution for urate than for inulin (estimated by delayed excretion), was primarily accessible along the distal tubule and/or collecting duct. Interpretation of this finding was previously discussed (2). Following intravenous infusion of benzbromarone and benziodarone urate reabsorption was markedly decreased along the proximal tubule. The drug effect was presumably located on the luminal membrane since direct recovery was increased. Similar results were obtained in rats pretreated with probenecid and pyrazinamide (2). A distinct difference between the drug action was found after distal injections. The volume of distribution for urate was markedly decreased following infusion with benziodarone in striking contrast with the lack of significant change in rats pretreated with benzbromarone.
These findings in benziodarone-treated rats may be suggestive of changes in handling of urate along the distal tubule and/or collecting duct in addition to the inhibition of proximal reabsorption. Their precise significance remains to be further investigated under appropriate experimental conditions.

ACKNOWLEDGEMENTS

This study was supported by a Grant-in-Aid from the Fonds National de la Recherche Scientifique of Belgium.
Dr. Kramp is Chargé de Recherches of the F.N.R.S.
Solutions of benziodarone and benzbromarone were kindly provided by the laboratories of Labaz, Société belge de l'Azote, Brussels, Belgium.

REFERENCES

1. Gutman A.B. and Yü T.F. - A three component system for regulation of renal excretion of uric acid in man. Trans. Ass. Amer.
   Phycns 74 : 353, 1961.

2. Kramp R.A., Lassiter W.E. and Gottschalk C.W. - Urate-2-^{14}C
   transport in the rat nephron. J. clin. Invest. 50 : 35, 1971.

3. Sternon J., Kocheleff P., Couturier E., Balasse E. and
   Van den Abeele P. - Effet hypouricémiant de la benzbromarone.
   Etude de 24 cas. Acta Clin. Belg. 22 : 285, 1967.

4. Podevin R., Paillard F., Amiel C. and Richet G. - Action de la
   benziodarone sur l'excrétion rénale de l'acide urique. Rev.
   franç. Etud. clin. biol. 12 : 361, 1967.

5. Gottschalk C.W., Morel F. and Mylle M. - Tracer microinjection
   studies of renal tubular permeability. Amer. J. Physiol. 209 :
   173, 1965.

6. Gutman Y., Gottschalk C.W. and Lassiter W.E. - Micropuncture
   study of inulin reabsorption in the rat kidney. Science 147 :
   753, 1965.

7. Morel F., Mylle M. and Gottschalk C.W. - Tracer microinjection
   studies of effect of ADH on renal tubular diffusion of water.
   Amer. J. Physiol. 209 : 179, 1965.

8. Arrizurieta-Muchnik E.E., Lassiter W.E., Lipham E.M. and
   Gottschalk C.W. - Micropuncture study of glomerulotubular
   balance in the rat. Nephron 6 : 418, 1969.

THE EFFECT OF GLUCOSE UPON REABSORPTIVE TRANSPORT OF URATE BY THE KIDNEY

Geoffrey Boner* and Richard E. Rieselbach

Nephrology Program, University of Wisconsin Medical

School, Madison. Wisconsin 53706, U.S.A.

The mechanism of the increase in urinary excretion of urate following hyperglycemia has been variously interpreted by different workers (1,2,3,4,5). Hyperglycemia and/or its consequences, in all probability inhibit the tubular reabsorption of urate. This inhibitory effect may be the result of:
1) the elevation of blood sugar per se,
2) an increase in filtered load of glucose with an associated acceleration of reabsorptive transport in that segment of the proximal tubule where glucose is primarily reabsorbed, or
3) the presence of glucose in tubular fluid in a distal segment ot the proximal tubule in a concentration substantially greater than that usually present.

In order to examine the foregoing possibilities, a series of studies were performed utilizing volunteers for various experimental protocols.

MATERIAL AND METHODS

All experiments were performed in well-hydrated subjects after an overnight fast. Glomerular filtration rate (GFR) was measured using Inulin or Iothalamate I125 clearance (6), the latter being especially useful in the presence of high blood and urine levels of glucose. The GFR was always expressed as ml/min/1.73 M2

---

*Present address - Nephrology Unit, Hillel Yaffe Government Hospital, Hadera, Israel.

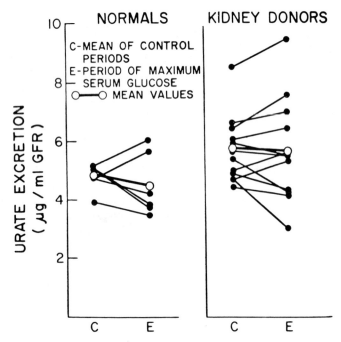

Fig. 1   Urinary urate excretion (μg/ml GFR) in 6 normal individuals
         and 12 kidney donors before and after an oral glucose load.

body surface area. Glucose was measured by the autoanalyzer or
hexokinase methods (7) and serum and urine urate by the uricase
method (8). Urate excretion was expressed as μg/ml GFR. All results
were expressed ± standard error of mean. Standard clearance methods
were used in all the studies (9).

RESULTS

     In the first series of studies 150 Gm of oral glucose was
given to 6 normal subjects and 12 kidney donors, who had undergone
nephrectomy a mean period of 22 months previously and had achieved
stable renal function. The kidney donors had a mean GFR of
77.5±3.4 ml/min/1.73 M2, an increase of 45% over their prenephrec-
tomy one kidney GFR. After the oral glucose load the mean serum
glucose of all the subjects increased from 92.2±2 to a maximum of
165±7 mg/100 ml without inducing glucosuria, there being no
difference in serum glucose between normals and donors. Fig. 1
shows that there was no change in the excretion of urate in either
group. The control excretion of urate in the normals was
4.8±0.2 μg/ml GFR and in the period of maximal serum glucose was

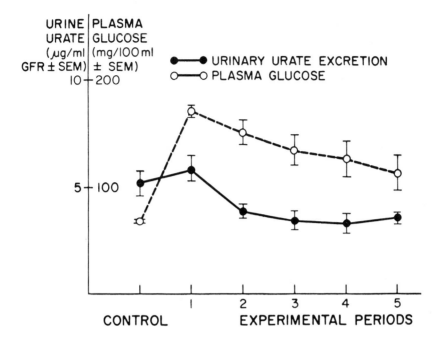

<u>Fig. 2</u>   Plasma glucose (mg/100ml) and urinary urate excretion
(μg/ml GFR) in 4 normal individuals before and after an
intravenous glucose load.

4.5±0.4 μg/ml GFR the respective values in donors being 5.7±0.3
and 5.6±0.5 μg/ml GFR. Thus an increase of 79% in the tubular
reabsorptive transport of glucose had no effect on urate excretion.
Moreover, the donors had a 45% increase in GFR/nephron and thus
had an increased reabsorptive transport of glucose both before
and after glucose load, with no significant effect on urate
excretion.

     In order to rule out the possibility that the previously
observed uricosuric effect occurred as a result of intravenous
glucose administration, I.V. glucose was administered to four
normal volunteers in a sufficient quantity so as to induce a
similar increase in the tubular reabsorptive transport of glucose
as obtained with oral administration. Fig. 2 demonstrates the mean
increase in serum glucose as well as the urinary excretion of urate.
The difference between the baseline urate excretion of 5.3±0.5 and
the subsequent excretion of 5.9±0.6 μg/ml GFR in the first period
is not significant. Further more, the first period was the period
of maximal serum glucose levels and all four subjects demonstrated
mild glycosuria. These four studies demonstrated the absence of a
uricosuric response to intravenously administered glucose.

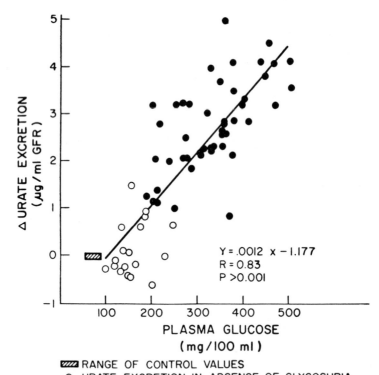

Fig. 3   Changes in urinary urate excretion in 10 normal individuals
         at control and increasing plasma glucose levels. (Plasma
         glucose levels ranged from 60 to 500 mg/100ml.)

     In the next series of studies, progressively increasing
hyperglycemia was induced in 10 normal volunteers by administering
intravenously, increasing concentrations of glucose and thus,
glycosuria of increasing magnitude was obtained. Fig. 3 depicts
the changes in urinary urate as compared to serum glucose levels.
The hatched area shows the range of control values, the mean urate
excretion being 3.9±0.4 μg/ml GFR. The open circles depict the
changes in urate excretion with hyperglycemia but without
glycosuria. The mean value is 4.3±0.4 μg/ml GFR; this value was
not significantly different than the control value. However, with
the appearance of glycosuria, as depicted by the closed circles,
there was a rapid increase in the urate excretion. This increase
over control values was directly related to blood glucose levels
(R=0.83; P<0.001). This series of studies demonstrated the
increase in urate excretion after the appearance of glycosuria.

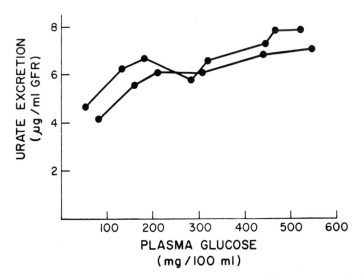

Fig. 4   Urinary urate excretion in mother and her son, with renal glycosuria, at increasing levels of plasma glucose.

This finding was further substantiated in two additional studies in a mother and her son with renal glycosuria. Both these subjects demonstrated glycosuria at low plasma glucose levels (162 and 133 mg% respectively). As shown in Fig. 4 they both showed an increase in urinary urate at substantially lower plasma glucose levels than normals. Thus, in these subjects with a decrease in tubular reabsorptive capacity for glucose, the uricosuric response to glucose was related to the urinary glucose concentration rather than that of plasma.

DISCUSSION

The uricosuric effect of glucose was first reported on by Talbott in 1944 (1). Bonsnes and Dana (2) clearly demonstrated, two years later, that the uricosuric effect of glucose is present only when the glucose concentration in tubular fluid is maintained at relatively high levels. However, Christensen and Steenstrup in 1958 (3) showed a linear relationship between urate excretion and plasma glucose suggesting that the inhibition of urate reabsorption is dependent on either plasma glucose or tubular reabsorptive transport of glucose. Padova et al (4) demonstrated the uricosuric effect of induced hyperglycemia in five diabetics. However, no mention was made of the presence of glycosuria and no explanation was offered for this phenomenon. Herman and Keynan (5) examined the effect of glucose infusion on urate excretion in seven volunteers. Six subjects who became glycosuric showed the uricosuric effect. The remaining subject did not demonstrate glycosuria nor did he demonstrate the uricosuria of hyperglycemia.

Most of these studies as well as our study support the contention that the uricosuric effect of glucose is dependent on the presence of glucose in the more distal parts of the renal tubules. Since Skeith et al (10) have shown that the osmotic diuresis following mannitol administration was only minimally uricosuric it would appear that glucose must have a direct inhibitory effect on the reabsorption of urate. However, as urate is assumed to be reabsorbed in the proximal tubule as is glucose, it is difficult to explain the ineffectiveness of an increased glucose load in inhibiting urate reabsorption. It has been recently suggested by Steele and Boner (11) that urate in addition to being partially reabsorbed and then secreted may also be reabsorbed at a second site, distal to the secretory site. It could then be postulated that the presence of glucose distal to the tubular site of its reabsorption inhibits urate reabsorption. Thus, we postulate that urate transport at this distal reabsorptive site is diminished by the presence of an elevated concentration of glucose in that segment of the proximal tubule.

In summary our data suggest that
1) the presence of an elevated concentration glucose in tubular fluid inhibits urate reabsorption only at a site (or sites) where it, i.e. glucose, is normally virtually absent, and
2) the mechanisms for urate reabsorption at this site (or sites) possess different characteristics than the more proximal site, which does not appear to be inhibited by increased tubular reabsorptive transport of glucose.

REFERENCES

1) TALBOTT, J.H.: Gout., Oxford Medicine, Vol. IV Oxford University Press, New York, 1943, p.108.
2) BONSNES, R.W. and DANA, E.S.: On the Increased Uric Acid Clearance Following the Intravenous Infusion of Hypertonic Glucose Solution.
    J. Clin. Invest., 25:386, 1946.
3) CHRISTENSEN, P.J. and STEENSTRUP, O.R.: Uric Acid Excretion with Increasing Glucose Concentration (pregnant and non-pregnant cases).
    Scan. J. Clin. & Lab. Invest., 10:182, 1958.
4) PADOVA, J., PATCHEFSKY, A., ONESTI, G., FALUDI, G. and BENDERSKY, G.: The Effect of Glucose Loads on Renal Uric Acid Excretion in Diabetic Patients.
    Metab., 13:507, 1964.

5) HERMAN, J.B. and KEYNAN, A.: Hyperglycemia and Uric Acid.
    Israel J. Med. Sci., 5:1048, 1969.
6) MAHER, F.T., NOLAN, N.G. and ELVEBACK, L.R.: Comparison of
    Simultaneous Clearances of I125 Labelled Iothalamate
    (Glofil) and Inulin.
    Mayo Clin. Proc., 46:690, 1971.
7) SLEIN, M.W.: Methods of Enzymatic Analysis, Bergmeyer, H.V.
    Ed., Academic Press, New York 1963, p.117.
8) LIDDLE, L., SEEGMILLER, J.E. and LASTER, L.: The Enzymatic
    Spectrophotometric Method for Determination of Uric Acid.
    J. Lab. & Clin. Med., 54:903, 1959.
9) SMITH, H.W.: Principles of Renal Physiology, Oxford
    University Press, New York, 1956, p.196.
10) SKEITH, M.D., HEALEY, L.A. and CUTLER, R.E.: Urate excretion
    During Mannitol and Glucose Diuresis.
    J. Lab. & Clin. Med., 70:213, 1967
11) STEELE, T.H. and BONER, G.: Origins of the Uricosuric
    Response.
    J. Clin. Invest., In press.

MECHANISMS OF RENAL EXCRETION OF URATE IN THE RAT

F. ROCH-RAMEL,
D. De ROUGEMONT, G. PETERS
Institut de Pharmacologie
de l'Université de Lausanne
Lausanne, Switzerland

In free flow micropuncture experiments performed in rats undergoing moderate mannitol diuresis and infused with small amounts of urate (plasma urate : 1.1 ± 0.1 mg /100 ml) , the concentration factor for urate (TF/P_{urate}) in 15 samples of proximal tubule fluid was 1.7 ± 0.2 (mean $\pm$ SE) and the fraction of filtered urate recovered (TF/P_{urate} : TF/P_{inulin}) 1.3 ± 0.1, indicating net secretion of urate.

The distal values were 1.2 ± 0.4 for TF/P_{urate} (n=5) and 0.39 ± 0.02 for TF/P_{urate} : TF/P_{inulin}, indicating extensive reabsorption of urate between late proximal and early distal tubules. Along distal tubules or collecting ducts, no urate appeared to be reabsorbed, the U/P_{urate}: U/P_{inulin} ratio in pelvic urine being 0.47 ± 0.04 (22).

These findings obtained with a fluorescent ultramicro-determination of urate , [1] confirm those of Greger et al [2] who used an ultramicroadaptation of a colorimetric method.

The net secretion observed in the rat proximal tubules contrasts with the net reabsorption reported in those of the Cebus monkey [1], where the mean proximal TF/P_{urate} : TF/P_{inulin} was 0.62 ± 0.05 (n=20).

At the end of the proximal tubules, this ratio reached approximately 0.2. Along distal tubules, the mean TF/P_{urate} : TF/P_{inulin} was 0.18 ± 0.03 (n= 10). Apparently some reabsorption occured in the collecting ducts, since

in pelvic urine U/P_{urate}: U/P_{inulin} was 0.05 ± 0.005
(n=7). Alernatively the deeper nephrons, not accessible
to micropuncture could reabsorb more urate than superficial nephrons.
In the Cebus, reabsorption of urate, thus appears to
predominate over secretion.

As xanthine oxidase activity has been demonstrated to
occur in the rat kidney but not in monkeys [3] the net
secretion observed in rat proximal tubules could be due
to cellular rather than to transtubular secretion. The
influence of urate synthesis on urate excretion was
studied by clearance methods in rats anesthetized with
pentobarbital. The animals were infused with hypoxanthine (0.5 ug/min) a metabolic precursor of urate, or
allopurinol, (5 ug/min), a xanthine-oxidase inhibitor.
In control rats C_{urate}: C_{inulin} decreased from 0.65 to
0.16 as plasma urate concentration increased from 0.2
to 0.8 mg per 100ml of plasma (physiological range). At
plasma levels between 0.2 to 0.4 mg/100 ml, the clearance ratio was 0.52 ± 0.07 (n=5) in allopurinol treated
rats. At plasma levels 0.4 to 0.8 mg/100 ml, it was 0.22 ± 0.04 (n=4) in controls, but 0.65 ± 0.07 (n=4) in hypoxanthine infused animals. The clearance ratio in animals
treated with both hypoxanthine and allopurinol was as
low as in animals treated with allopurinol alone. These
differences in urate excretion observed at similar
plasma urate concentrations are interpreted as resulting
from differences in urate synthesis, newly synthetized
urate entering the tubular lumen from the tubular cells.

1) F. Roch-Ramel and I.M. Weiner, Am.J.Physiol. in Press
   (June 1973)
2) R. Greger, F. Lang and P. Deetjen, Pflügers Arch.
   324, 279 (1971)
3) E.J. Morgan, Biochem. J. 20,1926

Supported by Fonds National Suisse de la recherche scientifique grant Nr. 3.9000.72

METHODOLOGY

IMMUNOADSORBENT CHROMATOGRAPHY OF HYPOXANTHINE-GUANINE PHOSPHORI-BOSYLTRANSFERASE

W. J. Arnold, R. B. Jones and W. N. Kelley

Duke University Medical Center, Durham, North Carolina

27710

Recent evidence indicates that the virtual absence of hypoxan-thine-guanine phosphoribosyltransferase (HGPRT) in patients with the Lesch-Nyhan syndrome is due in most if not all instances to a mutation(s) on the gene coding for the HGPRT protein (Kelley and Meade, 1971; Rubin, et al., 1971; Arnold, Meade and Kelley, 1972). This mutation(s) results in the synthesis of a normal amount of a cat-alytically defective yet immunoreactive HGPRT enzyme protein. We have purified normal human HGPRT from erythrocytes and have prepared a highly specific antiserum in rabbits (anti-HGPRT) which displays a single precipitin line on immunodiffusion and immunoelectrophoresis with both normal hemolysate and hemolysate from patients with the Lesch-Nyhan syndrome. We have succeeded in coupling a partially-pur-ified preparation of anti-HGPRT to cyanogen bromide activated sepharose and will present data which indicates that this technique is a promising method for the rapid isolation of both normal and mutant HGPRT.

Crude anti-HGPRT serum from rabbits was partially-purified by three successive 0-40% ammonium sulfate precipitations at 4° and dialyzed for 12-18 hrs. at 4° against 1000 volumes of the buffer to be used during coupling to the Sepharose derivatives. The preparation of the Sepharose derivatives was done exactly as described by Cuatrecasas (1970) and those which we have used are listed in Table 1. As can be seen the less reactive but potentially more useful Bromo-acetyl Sepharose derivative would not couple anti-HGPRT at pH 6.5 or 9.0 even if the coupling step were lengthened to 48 hrs. When the cyanogen bromide-activated derivative was used, coupling of as much as 20 mg of anti-HGPRT protein to one milliliter of the gel could be demonstrated at either pH 9.0 or pH 6.5. However, the anti-HGPRT Sepharose complex would retain its ability to bind

TABLE 1

COUPLING OF PURIFIED ANTI-HGPRT TO SEPHAROSE DERIVATIVES

Derivative	Anti-HGPRT in Supernatant	HGPRT Binding to Gel	APRT Binding to Gel
		% of Applied Activity	
I. Bromo-acetyl Sepharose			
1) 10 mM NaP pH 6.5, 24 hr coupling	+	0	0
2) 10 mM NaP pH 9.0, 48 hr coupling	+	0	0
II. Cyanogen Bromide Activated Sepharose			
1) 0.1 M NaHCO$_3$ pH 9.0, 18 hr coupling	0	10%	10%
2) 0.02 M NaP pH 6.5, 72 hr coupling	0	100%	10%

HGPRT specifically only if the coupling step were done at pH 6.5. This loss of biologic activity when coupling is done at pH 9.0 is apparently due to the exposure of excessive numbers of epsilon-amino groups of lysine and coupling of the antibody at too many sites on the Sepharose bead. Coupling time at pH 6.5 was extended to 72 hours to ensure that all reactive groups on the Sepharose were inactivated before the column was used. When the coupling step was performed at pH 6.5 the anti-HGPRT retained at least 60% of its calculated binding capacity and the sample size could be adjusted so that essentially 100% of the HGPRT activity subsequently applied to the column was bound. The small amount of APRT activity bound to the column indicates the specificity of the binding for HGPRT.

After the coupling step the anti-HGPRT Sepharose gel was washed at 4° with 1000 ml of 50 mM Tris buffer pH 7.4 followed by 100 ml of 6 M Urea, then equilibrated in 50 mM sodium phosphate pH 7.4, 0.15 N NaCl, and 0.1% sodium azide for storage at 4°. Just prior to use, the column was washed with 100 ml of 50 mM sodium phosphate pH 7.4, 0.15 N NaCl. Table II summarizes the results of one experiment in which a de-hemoglobinized preparation of normal hemolysate was applied to the column and eluted at 20 ml per hour. Before and after each of the buffer washes, two 20

TABLE II

ELUTION OF ENZYMATICALLY-ACTIVE HGPRT FROM ANTI-HGPRT IMMUNOABSOR-
BENT

	Buffer Wash	HGPRT Activity in Eluate	HGPRT Remaining Bound to Gel
		(% of Applied Activity)	(IMP-C^{14}CPM/20 μl Gel)
1.	50 mM NaP pH 7.4 0.15 N NaCl	14%	8215
2.	50 mM NaP pH 7.4 1 M NaCl	24%	6050
3.	50 mM NaP pH 7.4 3 M Urea	42%	2800

μl samples of the gel were removed and used as the enzyme source in
the standard radiochemical assay for HGPRT and adenine phosphorbo-
syltransferase (APRT) (Kelley, et al., 1967). HGPRT activity was
shown to be bound by direct assay of HGPRT activity on the anti-HGPRT
Sepharose gel while only a small amount of APRT activity could be
detected (80-100 cpm AMP-C^{14}/20 μl). No HGPRT activity was assayable
on the gel prior to the application of the sample.

Figure 1 illustrates the elution pattern of HGPRT activity
when the column was washed with 3 M urea. Similar results were
obtained when 0.8 M NH$_4$OH was used as the eluant   Those fractions
in the 3 M Urea eluate containing 42% of the HGPRT activity applied
to the column were pooled and concentrated by Amicon ultrafiltration
(UM-10) at 4° then dialyzed for 2 hr. against 1000 volumes of 50 mM
Tris buffer pH 7.4 Aliquots of this sample were applied to analytical polyacrylamide gels and electrophoresis was performed as previously described (Arnold and Kelley, 1971). Fig. 2 illustrates that,
although several minor contaminant bands were present, the major
protein band present in the eluate from 3 M migrated with a sample of HGPRT purified by another technique (Arnold and Kelley, 1971).
The final purification of HGPRT obtained was calculated at 1000-
fold from hemolysate.

Beaudet and Caskey (1973) have reported the binding of HGPRT
derived from cultured Chinese hamster fibroblasts to anti-HGPRT Sepharose activated by cyanogen bromide. However, the small amount of
antisera available precluded isolation of a sufficient quantity of
HGPRT for analysis of purity. We believe that the successful application of this technique will allow a rapid and quantitative isola-

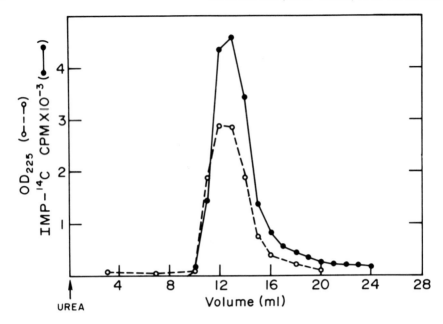

Fig. 1. Elution of HGPRT from antibody-Sepharose column with 3 $\underline{M}$ urea.

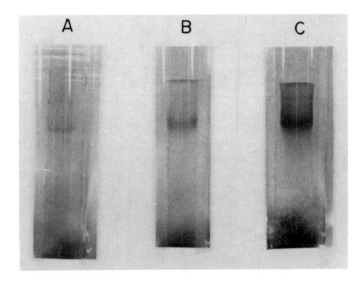

Fig. 2. Acrylamide gel electrophoresis of HGPRT purified by anti-HGPRT immunoabsorbent. A. 3 M urea eluate from anti-HGPRT sepharose. B. HGPRT purified by usual technique (Arnold and Kelley, 1971). C. Mixture of A and B.

tion of HGPRT protein from relatively small samples such as would be obtainable from individual patients with the Lesch-Nyhan syndrome.

REFERENCES

Arnold, W. J. and Kelley, W. N. 1971. Human hypoxanthine-guanine phosphoribosyltransferase: Purification and subunit structure. J. Biol. Chem. 246: 7398-7404.

Arnold, W. J., Meade, J. C. and Kelley, W. N. 1972. Hypoxanthine-guanine phosphoribosyltransferase: Characteristics of the mutant enzyme in erythrocytes from patients with the Lesch-Nyhan syndrome. J. Clin. Invest. 51: 1805-1812.

Beaudet, A. L., Roufa, D. J. and Caskey, C. T. 1973. Mutations affecting the structure of hypoxanthine-guanine phosphoribosyl-transferase in cultured Chinese hamster cells. Proc. Nat. Acad. Sci. 70: 320-324.

Cuatrecasas, P. 1970. Protein purification by affinity chroma-tography derivatizations of agarose and polyacrylamide beads. J. Biol. Chem. 245: 3059-3065.

Kelley, W. N., Rosenbloom, F. M., Henderson, J. F. and Seegmiller, J. E. 1967. A specific enzyme defect in gout associated with overproduction of uric acid. Proc. Nat. Acad. Sci. 57: 1735-1739.

Kelley, W. N. and Meade, J. C. 1971. Studies on hypoxanthine-gua-nine phosphoribosyltransferase in fibroblasts from patients with the Lesch-Nyhan syndrome. Evidence for genetic heterogeneity. J. Biol. Chem. 246: 2953-2958.

Rubin, C. S., Dancis, J., Yip, L. C., Nowinski, R. C. and Balis, M. E. 1971. Purification of IMP: Pyrophosphate phosphoribosyl-transferases, catalytically incompetent enzymes in Lesch-Nyhan disease. Proc. Nat. Acad. Sci. 68: 1461-1464.

THE USE OF HIGH PRESSURE LIQUID CHROMATOGRAPHY TO MONITOR NUCLEOTIDE LEVELS IN CELLS

Phyllis R. Brown and R. E. Parks, Jr.

Division of Biological and Medical Sciences
Brown University, Providence, R. I. 02912, U.S.A.

High pressure liquid chromatography has proved to be a valuable tool in measuring the nucleotide levels in cell extracts (1, 2). The analyses are accomplished with high speed, sensitivity, resolution and accuracy. A decided advantage of this technique is that the major naturally occurring nucleotides can be monitored at one time concurrently with nucleotides of purine and pyrimidine analogs.

HPLC chromatographs are relatively simple instruments composed of a hydraulic system, a column, a detector and a recorder. Many commercial instruments are available, and instruments can be readily built from components. For nucleotides, pellicular resins are usually used. This material consists of microglass beads coated with a thin layer of an anion exchange resin. The detector is a micro UV detector with a wavelength at 254 nm.

The high pressure liquid chromatograph is a very versatile instrument because many operating conditions can be varied. To optimize separations, it is possible to change any of the following operating parameters:

1) The resin properties (particle size, structure of polymer, shell thickness, etc.);

2) Column length, diameter and design;

3) Chemical composition of the eluents;

4) Concentration of the eluents;

799

5) pH of the eluents;

6) Column temperature;

7) Flow rates;

8) Elution mode.

For the analysis of nucleotides, the optimum conditions
using a 1 mm x 3 m stainless steel column packed with a pellicular anion exchange resin are:

Eluents: Low concentration - 0.015 M KH_2PO_4
           High concentration - 0.25 M KH_2PO_4 in 2.2 M KCl;

pH of Eluents: 4.5;

Flow rates: Gradients into the column - 12 ml/hr
             High concentration eluent into the low con-
                centration eluent - 6 ml/hr;

Elution mode: Linear gradient.

Factors affecting quantitation were investigated to insure
accuracy of results. In order to determine stability of column
and instrument conditions, standard solutions of adenine and
guanine nucleotides were run routinely. Chromatograms of standard solutions of either naturally occurring nucleotides or
nucleotides of purine analogs were obtained to be used as references. The reliability of quantitative results was checked by
comparing the results of total adenine nucleotide content
obtained on the chromatograph and by enzymatic analysis. According to results obtained by Burtis and Gere (3), the precision
that can be achieved using digital integration is 0.5%. Because
of the high sensitivity of HPLC, modified Beer's Law plots
were obtained (2) and the stability of the extracts on storage
(4) and sample preparation procedures were examined (5). One
of the most important steps in this analytical procedure is the
preparation of the sample. Care must be taken in multi-step
preparations to prevent errors which can be cumulative and cause
large errors in the final data.

In analyzing cell extracts by HPLC the peaks in the chromatogram must be identified. Methods used are:

1) Internal standards;

2) Retention times of standard solutions;

3) Characterization of collected fractions by chemical,
    spectral or other chromatographic techniques;

4) Isotopic techniques;

5) Derivatization;

6) The enzymic peak-shift technique.

The enzymic peak-shift technique utilizes the specificity
of enzyme reactions with a nucleotide or a class of nucleotides.
It has proved useful not only in the identification of peaks but
also in the removal of a peak so that concentrations of compounds
represented by smaller peaks that are hidden by larger peaks can
be measured.

Baseline studies of the nucleotide patterns of the formed
elements of normal human peripheral blood were obtained and the
concentrations of the major nucleotides calculated (6). The
nucleotide patterns are remarkably reproducible for each of the
respective elements. The profiles of erythrocytes, leukocytes
and platelets, however, are quite distinctive (see Figure 1).
The profiles of normal erythrocytes are relatively simple and
contain mainly adenine nucleotides. Only small amounts of other
nucleotides are present as is shown in Figure 1A. In leukocytes,
guanine nucleotides and UTP are noted in the chromatogram (Figure
1B). In the platelets (Figure 1C) the distinctive feature is the
ATP:ADP ratio which is approximately 1.5:1. The ATP:ADP ratio
in most other tissues is in the neighborhood of 10:1. Because
quantitative analyses can be made on milligram amounts of tissue
and a complete nucleotide profile obtained in one hour, this is
a powerful tool in studying nucleotide changes that may result
from disease or chemotherapy. Therefore, HPLC has potential for
diagnosing or predicting disease states and the effects of drug
therapy.

The species differences in the nucleotide profiles of whole
blood were determined (7, 8) to investigate evolutionary development of metabolic systems and to indicate the species best
suited for pharmacological studies. The chromatograms within a
species are highly reproducible. Although chromatograms of
various species are similar, each species has distinctive characteristics. The nucleotide profiles of whole blood of cow, horse
and rooster are shown in Figure 2. The nucleotide profile of
eel is distinctive because the concentrations of guanine nucleotides are higher than those of adenine nucleotides. The profile
of eel and that of human whole blood are shown in Figure 3.

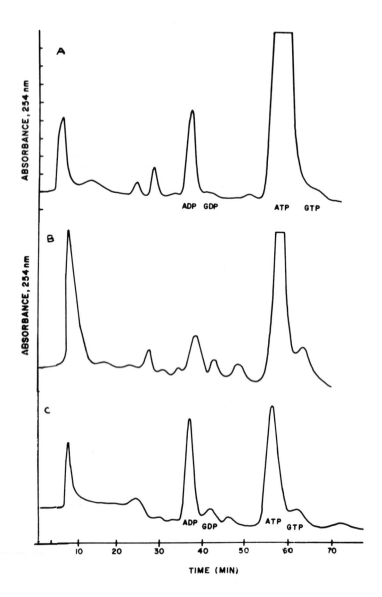

FIGURE 1.   Nucleotide profiles of formed elements of human
blood.   A. Erythrocytes.   B. Leukocytes.   C. Platelets.   One
volume of a suspension of erythrocytes, leukocytes or platelets
was added to two volumes of cold 12% trichloracetic acid (TCA).
The suspensions were mixed on a vortex-type stirrer and centrifuged.   The TCA was extracted from an aliquot portion of the
supernatant fluid with water-saturated diethyl ether.   20 μl
samples were analyzed by HPLC (6).

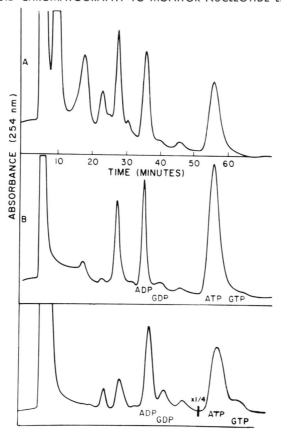

FIGURE 2.   Nucleotide profiles of the whole blood of cow,
horse and rooster.   The samples of whole blood of cow (A), horse
(B) and rooster (C) were prepared by the procedure described in
Fig. 1.   20 μl samples of the extracts from each species were
analyzed by HPLC (7).

In pharmacological studies, the metabolism of purine antimetabolites in normal human erythrocytes was followed simultaneously with the effect of these drugs on the naturally occurring
nucleotide pools (9, 10).   Studies were also carried out with
mercaptopurine analogs in Sarcoma 180 cells (11, 12).   The
effect of preincubation with 6-methylmercaptopurine ribonucleoside
on the treatment of infected mice with 6-thioguanine was investigated.   As is shown in Figure 4, there was a decrease in the ATP
concentration and an increase in the UTP concentration when the
infected mice were pretreated with 6-methylmercaptopurine ribonucleoside for 12 hours.   In investigating possible drugs for the
treatment of schistosomiasis, the metabolism of the purine analog,
tubercidin, in the blood fluke, Schistosoma mansoni, was studied.

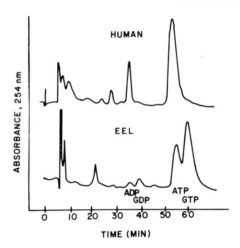

FIGURE 3.   Nucleotide profiles of the whole blood of human and eel.   The extracts of human (upper chromatogram) and eel (lower chromatogram) were prepared according to the procedure described in Fig. 1 (8).

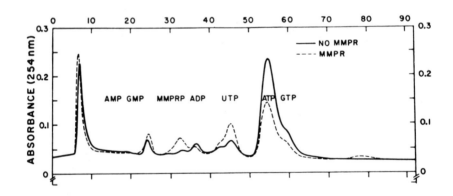

FIGURE 4.   The effect of pretreatment with 6-methylmercapto-purine ribonucleoside on the nucleotide profiles of Sarcoma 180 cells treated with 6-thioguanine.   Sarcoma 180 cells were removed from mice 1 hour after incubation with 6-thioguanine.   The dotted line indicates cells from mice that had been pretreated for 12 hours with 6-methylmercaptopurine ribonucleoside, and the solid line indicates cells from mice that had not been pretreated (12).

With HPLC, the formation of the triphosphate nucleotide of tubercidin in the blood fluke was followed simultaneously with the effects of the drug on the ATP level (13).

In summary, high pressure liquid chromatography can be used not only in the analysis of nucleotides, but in the analysis of almost any biologically active compound. It is especially useful for the separation of those thermally-labile, non-volatile, polar cell components and drugs that are difficult to measure by other methods (14). In addition, drugs and their metabolites can be monitored concurrently with the effects of these drugs on the naturally occurring cell constituents. Therefore, this instrument is a valuable addition to any well equipped biomedical laboratory for use by itself or in combination with other available techniques.

ACKNOWLEDGEMENTS

The authors wish to thank Ms. Sandra Bobick and Mr. Jonathan Gell for their expert technical assistance, Ms. Andrea Bullard for the excellent illustrations and the Hematologic Research Division of the Pawtucket Memorial Hospital for supplying us with fresh blood.

This work was supported by Grant # GM-16538 from the United States Public Health Service.

BIBLIOGRAPHY

1. C. G. Horvath, B. Preiss and S. Lipsky, Anal. Chem. 39, 1422 (1967)

2. P. R. Brown, J. Chromatogr. 52, 257 (1970)

3. C. A. Burtis and D. Gere, "Nucleic Acid Constituents by Liquid Chromatography", Varian Aerograph, Walnut Creek, California (1970)

4. P. R. Brown, Anal. Biochem. 43, 305 (1971)

5. P. R. Brown and R. P. Miech, Anal. Chem. 44, 1072 (1972)

6. E. M. Scholar, P. R. Brown, R. E. Parks, Jr. and P. Calabresi, Blood (In Press)

7.   P. R. Brown, R. P. Agarwal, J. Gell and R. E. Parks, Jr.,
     Comp. Biochem. Physiol. 43B, 891 (1972)

8.   R. E. Parks, Jr., P. R. Brown, Yung-Chi Cheng, K. C. Agarwal,
     Chong M. Kong, R. P. Agarwal and Christopher C. Parks,
     Jr., Comp. Biochem. Physiol. (In Press)

9.   R. E. Parks,Jr. and P. R. Brown, Biochem. (In Press)

10. P. R. Brown and R. E. Parks, Jr., Anal. Chem. (In Press)

11. E. M. Scholar, P. R. Brown and R. E. Parks, Jr., Cancer
     Res. 32, 259 (1972)

12. J. A. Nelson and R. E. Parks, Jr., Cancer Res. 32, 2034
     (1972)

13. R. Stegman, A. W. Senft, P. R. Brown and R. E. Parks, Jr.,
     Biochem. Pharmacol. 22, 459 (1973)

14. P. R. Brown, "High Pressure Liquid Chromatography, Biochemical
     and Biomedical Applications", Academic Press, New York,
     1973.

ELECTROPHORETIC SEPARATION OF NORMAL AND MUTANT

PHOSPHORIBOSYLPYROPHOSPHATE SYNTHETASE

P. Boer, O. Sperling and A. de Vries

Rogoff-Wellcome Medical Research Institute and
the Metabolic Unit of Department of Medicine D,
Tel-Aviv University Medical School, Beilinson
Hospital, Petah Tikva, Israel

Phosphoribosylpyrophosphate (PRPP) synthetase
(E.C. 2.7.6.1) catalyzes the formation of PRPP from
ribose-5-phosphate and ATP in the presence of Mg^{++} and
inorganic phosphate. The product, PRPP, is a substrate
of the first rate limiting step of the de novo synthesis
of purine nucleotides and its availability has been
shown to regulate this pathway in human tissue (1). A
superactive mutant erythrocyte PRPP synthetase with
decreased sensitivity to feedback inhibition has recently
been found by us in a gouty family (2,3).

In the present communication we report a method
for the electrophoretic separation of PRPP synthetase
on cellulose acetate, and its application to the study
of this enzyme derived from various normal human tissues
and from the erythrocytes of two gouty siblings affected
by the enzyme mutation.

Tissues were homogenized in phosphate buffer and
sonicated, and the supernatant was used for electrophoresis. Samples were run in phosphate buffer, pH 8.5,
on cellulose acetate paper for $2\frac{1}{2}$ hours at 4° and at
0.5 mA/cm. PRPP synthetase was located on the paper by
a radiochemical assay: formation of PRPP from ribose-5-
phosphate and ATP was coupled to inosinic acid (IMP)
synthesis by the addition to the reaction mixture of
labelled hypoxanthine and partially purified hypoxanthine-guanine phosphoribosyltransferase (HGPRT).

Hemolysates obtained by dilution of human erythro-cytes with phosphate buffer and also when further subjected to rapid freezing and thawing, contained enzyme activity which remained at the application point. Storage of such hemolysates at $-20°$ for two weeks resulted in migration of the peak towards the anode, moving slightly faster than hemoglobin. Sonication of the hemolysate caused electrophoretic mobilization of the enzyme activity, migrating somewhat faster than the stored enzyme. Under both these conditions, the mobile enzyme activities gave a single, narrow and symmetric peak, indicating that the erythrocyte PRPP synthetase activity resides in a single protein.

The various other human tissues studied leukocytes, platelets, spleen, kidney and heart, gave a single peak, identical to that exhibited by the sonicated erythrocytes, while liver gave an additional slower moving peak, appearing only in concentrated extracts. Application of increasing amounts of extract resulted in increasing height of the slow peak, at the same time the height of the fast peak rising only little. Among all tissues studied liver extracts exhibited on electrophoresis the highest amount of enzyme activity and the appearance of the slower peak could have therefore been a technical artifact. Nevertheless, in view of the tendency of PRPP synthetase to aggregate in the presence of substrates (4), and also as a function of enzyme concentration (5), the appearance of the second peak might possibly repre-sent a heavier aggregate of the enzyme.

The results are taken to indicate the presence of only one genetic type of PRPP synthetase in the various human tissues. If valid, this conclusion would justify the conclusion that PRPP synthetase is superactive in the liver of patients with a superactive mutant erythro-cyte enzyme.

The mutant enzyme from the erythrocytes of our patient exhibited normal electrophoretic mobility. Thus, this electrophoretic method appears not to be practical for the screening of a gouty population for a PRPP synthetase mutant of this kind. On the other hand, it can not be excluded as yet that this method may show to be suitable for the detection of other mutant PRPP synthetase enzymes.

References

1. Fox I.H. and Kelley W.N. Ann. Intern. Med.
    74:424, 1971.

2. Sperling O., Boer P., Persky-Brosh S., Kanarek E.
    and De Vries A. Europ. J. Clin. Biol. Res.
    17:703, 1972.

3. Sperling O., Persky-Brosh S., Boer P. and De Vries A.
    Bioch. Medicine, In press, 1973.

4. Fox I.H. and Kelley W.N. J. Biol. Chem. 246:5739,
    1971.

5. Sperling O., Boer P., Persky-Brosh S. and De Vries A.
    FEBS letters, 27:229, 1972.

DEVELOPMENT OF A MICRO HG-PRT ACTIVITY ASSAY: PRELIMINARY COMPLE-

MENTATION STUDIES WITH LESCH - NYHAN CELL STRAINS

P.Hösli, Department of Human Genetics, Sarphatistraat 217,
    Amsterdam, The Netherlands
C.H.M.M. de Bruyn and T.L. Oei, Department of Human Genetics,
    Faculty of Medicine, University of Nijmegen,
    Nijmegen, The Netherlands

The main purpose of the present study was to look with the hetero-
polykaryontest (ref. 1,2) for genetic complementation between different HG-PRT deficient cell strains, which had been fused and cultured in Plastic Film Dishes (PFD's ref. 3,4) A radiochemical HG-
PRT microassay, based on the use of the Parafilm Micro Cuvette
(PMC; ref. 3,4) was developed, which permits the quantification of
the enzyme activity at the single cell level.

Methods and Materials

The technical equipment necessary for the described techniques
is commercially available from Tecnomara AG (ref. 3).

Normal or mutant fibroblasts, respectively, were fused in a 1:1
ratio with B-propiolactone inactivated Sendai virus and seeded into PFD's. 72 hours later single homo- or heteropolykaryons, each
containing 10 nuclei, were marked on the plastic film bottom of the
PFD's. The cultures were washed and lyophilized. Subsequently, little plastic film leaflets, each carrying one marked polykaryon, were
cut out. The leaflets were transferred into PMC's, each containing
0.3 µl of incubation mixture. The final concentrations of reagents
in the incubation mixtures were as follows: 0.17M Tris HCl buffer
pH 7.4; 17mM $MgCl_2$; 1.7mM PRPP; 0.13 mM $^{14}C-$ labeled guanine or hypoxanthine; 0.5 % BSA. After incubation for 2-24 hours at 37° the
whole contents of the PMC's were pushed onto Whatman 3MM paper
strips and substrate and products were separated by chromatography
The identified radioactive spots were cut out and radioactivity
was measured quantitatively with a liquid scintillation counter. To
study the HG-PRT activity of single skin- or amniotic fluid derived
fibroblasts, essentially the same procedure as described for the
heteropolykaryon test was used. Details of the micro-assay are reported elsewhere.

811

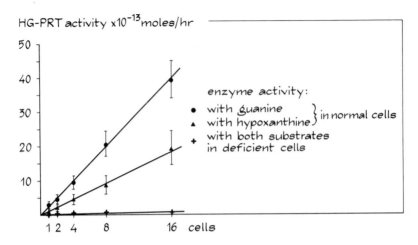

Fig.1: Relation between cell number and HG-PRT activity in normal and enzyme deficient skin fibroblasts.

Results.

        In Fig. 1 the mean HG-PRT activities of 1,2,4,8 and 16 cells (normal and HG-PRT deficient) are given. Each point in the graph represents the mean of 15 isolations. The ranges indicated are not due to methodological errors, but reflect the cell-cycle-dependent variation of asynchronously growing cells. It is obvious, that HG-PRT can be assayed in the most sensitive way by using guanine as substrate and that deficient and normal cells can be distinguished without any overlap. The whole experiment of figure 1 has been re-peated several times, leading to identical results, which demonstra-tes the accuracy and reproducibility of the method.
        Fig.2 represents the HG-PRT measurements with guanine of 6 different fibroblastic cell strains: 3 were derived from normal skin and 3 from normal amniotic fluid. For each cell strain, 15 plastic leaflets with 5 or 10 cells were tested. The data demon-strate that there is no difference in HG-PRT activity between skin-and normal amniotic fluid derived fibroblasts.
        In Fig.3 the preliminary results of crossings of 3 different HG-PRT deficient cell strains are summarized. Each number repre-sents the mean of 15 independently isolated homo- or heteropoly-karyons. The values in brackets are indicative of HG-PRT activities caused by weak genetic complementation; they are corrected for the expected additive gene dosage effects of the parental strains. These experiments have been repeated twice (with hypoxanthine and guanine as substrates) and seemed to lead to comparable results. In contrast to the results shown in fig.1 and 2 these data should be considered as semiquantitative because of technical reasons.

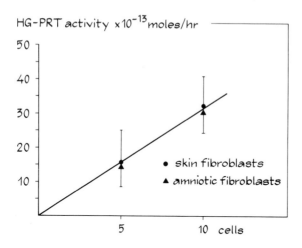

Fig.2: Comparison between HG-PRT activity in skin fibroblasts and amniotic fibroblasts.

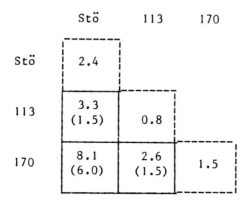

Fig.3: HG-PRT activities of heteropolykaryons from crossings between 3 unrelated deficient patients, expressed in pmoles/heteropolykaryon/24 hour.

814   P. HÖSLI, C.H.M.M. DE BRUYN, AND T. L. OEI

Discussion

The PMC-technique permits miniaturization of most pre-existing macro enzyme activity assays. It allows incubations in exceedingly small volumes (lowering of the blanc, improving the signal/noise ratio) for as many hours as the enzyme is stable (increased sensitivity). This is a reliable way to assay enzymes routinely 1. at the single level, 2. with very low costs.

The present HG-PRT micro assay introduces radioactive substrates to the PMC-technique. Therefore, is becomes possible with the PMC to use natural substrates with high specificity and sensitivity. Moreover radioactive substrates and products can be separated in a straightforward way to check possible side reactions.

As shown in figure 2, skin-derived fibroblasts serve reliably as standards for prenatal diagnosis of severe or partial HG-PRT deficiencies. The heteropolykaryon test has principal advantages compared to the more conventional macro-technique with hybrid cell strains isolated by selective procedures:
1. The heteropolykaryon test works without vital hybrid cell clones, which are sometimes very difficult to establish.
2. Any heteropolykaryon can be isolated without selective media. Therefore, the heteropolykaryon test permits complementation of any mutants.
3. Chromosome losses are known to occur in the early phases of the establishment of the hybrid clones even in intraspecies crosses. With the conventional techniques this can obviously lead to false negative results.
4. Multiple crosses can be studied in 3 to 5 days by means of the heteropolykaryon test, which thus approaches the speed and simplicity of the classical techniques used in microbial genetics.

We presently use the micro HG-PRT activity assay for rapid prenatal diagnosis in early pregnancy and more extensive complementation studies. The radiochemical micro assays of visually selected cells should, however, serve many other experimental and diagnostic purposes.

Acknowledgements

This investigation was supported by grants of the Department of Health of the Dutch Government and FUNGO (Foundation for Medical Scientific Research in the Netherlands).

The authors thank Miss E. Vogt, Mr. C. van Bennekom and Mr. A. Janssen for excellent technical assistance.

References

1. P. Hösli: (1972) Microtechniques for the detection of inter-
   genic and interallelic complementation in man. Bull. Europ.
   Soc. Hum. Genet. p.32.
2. P. Hösli: (1973) Inborn lysosomal disease: detection of genetic
   heterogeneity by complementation studies. Proceedings of the
   "Third International Research Conference on Lysosomes in Cell
   Pathology",Louvain, in press.
3. P. Hösli: (1972) Tissue cultivation on plastic films. Tecnomara
   A.G., Rieterstrasse 59, Postfach CH 8059, Zürich, Zwitserland.
4. P. Hösli: (1972) Micro techniques for rapid prenatal diagnosis
   of inborn errors of metabolism in early pregnancy. Prenatal
   diagnosis newsletter 1 (1972) 10.

COMMENT ON THE ASSAY OF PURINE PHOSPHORIBOSYLTRANS-FERASES IN CULTURED HUMAN FIBROBLASTS

E. Zoref, O. Sperling and A. de Vries

Tel-Aviv University Medical School, Department
of Pathological Chemistry, Tel-Hashomer and
the Rogoff-Wellcome Medical Research Institute,
Beilinson Hospital, Petah Tikva, Israel

Cultured human skin fibroblasts are commonly
utilized in the detection of hemizygosity and hetero-
zygosity for hypoxanthine-guanine phosphoribosyltransferase (HGPRT) deficiency. Two obstacles are encountered
in the determination of HGPRT and adenine phosphoribosyltransferase (APRT) in extracts of cultured skin fibroblasts: the sensitivity of these enzymes in dilute cell
suspension to freezing and thawing (1), and the presence
of nucleotidase activity (2).

The loss of HGPRT and APRT activities by freezing
and thawing of the cells has been dealt with previously
and can be prevented by the use of concentrated cell
suspensions and the addition of PRPP (1). The presence
of nucleotidase activity in the cell extracts causes
the rapid degradation of the nucleotides, produced in
the assay, to their corresponding nucleosides.

The efficacy of some modifications in correcting
the assay were studied. Skin biopsies were removed and
cultured by standard methods (3). Cell extracts were
prepared by freezing and thawing of concentrated cell
suspensions ($>6 \times 10^6$/ml) in the presence of 1 mM PRPP.
HGPRT and APRT were determined in the dialyzed super-
natants of the cell extracts by a radiochemical method
in which the labelled purine bases are converted to
their respective nucleotides by reacting with PRPP,

catalyzed by the corresponding phosphoribosyltransferases (4). The reaction products were separated by
thin layer chromatography on microcrystalline cellulose.
The nucleotide fractions were separated by development
in butanol:methanol: H_2O: NH_4OH 25% (60:20:20: 1 v/v).
In addition, nucleosides and nucleotides fractions were
separated by development in saturated ammonium bicarbonate, and their sum total determined.

Further degradation of the nucleosides to the purine
bases by the catabolic activity of nucleoside phosphorylase has not been detected in the assay system used.
While the sum of the nucleotides and nucleosides formed
accurately reflects the enzyme activities, inhibition
of the conversion of nucleotides to nucleosides by
thymidine-5'triphosphate (TTP) (5) facilitates the assay
by measuring the nucleotides only. Indeed, for normal
cells, the accumulation of radioactivity was found to be
linear with time and protein concentration, in the
nucleotide fraction only when TTP was present, while in
the sum of the nucleotide and nucleoside fractions in
both the presence and the absence of TTP.

The specific activities of HGPRT and APRT were
determined in extracts of cultured skin fibroblasts
from both normal and HGPRT-deficient subjects, comparing
the results obtained with and without TTP (Table 1).
It is noteworthy that for the Lesch-Nyhan syndrome
fibroblasts, the assay in the presence of TTP furnished
the bulk of the radioactivity in the nucleoside fraction
rather than in the nucleotide fraction, in contrast to
normal cell extracts in which the bulk of the radioactivity is found in the nucleotide fraction. Two
possible explanations may be entertained to explain
this observation. First, in the HGPRT-deficient system
the nucleotidase activity is far in excess over the
HGPRT activity, and presumably the residual nucleotidase activity, not inhibited by TTP, is sufficient to
degrade the relatively small amount of nucleotide formed.
Secondly, the possibility should be considered that the
accumulation of radioactivity in the nucleoside fraction
reflects the anabolic activity of nucleoside phosphorylase reacting the purine base with ribose-1-phosphate
to form the nucleotide by an alternative pathway. However, this latter explanation seems to be invalid in view of
the absence of the suitable substrate, ribose-1-phos-
phate, from the incubation system, and by the linearity

TABLE 1. SPECIFIC ACTIVITIES OF HGPRT AND APRT IN NORMAL
AND HGPRT DEFICIENT CULTURED SKIN FIBROBLASTS, ASSAYED
WITH AND WITHOUT THYMIDINE-5'-TRIPHOSPHATE

Subject	HGPRT nmoles/mg protein/h			APRT nmoles/mg protein/h		
	GMP (+TTP)	GMP+guanosine (+TTP)	(-TTP)	AMP (+TTP)	AMP+adenosine (+TTP)	(-TTP)
Normal*	129	126	141	175	166	237
HGPRT deficient						
Partial deficiency**	13.4	14.4	16.0	190	167	252
Lesch–Nyhan syndrome***	0.6 (0–2.2)	2.82	3.68	151	128	192

* The results on normal subjects were obtained by the
determination of the specific activity in cell
extracts, for APRT from 10 individuals (20 experiments)
and for HGPRT from 5 individuals (6 experiments)

** One subject

***One subject, mean value (range in parenthesis)

of the relationship between the formation of the nucleoside and the time of incubation up to 2 hrs. Furthermore, it is noteworthy that also for the HGPRT-deficient
Lesch–Nyhan fibrobalsts, the accumulation of the radioactivity in the nucleoside fraction is sensitive to inacti-
vation by freezing and thawing and to protection by PRPP,
a property characteristic of normal HGPRT (1).

References

1. Zoref E., Sperling O. and De Vries A. Proceedings
    of the International Symposium on Purine Metabolism
    in Man, Tel-Aviv, June 17-22, 1973. Plenum Pub.

2. Fujimoto W.Y. and Seegmiller J.E. Proc. Nat. Acad.
    Sci. 65:577, 1970.

3. Kelley W.N. and Wyngaarden J.B. J. Clin. Invest.
    49:602, 1970.

4. Sperling O., Frank M., Ophir R., Liberman U.A.,
    Adam A. and De Vries A. Europ. Clin. Biol. Res.
    15:942, 1970.

5. Murray A.W. and Friedrichs B. Biochem. J.
    111:83, 1969.

SIGNIFICANCE OF STAINING URIC ACID CRYSTALS WITH NATURAL AND SYNTHETIC DYES

J. Kleeberg,* Esther Warski,+ and J. Shalitin+

*Research Department, Rothschild University Hospital
 Haifa, Israel
+Biochemistry Department, Israel Institute for
 Technology, Haifa, Israel

In the urinary sediment of healthy and of sick people only crystals of uric acid appear stained (yellow-brown). No other colored metabolites or calciumsalts in crystallized form are excreted. There is no uniform opinion about the nature of this yellow-brown pigment (1;2;3;); it is probably a mixture of urobilin – urobilinogen. Among the many pigments appearing in the urine of healthy individuals, two are chemically and biochemically well-known: urobilin and urorosein. Experiments with urobilin would have required the production of this bilirubin derivative in sufficient quantities; this task was beyond our technical conditions at that time. Studies to try staining with urorosein, on the other hand, were easy because its synthesis from its basic substance, indole-acetic-acid (4;5;) is easy (6;7;8;). It seemed, furthermore, obvious to experiment with another natural pigment of strong staining property, beetroot extract. Since both kinds of dye studies proved successful it was but a simple, logical step to test a synthetic dye also: methylene blue.

For reprints: Professor J. Kleeberg, M.D. Rothschild University Hospital, Haifa, Israel.

o Part of a thesis for M. Sc.

MATERIAL AND EXPERIMENTS

Pure uric acid (powder) was purchased from SIGMA (U.S.A.); indole-acetic-acid from E. MERCK (Darmstadt); fresh beetroots from the market and beetroots in powder form from C. Roth (Karlsruhe); methylene blue from SIGMA and the same dye for medical purposes from E. MERCK (Darmstadt).

Preparation of beetroot extract: fresh beetroots were cleaned, put in hot water for 5 minutes, peeled and then cut into small pieces. These were boiled with distilled water, half a liter for half a kilo vegetable or 1 liter for 1 kilo. After boiling for half an hour to ensure sufficient evaporation, a concentrated deep red-purple extract was obtained. This was filtered with a sieve or a piece of cloth. The staining power was intensive during the 24 hours after preparation, becoming less after that date. To prevent fungus contamination 2 ml of a 2% thymol-ethanol solution per 100 ml extract were added. Such preparations could be kept effective for 2 weeks, especially if preserved at $4^{\circ}C$.

We also worked with a dry but hygroscopic powder, durable in sealed tins for many months. A 5% solution was prepared by boiling the powder with distilled water over a small flame for a maximum of 10 minutes. To avoid excessive foam, the fluid was stirred. After filtration the extract was kept and used like the fresh one.

Methylene blue dye: the material as a powder may be kept almost indefinitely in a well-closed container. We prepared a 0,5% aqueous solution which remained effective 5-6 months. The dye-solution resulting from the SIGMA material had a somewhat stronger coloring effect than the methylene blue labelled "for medical purposes".

I STAINING WITH UROROSEIN

1.) About 20 ml of pure uric acid were boiled in a test tube with 2 ml of a 3% or 4% sodium hydroxide solution for half a minute. After cooling, 30 mg of 3-indol-acetic-acid were added, gently shaken, warmed, mixed with 5 - 6 drops of a freshly prepared 0.5% sodium nitrite solution and then quickly acidified with 3 - 4 ml of a 10% hydrochloric acid. The fluid immediately imparted a pink color, changing rapidly into pink-violet and, at the same time, presenting precipitated glittering crystals. The test tube was then filled almost to the rim with distilled water, turned upside down gently once or twice and kept in a rack for 15–20 minutes. After discarding the supernatant fluid, the sediment was ready for examination under the microscope.

2.) in a 2nd series of experiments the same procedures were carried out but with two differences: the material to be stained were the powders (25–30mg) of 4 pure and of 4 uric-acid-oxalate stones and the boiling time was 1½–2 minutes.

3.) 20 mg uric acid were dissolved in 100 ml boiling distilled water and the fluid cooled only to 80°C mixed with an acid urorosein solution; it was prepared as follows: 30 mg indole-acetic-acid were dissolved in 4 ml ethylether, 6–7 drops of a freshly prepared 0,5% sodium nitrite solution and 8 ml distilled water added. After shaking the fluid was acidified with 6–8 ml of a 10% hydrochloric acid and the mixture vigorously shaken again. This deep-pink-violet fluid was filtered and then used. After coding to room temperature the precipitate was examined under the microscope.

4.) Urine (150–200 ml) from healthy people was made alcaline with sodium hydroxide to pH 8.0–8.5, mixed with 100 mg di-sodium-urate (100 mg uric acid dissolved in 8–10 ml of a boiling 4% sodium-hydroxide solution), 10 ml of the just described acid urorosein solution were added. Sedimentation of the crystals took sometimes up to 10 hours.

II STAINING WITH BEETROOT EXTRACT

1 and 2.) Alkaline solutions of pure sodium urate and of pure and mixed uric acid calculi (prepared as described under I,1) were mixed with about 3–4 times the amount of beetroot extract, e.g. to 2 ml sodium urate solution were added 6–8 ml of beetroot extract, mixed and the dark-brown-red fluid carefully acidified with a 5% or 10% sulphuric acid to a pH of 3.0 – 2.5 or with a 10% acetic acid. Partial precipitation took place within 15–20 minutes but complete sedimentation required several hours.

3.) The experiment of I.3 was repeated, the staining agent being 5 ml beetroot extract.

4.) The experiment of I.4 was repeated with two changes: a) staining was carried out with 10–12 ml beetroot extract and acidification to pH 3.5–3.0 by 5% sulfuric acid. Complete sedimentation took also up to 10 hours

III STAINING WITH METHYLENE BLUE

1. and 2.) In test tubes amounts of 25–30 mg of either pure uric acid or of pure uric acid stones and of uric acid–oxalate calculi were transformed to their respective sodium urates by boiling the powders for 1–2 minutes with 2 ml of a 4% sodium hydroxide solution. To each test tube 3–4 drops of the methylene blue stain and 2–3 ml distilled water were added and the fluids gently shaken. They were then acidified by 10% hydrochloric acid, drop by drop, until a greyish–blueish precipitate had formed (12–20 drops usually). Examination under the microscope was made after 10 and 60 minutes and after 4 and 24 hours.

3.) The procedure I.3 was repeated, the dye being 10 drops of the methylene blue solution.

4.) Experiment I.4 was repeated, the staining agent was methylene blue (20–25 drops); acidification was carried out with 5% or 10% hydrochloric acid to pH 3.0 or 2.5.

RESULTS WITH UROROSEIN

Experiment I.1 Large, thin, transparent rectangular plates appeared, pale pink colored.

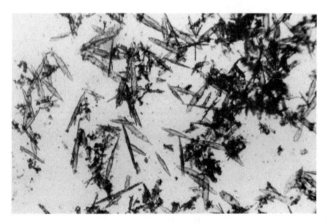

Fig. 1

Experiment I.2. Crystals from a pure uric acid stone powder were pink–violet stained in the form of long needles. They showed a peculiar shape at magnification of 225 (see Fig.2)

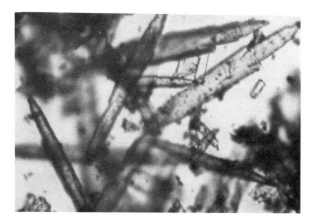

Fig. 2

The long needles of Fig. 1, at magnification 225.

Experiment I.3 revealed small, rectangular, transparent plates, hardly stained.

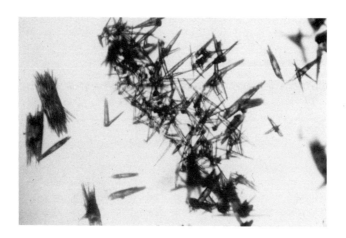

Fig. 3

Experiment I.4. These uric acid crystals actually originated from sodium urate dissolved in normal urine, acidified by hydrochloric acid to pH 2.0. They were of whetstone shape, appearing as isolated bodies and in bundles, all deep-yellow stained. This photograph was chosen as a model, because urorosein did change neither form nor color of the crystals.

RESULTS WITH BEETROOTS

Experiment II.1 The sodium urate-pigment solution acidified with 10% sulfuric acid to pH 2.5 resulted in pink-yellow stained crystals in the form of short crosses and stars of thin whetstones. If acetic acid was used (pH 3.5) the ensuing crystals were small, oval shaped of uniform size and yellow-orange stained.

Fig. 4

Experiment II.2 Powder of a pure uric acid stone produced aggregates of orange-red stained crystals in the form of butterflies or rosettes, at a pH of 1.6. Some fern-like patterns were also present. At a pH of 1.2 crystals were of the same shape but less brilliantly stained.

Fig. 5

Experiment II.3 Small, larger and large rhomboid shaped crystals orange to orange-pink stained had precipitated from a super saturated aqueous uric acid solution.

Fig. 6

Experiment II.4 Crystals in the shape of whetstones, prisms and plates had precipitated, stained brownish red.

RESULTS WITH METHYLENE BLUE

Experiment III.1 resulted in the appearance of long, transparent blue-stained rectangular plates.

Fig. 7

Experiment III.2 Pure uric acid stone powder yielded long plates, whetstones, stars of whetstones and long feathered leaves and swords; all crystals were stained blue. Blue stained crystals of other pure uric acid stones appeared in the form of X.

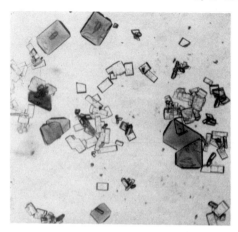

Fig. 8

Experiment III.3 The picture shows small, unstained, transparent
plates mixed with larger, blue stained transparent plates, precipitated from a super saturated aqueous solution.
(By courtesy of "Ztschr. Urol. 65. 620. 1972)

Fig. 9

Experiment III.4 The urinary sediment showed thick, blue-green
stained whetstones and barrels. Most whetstones appeared in
clusters.

DISCUSSION

Although the staining methods described proved to be effective
and the results were always reproducible, modifications of the
procedures may also give satisfactory results. The nature and
concentration of the acids as well as type and concentrations
of the pigments (9) determine size and shape of the uric acid
crystals (10). From organic acids lactic acid was only slightly effective, acetic acid useful for beetroot experiments only.
One reason is that uric acid needs a pH of at least 5.5 to precipitate, for complete and rapid precipitation a pH of 3.0 or
2.5, values which can be obtained by mineral acids only. Values
of pH of 2.0 or 1.0 on the other hand, produced sometimes long
thin needles (forming stars) of acid sodium-urate or small
prisms of acid ammonium urate (from stones) which would also
stain, but less intense; in beetroot extract experiments their
color was yellow. Yet, in beetroot studies, acidification by
acetic acid (pH 3.5 - 3.0) small, oval uniform pink red stained crystals ensued, while sulfuric acid had produced "rosettes"
and crosses of larger size; this proves that the nature of the
acid exerts a specific influence. Crystals, precipitated from
supersaturated aqueous uric acid solutions presented themselves as transparent small plates of an almost uniform size and
form, crystals obtained from aqueous sodium urate solutions,
by acidification, were much larger, but still of a nearly uniform shape. However, in aqueous solutions, in the presence of
the described pigments, larger crystals and in a multitude of
forms appeared, with a great tendency to produce large aggregates.

Two important facts emerged from our experiments: 1. no
crystals already in existence could be stained by any of the
three dyes. Pigments have to be present in alkaline aqueous
urate solutions when on acidification crystals would form —
staining in statu nascendi. 2. Only tri-oxy-purine crystals
could be stained by these pigments. Neither cystine, creatinine nor calcium-oxalate-phosphate- or culfate would absorb
the dyes. It is evident, therefore, that these three pigments
are specific and characteristic stains for uric acid crystals,
whether they originate from pure chemicals, urine sediment or
urinary stones. The only exception to this rule was the coloring of calcium-carbonate with beetroot extract if the substrate was alkaline. However, this result in no way affects
the previous statement, because all our staining was carried
out in strongly acid solution where no carbonate exists.

The three dyes produced fast colors though not unfading,
because repeated washings with boiling water or boiling 50%
and 75% ethanol took out most of the urorosein and the beet-

root-red, less of the methylene blue. If the staining was performed in the presence of a colloid such as agar or gelatine (10) the blue stained crystals kept their form and color even if they remained in water for several weeks.

Double refraction of the various crystal forms was unaltered after staining. X-ray-diffraction (X.R.D.) analysis revealed that the single sword-shaped crystal was tri-oxy-purine di-hydrate with and without methylene blue. The murexide test could be carried out even with stained crystals.

The mechanism of the staining might be explained on the basis of the similar molecular structure of urorosein, of betanine or betanidine (the coloring matter of beetroot) and the substrate uric acid. Urorosein is, according to von Dobeneck et al (6;7;8;) an alpha-beta-di-indolyle-methene. This polymer has as its basic structure a benzene-pyrrol ring. Betanin (betanidin) has a very similar molecular pattern, a 6- and 5 benzene-pyrrol ring. Even tri-oxy-purine may be regarded as a compound of a 6- and a 5-ring, where 3 nitrogen atoms have replaced the carbon. We cannot be sure that betanine or betanidine were the staining factors, because our methods of producing and using the extract may have converted it to isomers or to neo-betanidines (Dreiling et al 11; 12. Marby and Dreiling 13). However, even these derivatives keep their principle indole structure. Thus both pigments fit into the lattice of the purine metabolite. True, this explanation cannot be applied to the action of methylene blue, a three-ring fused phen-thiazine.

Our experiments with bilirubin or chlorophyll as natural pigments for uric acid crystals were unsuccessful. Within the last five years methylene blue has been tried as a therapeutic against idiopathic hyperoxaluria (14) and oxalate lithiasis (15;16). Good clinical results were reported with doses of 200 - 300 mg daily, administered for a period of several months. When we experimented in vitro with such low concentrations of the dye as they appear after the intake of doses of 300 mg, we found no change in size, form or color of excreted tri-oxy-purine crystals. On the other hand, higher concentrations did increase both size of crystals and their tendency to form large stained aggregates (in vitro). Such aggregates may act as crystallisation centers in stone formation as far as uric acid is concerned. In our studies of pertinent literature no references to beetroot as a therapeutic against urinary lithiasis were found. We succeeded in dissolving 30% more uric acid powder in a boiling aqueous solution of beetroot extract than in pure water. Furthermore, precipitation of tri-oxy-purine crystals takes more time than in water or in methylene blue solutions. Whether betanin or methylene blue play a role in human

urinary uric acid lithiasis is one of our further clinical
and laboratory studies.

ABSTRACT

Crystals of uric acid whether originating from pure chemicals, urinary sediment or stones, could be stained by urorosein, beetroot extract or methylene blue. Staining became possible only at the moment of crystal formation – in statu nascendi – i.e. if in the presence of those pigments crystals
were formed out of aqueous urate solutions on acidification.

Form and size of the stained uric acid crystals depended
on the nature and concentration of the three stains as well as
on the type and concentration of the acid used.

The staining: pink, red-violet and yellow red or blue in
acid aqueous solutions indicate uric acid crystals specifically
and has, therefore, diagnostic value.

The relationship between molecular structure of the pigments and of tri-oxy-purine has been discussed and also their
possible role in stone formation, as far as uric acid lithiasis is concerned.

REFERENCES

1. PAGE, L.B. and CULVER P.J.
   A syllabus of laboratory examinations in clinical Diagnosis.
   Cambridge (Mass), Harvard University Press 1961. pp.293,326.

2. HAWK's Physiological Chemistry 14th ed. by B.L. OSER
   New York. London
   Blackiston Division. Mc. Graw Hill Book Comp. 1965

3. VARLEY
   Practical clinical biochemistry 4th ed.
   London. William Heinemann Medical Books 1967

4. HERTER, C.A.
   On Indol-acetic-acid as the Chromogen of the Urorosein in
   the Urine.
   J. biol. Chem. 4. 239. 1908

5. NENCKI, M. und SIEBER, N.
   Ueber das Urorosein, einen neuen Harnfarbstoff
   J. f. prakt. Chem. 26. 333. 1882.

6. von DOBENECK, H., LEHNERER, Wo. und MARESCH, G.
Ueber das Urorosein
Ztschr. phys. Chem. 304. 26. 1956

7. von DOBENECK, H.
Die Urorosein Reaktion.
Ztschr. f. klin. Chem. 4. 141. 1966

8. von DOBENECK, H., WOLKENSTEIN, D. und BLANKENSTEIN, G.
Alpha-beta-di-Indolyl-Methan und Methene; der Urorosein
Chromophor.
Chem. Ber. 102. 1347. 1969

9. KLEEBERG, J.
Faerbung von Harnsaeurekristallen mit Methylenblau
Ztschr. Urol. 63. 619. 1972

10. WARSKI, Esther
Formation and Dissolution of Urinary Calculi
Thesis in Bio-Medical Engineering
Israel Institute of Technology, Haifa. 1972

11. DREIDING, A.S.
Recent Development in the Chemistry of natural Phenol
Compounds
ed. by G. ALLIS. Chapter 11
London. New York. Pergamon Press 1961

12. WYLER, H., MABRY, T.J. und DREIDING, A.S.
Ueber die Konstitution des Randenfarbstoffs Betanin.
Helvet. Chim. Acta 46. 1745. 1963

13. MABRY, T.J. and DREIDING, A.S.
The Betanines (in "Recent Advances in Phytochemistry ed.
by MABRY - ALSTON - RUNEKLES). Chapter 4.
New York. Appleton Century Crofts. 1968

14. SUTOR, Jane
The possible Use of Methylene Blue in the Treatment of Primary Hyperoxaluria
Brit. J. Urol. 42. 398. 1970

15. BOYCE, W.M., KIMNEY, W.M., LONG, N.T. and DRACH, G.W.
Oral Administration of Methylene Blue to Patients with Renal Calculi
J. Urol. (U.S.A.) 97. 783. 1967

16. CHOW, F.H.C., BRASE, J.L., DWAYNE, D.H. and UDELL, R.H.
Effects of Dietary Supplements and Methylene Blue on Urinary Calculi.
J. Urol. (U.S.A.)   104. 315. 1970

SUBJECT INDEX

Pages 1 through 366 will be found in Volume 41A,
pages 367 through 833 in Volume 41B